CHEMICAL PROCESS ENGINEERING

CHEMICAL INDUSTRIES

A Series of Reference Books and Textbooks

Founding Editor

HEINZ HEINEMANN

ADDITIONAL VOLUMES IN PREPARATION

CHEMICAL PROCESS ENGINEERING
Design and Economics

Harry Silla
Stevens Institute of Technology
Hoboken, New Jersey, U.S.A.

MARCEL DEKKER, INC.   NEW YORK · BASEL

Library of Congress Cataloging-in-Publication Data
A catalog record for this book is available from the Library of Congress.

ISBN: 0-8247-4274-5

This book is printed on acid-free paper.

Headquarters
Marcel Dekker, Inc., 270 Madison Avenue, New York, NY 10016, U.S.A.
tel: 212-696-9000; fax: 212-685-4540

Distribution and Customer Service
Marcel Dekker, Inc., Cimarron Road, Monticello, New York 12701, U.S.A.
tel: 800-228-1160; fax: 845-796-1772

Eastern Hemisphere Distribution
Marcel Dekker AG, Hutgasse 4, Postfach 812, CH-4001 Basel, Switzerland
tel: 41-61-260-6300; fax: 41-61-260-6333

World Wide Web
http://www.dekker.com

The publisher offers discounts on this book when ordered in bulk quantities. For more information, write to Special Sales/Professional Marketing at the headquarters address above.

Current printing (last digit):

10 9 8 7 6 5 4 3 2 1

PRINTED IN THE UNITED STATES OF AMERICA

ac

Preface

Chemical engineers develop, design, and operate processes that are vital to our society. Hardigg[*] states: "I consider engineering to be understandable by the general public by speaking about the four great ideas of engineering: structures, machines, networks, and processes." Processes are what distinguish chemical from other engineering disciplines. Nevertheless, designing chemical plants requires contributions from other branches of engineering. Before taking process design, students' thinking has been compartmentalized into several distinct subjects. Now, they must be trained to think more globally than before. This is not an easy transition. One of my students said that process design is a new way of thinking for him. I have found it informative to read employment ads to keep abreast of skills required of process engineers. An ad from General Dynamics[†] in San Diego, CA, states, "We are interested in chemical engineers with plant operations and/or process engineering experience because they develop the total process perspective and problem-solving skill we need."

The book is designed mostly for a senior course in process design. It could be used for entry-level process engineers in industry or for a refresher course. The book could also be used before learning to use process simulation software. Before enrolling in process design, the student must have some knowledge of chemical engineering prerequisites: mass and energy balances, thermodynamics, transport

[*] Hardigg, V, ASEE Prism, p.26, April 1999.
[†] Chemical and Engineering News, January 29, 1990.

phenomena, separator design, and reactor design. I encourage students to refer to their textbooks during their process design, but there is need for a single source, covering the essentials of these subjects. One reason for a single source is the turnover in instructors and texts. Besides, it is difficult to teach a course using several texts, even if the students are familiar with the texts. Another objective of a process design course is to fill the holes in their education. This book contains many examples. In many cases, the examples are familiar to the student. Sources of process-design case studies are: the American Institute of Chemical Engineers (AIChE) student contest problems; the Department of Chemical Engineering, Washington University, at St. Louis, Missouri; and my own experience.

I am fortunate to have worked with skilled engineers during my beginning years in chemical engineering. From them I learned to design, troubleshoot, and construct equipment. This experience gave me an appreciation of the mechanical details of equipment. Calculating equipment size is only the beginning. The next step is translating design calculations into equipment selection. For this task, process engineers must know what type and size of equipment are available. At the process design stage, the mechanical details should be considered. An example is seals, which impacts on safety. I have not attempted to include discussion of all possible equipment in my text. If I had, I would still be writing.

The book emphasizes approximate shortcut calculations needed for a preliminary design. For most of the calculations, a pocket calculator and mathematics software, such as Polymath, is sufficient. When the design reaches the final stages, requiring more exact designs, then process simulators must be used. Approximate, quick calculations have their use in industry for preparing proposals, for checking more exact calculations, and for sizing some equipment before completing the process design. In many example problems, the calculated size is rounded off to the next highest standard size. To reduce the completion time, the approach used is to purchase immediately equipment that has a long delivery time, such as pumps and compressors. Once the purchase has been made the rest of the process design is locked into the size of this equipment. Although any size equipment – within reason – could be built, it is less costly to select a standard size, which varies from manufacturer to manufacturer. Using approximate calculations is also an excellent way of introducing students to process design before they get bogged down in more complex calculations.

Units are always a problem for chemical engineers. It is unfortunate that the US has not converted completely from English units to SI (Système International) units. Many books have adopted SI units. Most equipment catalogs use English units. Companies having overseas operations and customers must use SI units. Thus, engineers must be fluent in both sets of units. It could be disastrous not to be fluent. I therefore decided to use both systems. In most cases, the book contains units in both systems, side-by-side. The appendix contains a discussion of SI units with a table of conversion factors.

Chapter 1, The Structure of Processes and Process Engineering, introduces the student to processes and the use of the flow diagram. The flow diagram is the way chemical engineers describe a process and communicate. This chapter contains some of the more common flow-diagram symbols. To reduce the complexity of the flow diagram, this chapter divides a process into nine process operations. There may be more than one process operation contained in a process unit (the equipment). This chapter also describes the chemical-engineering tasks required in a project.

Chapter 2, Production and Capital Cost Estimation, only contains the essentials of chemical-engineering economics. Many students learn other aspects of engineering economics in a separate course. Rather than placing this chapter later in the book, it is placed here to show the student how equipment influences the production cost. Chapter 2 describes cash flow and working capital in a corporation. This chapter also describes the components of the production cost and how to calculate this cost. Finally, this chapter describes the components of capital cost and outlines a procedure for calculating the cost. Most of the other chapters discuss equipment selection and sizing needed for capital cost estimation.

Chapter 3, Process-Circuit Analysis, first discusses the strategy of problem solving. Next, the chapter summarizes the relationships for solving design problems. The approach to problem solving followed throughout most of the book is to first list the appropriate design equations in a table for quick reference and checking. The numbering system for equations appearing in the text is to show the chapter number followed by the equation number. For example, Equation 5.7 means Equation 7 in Chapter 5. For equations listed in tables, the numbering system is to number the chapter, then the table and the equation. Thus, 3.8.12 would be Equation 12 in Table 8 and Chapter 3. Following this table another table outlines a calculating procedure. Then, the problem-sizing method is applied to four single-process units, and to a segment of a process consisting of several units.

Heat transfer is one of the more frequently-occurring process operations. Chapter 4, Process Heat Transfer, discusses shell-and-tube heat exchangers, and Chapter 7, Reactor Design, discusses jacket and coil heat exchangers. Chapter 4 describes how to select a heat-transfer fluid and a shell-and-tube heat-exchanger design. This chapter also shows how to make an estimate of heat-exchanger area and rate heat exchangers.

Transferring liquids and gases from one process unit to another is also a frequently occurring process operation. Heat exchangers and pumps are the most frequently used equipment in many processes. Chapter 5, Compressors, Pumps, and Turbines, discusses the two general types of machines, positive displacement and dynamic, for both liquids and gases. The discussion of pumps also could logically be included in Chapter 8, Design of Flow Systems. Instead, Chapter 5 includes pumps to emphasize the similarities in the design of pumps and compressors. This chapter shows how to calculate the power required for compressors and pumps. Chapter 5 also discusses electric motor and turbine drives for these machines.

Chapter 6, Separator Design, considers only the most common phase and component separators. Because plates and column packings are contained in vessels, this chapter starts with a brief discussion of the mechanical design of vessels. Although chemical engineers rarely design vessels, a working knowledge of the subject is needed to communicate with mechanical engineers. The phase separators considered are: gas-liquid, liquid-liquid, and solid-liquid. The common component separators are: fractionators, absorbers, and extractors. This chapter shows how to approximately calculate the length and diameter of separators. Flowrate fluctuations almost always occur in processes. To dampen these fluctuations requires installing accumulators at appropriate points in the process. Accumulators are sized by using a surge time (residence time) to calculate a surge volume. Frequently, a phase separator and a component separator include the surge volume. This chapter also discusses vortex formation in vessels and how to prevent it. Vortexes may form in a vessel, drawing a gas into the discharge line and forming a two-phase mixture. Then, the two-phase mixture flows into a pump, damaging the pump.

Chapter 7, Reactor Design, discusses continuous and batch stirred-tank reactors and the packed-bed catalytic reactor, which are frequently used. Heat exchangers for stirred-tank reactors described are the: simple jacket, simple jacket with a spiral baffle, simple jacket with agitation nozzles, partial pipe-coil jacket, dimple jacket, and the internal pipe coil. The amount of heat removed or added determines what jacket is selected. Other topics discussed are jacket pressure drop and mechanical considerations. Chapter 7 also describes methods for removing or adding heat in packed-bed catalytic reactors. Also considered are flow distribution methods to approach plug flow in packed beds.

Designing flow systems is a frequently occurring design problem confronted by the process engineer, both in a process and in research. Chapter 8 discusses selecting and sizing, piping, valves, and flow meters. Chapter 5 considered pump selection. Chapter 8 also describes pump sizing, using manufacturer's performance curves. Cavitation in pumps is a frequently occurring problem and this chapter also discusses how to avoid it. After completing the chapter, the students work on a two week problem selecting and sizing control valves and a pump from manufacturers' literature. Many of these problems are drawn from industrial experience.

Most things in life are not possible without the help of others. I am grateful to the following individuals:

the many students who used my class notes during the development of the senior course in process design, and who critiqued my class notes by the questions they asked

Otto Frank, formally Process Supervisor at Allied Signal Co., Morristown, NJ, who critiqued a draft of my book from an industrial point of view.

Prof. Deran Hanesian, Prof. of Chemical Engineering at New Jersey Institute of Technology, Newark, NJ, who also critiqued the draft but from an academic point of view

Charles Bambara, Director of Technology, Koch-Otto York Co., Parsippany, NJ, who contributed many flow-system design problems

 My wife, Christiane Silla, who guided me through the graphics software, Adobe Photoshop and Adobe Illustrator, and drew or edited many of the illustrations

and to BJ Clark, Executive Acquisitions Editor, for his help in the review process and Brian Black and Erin Nihill, Production Editors, who guided the book through the production process.

Harry Silla

Contents

CHEMICAL PROCESS
ENGINEERING

1

The Structure of Processes and Process Engineering

The activities of most engineering disciplines are easily identifiable by the public, but the activities of chemical engineers are less understood. The public recognizes that the chemical engineer is somehow associated with the production of chemicals, but often does not know the difference between chemists and chemical engineers. What is the distinguishing feature of chemical engineering? Briefly, chemical engineering is the development, design, and operation of various kinds of processes. Most chemical engineering activities, in one way or another, are process oriented.

The chemical engineer may work in three types of organizations. One is the operating company, such as DuPont and Dow Chemical, whose main concern is to produce products. These companies are also engaged in developing new processes. If a new plant for an old improved process, or a plant for a recently developed process is being considered, a plant construction organization, the second company type, such as the C.E. Lummus Corp. or the Forster Wheeler Corp., will

Table 1.1 Selected Process Types

Process	Example
1. Chemical Intermediaries	Ethylene
2. Energy	Gasoline
3. Food	Bread
4. Food Additive	Vitamin C
5. Waste Treatment	Activated Sludge Process
6. Pharmaceutical	Aspirin
7. Materials	
a) Polymer	Polyethylene
b) Metallurgical	Steel
8. Personal Products	Lipstick
9. Explosives	Nitrocellulose
10. Fertilizers	Urea

be contacted. Finally, numerous small and large companies support the activities of the operating and plant construction companies by providing consulting services and by manufacturing equipment such as pumps, heat exchangers, and distillation columns. Because many companies are involved in more than one activity, classifying them may be difficult.

PROCESS TYPES

There are numerous types of processes and any attempt to classify processes will meet difficulties. Nevertheless, attempts at classification should be made to achieve a better understanding of the process industries. Wei, et al. [1] discuss the structure of the chemical process industries. A classification is also given by Chemical Engineering magazine, and the North American Industry Classification System (NAICS) is provided by the U.S. Bureau of Budget. A selected list of process types, classified according to the product type, is given in Table 1.1, illustrating the variety and diversity of processes.

Chemical intermediates are listed first in Table 1.1. These are the chemicals that are used to synthesize other chemicals, and are generally not sold to the public. For example, ethlyene is an intermediate produced from hydrocarbons by cracking natural gas derived ethane or petroleum derived gas oil, either thermally using steam or catalytically. Ethlyene is then used to produce polyethylene (45%), a polymer; and ethylene oxide (10%), vinyl chloride (15%), styrene (10%), and other uses (20%) [2]. The number of chemicals that are classified as intermediates is considerable.

Examples of energy processes are the production of fuels from petroleum or electricity in a steam power plant. A steam power plant is not ordinarily considered a process, but, nevertheless, it is a special case of a process. The plant contains a combustion reactor, the furnace; pumps; fans; heat exchangers; a water treatment facility, consisting of separation and purification steps; and most likely flue gas treatment to remove particulates and sulfur dioxide. Because of the mechanical and electrical equipment used, mainly mechanical and electrical engineers operate power plants. However, all chemical plants contain more or less mechanical and electrical equipment. For example, the methanol–synthesis process, discussed later, contains steam turbines for energy recovery. Chemical engineers have the necessary background to work in power plants as well, complementing the skills of both mechanical and electrical engineers.

Bread making, an example of a food process, is almost entirely mechanical, but it also contains fermentation steps where flour is converted into bread by yeast [3]. Thus, this process can also be classified as a biochemical process. Another well known biochemical process that removes organic matter in both municipal and industrial wastewater streams is the activated sludge process. In this process, microorganisms feed on organic pollutants, converting them into carbon dioxide, water, and new microorganisms. The microorganisms are then separated from most of the water. Some of the microorganisms are recycled to sustain the process, and the rest is disposed of.

Aspirin, one of the oldest pharmceutical products, has been produced for over a hundred of years [4]. A chemist, Felix Hoffmann, who worked for the Bayer Co. in Elberfeld, Germany, discovered aspirin. He was searching for a medication for pain relief for his father who suffered from the pain of rheumatism. Besides pain relief, physicians have recently found that aspirin helps prevent heart attacks and strokes.

Vitamin C, classified as either a pharmaceutical [5] or a food additive [6], has annual sales of 325 million dollars, the largest of all pharmaceuticals produced [7]. Pharmaceuticals, in general, lead in profitability for all industries [6]. Although vitamin C can be extracted from natural sources, it is primarily synthesized. In fact, it was the first vitamin to be produced in commercial quantities [6]. Jaffe [8] outlines the synthesis. Starting with D-glucose, vitamin C is produced in five chemical steps, one of which is a biochemical oxidation using the bacterium Acetobacter suboxydans. D-glucose is obtained from cornstarch in a process, which will be described later.

The personal products industries, which also includes toiletries, is a large industry, accounting for $10.6 billion in sales in the United States in 1983 [9]. The operation required for manufacturing cosmetics is mainly the mixing of various ingredients such as emollients (softening and smoothing agents), surfactants, solvents, thickeners, humectants (moistening agents), preservatives, perfumes, colors, flavors and other special additives.

Over a period of many years polymeric materials have gradually replaced metals in many applications. Among the five leading thermoplastics; low and high density polyethylene, polyvinyl chloride, polypropylene, and polystyrene; polyethylene is the largest volume plastic in the world. Polyethylene was initially made in the United States in 1943. In 1997, the estimated combined worldwide production of both low and high-density polyethylene was 1.230×10^{10} kg (2.712×10^{10} 1b) [10]. Low density polyethylene is produced at pressures of 1030 to 3450 bar (1020 to 3400 atm) whereas high density polyethylene is produced at pressures of 103 to 345 bar (102 to 340 atm) [11].

Explosives are most noted for their military, rather than civilian uses, but they are also a valuable tool for man in construction and mining. Interestingly, as described by Mark [12], the first synthetic polymer, although it is only partially synthetic, was nitrocellulose or guncotton, a base for smokeless powder. Nitrocellulose was discovered accidentally in 1846 when a Swiss chemist, Christian Schoenbein, wiped a spilled mixture of sulfuric and nitric acids using his wife's cotton apron. After washing the apron, he attempted to dry it in front of a strove, but instead the apron burst into flames. Although the first application of modified cellulose was in explosives, it was subsequently found that cellulose could be chemically modified to make it soluble, moldable, and also castable into film, which was important in the development of photography. Nitrocellulose is still used today as an ingredient in gunpowder and solid propellants for rockets.

Nitrogen is an essential element for life, required for synthesizing proteins and other biological molecules. Although the earth's atmosphere contains 79% nitrogen, it is a relatively inert gas and therefore not readily available to plants and animals. Nitrogen must be "fixed", i.e., combined in some compound that can be more readily absorbed by plants. The natural supply of fixed nitrogen is limited, and it is consumed faster than it is produced. This led to a prediction of an eventual world famine until 1909 in Germany, when Badische Anilin and Soda Fabrik (BASF) initiated the development of a process for ammonia synthesis [13]. In 1910, the United States issued a patent to Haber and Le Rossignol of BASF for their process [14]. The first plant was started up in 1913 in Ludwigshafen, Germany, expanded in the 1960's, and only shut down in 1982 after seventy years of production [15]. This is certainly an outstanding engineering achievement. Although the fixed nitrogen supply is no longer limited by production from natural sources, they are still major sources. Agricultural land produces 38%; forested or unused land, 25%; combustion, resulting in air pollution, 9%; lightning, 4%; and industrial fixation, 24% [16]. The oceans produce an unknown amount.

Processes could be subdivided according to the type of reaction occurring, as illustrated by bread making and the activated sludge process, by also classifying them as biochemical processes. Similarly, we could also have electrochemical, photochemical, and thermochemical processes and so on, but this subclassification could lead to difficulties because in some processes more than one type of reaction occurs, such as in the vitamin C process.

CHEMICAL ENGINEERING ACTIVITIES

It is useful to delineate the various activities of a chemical engineer, from the conception of a project to its final implementation. Companies will assign a variety of job titles to these activities. In some companies, these activities will be subdivided, but in other companies many activities may be included under one job title, according to company policy. In this discussion, the engineering activity is of more concern than any particular job title assigned by a company. We will use the most frequently employed job title, keeping in mind that any particular company must be consulted for its definition of the job.

A project is initiated by determining if there is a market for a product, which may be a chemical, a processed food, a metal, a polymer or one of the many other products produced by the process industries. For example, a chemist first synthesizes a new drug in the laboratory, which after many tests is approved by the Food and Drug Administration (FDA) of the federal government. Then, chemical engineers develop and design the process for producing the drug in large quantities. The steps required to accomplish this task are outlined in Table 1.2. Under some circumstances, where knowledge of the process is highly developed and sufficient data exists, the research or pilot phase of the process, or both, may be omitted. In order to cover all aspects of a project, we will assume that a new chemical, which is marketable, has just been synthesized in the laboratory by a chemist.

Next, the technical, economic, and financial feasibility of proposed processes must be demonstrated. Unless the project shows considerable promise when matched against other potential projects, it may be abandoned. Any particular company will have several projects to invest in but limited financial resources so that only the most promising projects will be continued. The research engineer should estimate the capital investment required and the production cost of the product. No matter how crude or incomplete the process data may be, the research engineer must estimate the profitability of the process to determine if further process development is economically worth the effort. This analysis will also uncover those areas requiring further research to obtain more information for a more accurate economic evaluation.

If the project analysis shows sufficient uncertainty or the need for design data, the research engineer will plan experiments, design an experimental setup and correlate the resulting data. After completing the experiments, the research engineer, or more likely a cost engineer, revises the flow diagram and reevaluates the project. Again, he must show that the project is still economically feasible.

After completion of the research phase, it is usually found that further demonstration of the viability of the process and more design data is needed, but under conditions that will more closely resemble the final plant. It may also be required to obtain some product for market research. In this case, the development engineer will plan the development program and design the pilot plant. Whenever possible the equipment selected will be smaller versions of the plant size equipment, using the same materials of construction selected for the plant.

Table 1.2 Structure of a Project

Process Research
1. Process Evaluation
The objective is to evaluate the technical, economic, and financial feasibility of a process.
    a) Construct a preliminary process flow diagram
    b) Approximate equipment sizing
    c) Economic evaluation
    d) Locate areas requiring research
2. Bench Scale Studies
    The objective is to obtain additional design data for process evaluation.

a) Plan experiments	d) Revise flow diagram
b) Design experimental setup	e) Revise economic evaluation
c) Correlate data	f) Locate areas requiring development

Process Development
Objective: To obtain more design data and possibly product for market research.

a) Plan development program	e) Correlate data
b) Design pilot plant	f) Revise flow diagram
c) Supervise pilot-plant construction	g) Revise economic evaluation
d) Supervise pilot-plant operations	

Process Design
Objective: To establish process and equipment specifications.

a) Construct flow diagram	f) Conduct economic studies
b) Perform mass and energy balances	g) Conduct optimization studies
c) Consider alternative process designs	h) Evaluate safety and health
d) Size equipment	i) Conduct environmental impact
e) Design control systems	studies

Plant Design and Construction
Objective: To implement the process design.
    a) Specify equipment
    b) Design vessels (mechanical design of reactors, separators, tanks)
    c) Design structures
    d) Design process piping system
    e) Design data acquisition and control system
    f) Design electric-power distribution system
    g) Design steam-distribution system
    h) Design cooling-water distribution system
    i) Purchase equipment
    j) Coordinate and schedule project
    k) Monitor progress

Plant Operations
Objective: To produce the product.

a) Plant startup	d) Production
b) Trouble shooting	e) Plant engineering
c) Process improvement	

Marketing
Objective: To sell the product.
    a) Market research
    b) Product sales
    c) Technical customer service
    d) Product development

At the end of the pilot-scale tests, the process is again evaluated, but since the process-design phase of the project will require a substantial increase in capital investment, the calculations require improved accuracy. Table 1.2 lists the activities of the process-design engineer. Usually, there are several technically acceptable alternatives available for each process unit, so that the process-design engineer will have to evaluate these alternatives to determine the most economical design. Additionally, each process unit can operate successfully under a variety of conditions so that the engineer must conduct studies to determine the economically-optimum operating conditions. It is clear from the foregoing discussion that economics determines the direction taken at each phase of the project. Consequently, process economics will be discussed in the next chapter. It can also be seen from Table 1.2 that there are several social aspects of the process design that must be considered. The effects of any possible emissions on the health of the workers, the surrounding community, and the environment must be evaluated. Even aesthetics will have to be considered to a greater extent than has been done in the past.

The next phase of the project is plant design and construction, which employs a variety of engineering skills, mainly mechanical, civil, and electrical. The objective in this phase of the project is to implement the process design. Table 1.2 outlines the major activities of this phase. Most likely a plant design and construction company will conduct this phase of the project, commonly called outsourcing.

After the plant is constructed, the operations phase of the project begins, which includes plant startup. Rarely does this operation proceed smoothly. Troubleshooting, process modifications, and repairs are generally required.

Because of the need to get the plant on-stream as soon as possible, the process design, plant design, plant construction and plant startup must be completed as rapidly as possible. Electrical, mechanical or chemical systems, as well as any human activity need to be controlled or regulated to approach optimum performance. Similarly, project management, or more appropriately project control, is needed because of the complexity of process and plant design, and construction. Numerous activities must be scheduled, coordinated and progress monitored to complete the project on time. It is the responsibility of the project engineer to plan and control all activities so that the plant is brought on-stream quickly. It is poor planning to complete the tasks sequentially, i.e., completing one task before starting another task. To reduce the time from the initiation of a project to routine plant operation, the strategy is to conduct as many parallel activities as possible. Thus, as many tasks as possible are conducted simultaneously. This strategy, illustrated in Figure 1.1, shows that detailed plant design starts before completing the process design, construction before completing the plant design, and finally, startup begins

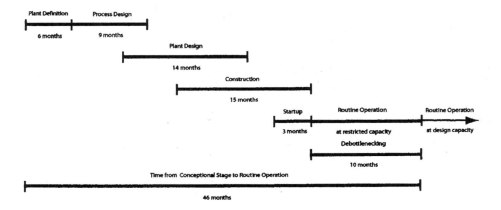

Figure 1.1 Sample of a process and plant-design schedule.
Source: Ref. 17, with permission.

before completing plant construction. Usually, from the start to the time a plant reaches design capacity may take anywhere from three to four years. [17].

Even after the plant has been successfully started, it will need constant attention to keep it operating smoothly and to improve its operation. This is the responsibility of the process engineer. Many of the skills that were used by the process-design engineer are also utilized by the process engineer. A major activity of the process engineer is the "debottlenecking" study to increase plant capacity, in which the process is analyzed to determine what process unit limits the plant capacity. When this unit is located, the process engineer will consider alternative designs for increasing plant capacity.

PROCESS DESIGN

Our main goal is to develop techniques for solving problems in process design. Process design generally proceeds in the following stages:

1. Developing process flow diagrams
2. Process circuit analysis
3. Sizing process units
4. Estimating production cost and profitability

Chemical engineers express their ideas by first constructing a process flow diagram to describe the logic of the process. At an early stage of the process design, several flow diagrams are drawn to illustrate process alternatives. Following this initial stage, a preliminary screening will reduce the many alternatives to a few of the most promising, which are studied in detail. Process-circuit analysis, which establishes specifications for the process, will be the subject of a later chapter. These specifications are quantities, such as flow rates, compositions, temperatures, pressures, and energy requirements. Once the process specifications are established, each process unit is sized. At the beginning of a process design, simple sizing procedures are sufficient to determine a preliminary production cost. In fact, it may be poor strategy to use more exact, and therefore more costly design procedures until the economics of the process demands it. The process design engineer will have a number of design procedures available, each one differing in accuracy. He will have to decide which procedure is the more appropriate one for the moment. To determine the economic viability of a process, the product manufacturing and capital costs are estimated first. Using simplified cost estimating techniques, the most costly process steps are located for a more detailed analysis.

The steps in a process design, listed above, do not have well defined boundaries, but overlap. New information is fed back continuously, requiring revision of previous calculations. Process design is a large-scale iterative calculation which terminates on a specified completion date.

PROCESS STRUCTURE

Because of the numerous process types, it is essential to be able to divide a process into a minimum number of basic logical operations to aid in the understanding of existing processes and in the development and design of new processes. The electrical engineer designs electrical circuits consisting of transistors, resistors, capacitors and other basic elements. Similarly, the chemical engineer designs process circuits consisting of reactors, separators, and other process units. Early in the development of chemical engineering the concept of unit operations and processes evolved to isolate the basic elements of a process. Unit operations consist of physical changes, such as distillation and heat transfer, and unit processes consist of chemical changes, such as nitration and oxidation. Thus, any process consists of a combination of unit operations and processes. Trescott [18] discusses the history of this concept.

A modification of the unit-operations, unit-process division is shown in Table 1.3, where a process is divided into nine basic process operations. According to this division, the unit operations are subdivided into several basic operations and conversion is substituted for all unit processes for a total of nine process

Table 1.3 Basic Process Operations

___

1. Conversion

Thermochemical
Biochemical
Electrochemical
Photochemical
Plasma
Sonochemical

2. Separations

Component (Examples)	Phase(Examples)
Distillation	Gas-Liquid
Absorption	Gas-Solid
Extraction	Liquid-Liquid
Adsorption	Liquid-Solid

3. Mixing

Component	Phase (Examples)
Dissolving	Gas-Liquid
	Gas-Solid
	Liquid-Liquid
	Liquid-Solid
	Solid-Solid

4. Material Transfer
Pumping Liquids
Compressing Gases
Conveying Solids

5. Energy Transfer
Expansion
Heat Exchange

6. Storage
Raw Materials
Internal
Products

7. Size reduction

8. Agglomeration

9. Size Separation

___

operations. The nine basic process operations will be discussed separately. More than one process operation can occur in a single piece-of-equipment, which is called a process unit.

Conversion of material from one form to another is a task of the chemical engineer. Table 1.3 lists a number of ways conversion can be accomplished, depending on what form of energy is supplied to the reactor. The most common form of energy is heat to carry out a reaction thermochemically.

Rarely do the reaction products have an acceptable degree of purity. Thus, separators are necessary process units. Together, conversion and separation constitute the heart of chemical engineering. In turn, separations consist of two parts, component and phase. In component separations, the components in a single phase are separated, usually by the introduction of a second phase. Molecules of different substances can be separated because their chemical potential in one phase differs from their chemical potential in a second phase. Thus, separation occurs by mass transfer, whereas phases separate because a force acting on one phase differs from a force acting on the other phase. Usually, it is a gravitational force. Examples are sedimentation and clarification, where a solid settles by the gravitational force acting on the solid. Generally, phase separation follows component separation. For example, in distillation vapor and liquid phases mix on a tray where component separation occurs, but droplets and possibly foam form. Then, the vapor is separated from the liquid drops and foam, by allowing sufficient tray spacing and time, for small drops to coalesce into large drops and the foam to collapse. The large drops and collapsing foam then settle on the tray by gravity.

Mixing, the reverse of component and phase separation also occurs frequently in processes. This operation requires energy to mix the two phases. For example, in liquid-liquid extraction, one of the liquid phases must be dispersed into small drops by mixing to enhance mass transfer and increase the rate of component separation. Thus, extractors must contain a method for dispersing one of the phases.

Material is transferred from one process operation to another by compression, pumping or conveying; depending on whether a gas, liquid or a solid is transferred. This operation also requires energy to overcome frictional losses.

Many of the process operations listed in Table 1.3 require an energy input. Energy must be supplied to the process streams to separate components and to obtain favorable operating temperatures and pressures. For example, it may be necessary to compress a mixture of gases to achieve a reasonable chemical conversion. This work is potentially recoverable by expanding the reacted gases through a turbine when the system pressure is eventually reduced downstream of the reactor. Similarly, a high-pressure liquid stream could be expanded through a hydraulic turbine to recover energy. Heat transfer and expansion of a gas or liquid through a turbine are energy transfer operations. In addition to elevating the gas pressure to obtain favorable reaction conditions, gases are also transferred from a previous process unit to the reactor. This material transfer operation requires work to overcome frictional losses. Both the material and energy transfer operations are combined and only one compressor is used. If the conversion is less than 100%, a recycle compressor will transfer the unreacted gases back to the reactor after separating out the products. Since the recycled gases are already at a high pressure, but at a lower pressure than at the reactor inlet because of frictional pressure losses, a compressor is needed to recompress the gases to the reactor inlet pressure. This step would be considered primarily material transfer.

Because raw-material delivery cannot be accurately predicated, on account of unforeseen events such as bad weather, strikes, accidents, etc., storage of raw materials is a necessity. Similarly, the demand for products can be unpredictable. Also, internal storage of chemical intermediates may be required to maintain steady operation of a process containing batch operations or to store chemical intermediates temporarily if downstream equipment fails. Production can continue when repairs are completed.

The last three process operations; size reduction, agglomeration, and size separation; pertain to solids. Examples of size reduction are grinding and shredding. An example of agglomeration is compression of powders to form tablets. Screening to sort out oversized particles is an example of size separation.

The first step in the synthesis, or development and design of a process, is to construct a flow diagram, starting with raw materials and ending with the finished product. The flow diagram is a basic tool of a chemical engineer to organize his thinking and to communicate with other chemical engineers. A selected list of flow-diagram symbols for the process operations discussed above are given in Figure 1.2. Other symbols are given by Ulrich [19] and by Hill [20] and have been collected and reviewed by Austin [21]. The various process operations discussed above, using the flow-diagram symbols in Figure 1.2, are used to describe a process for producing glucose from cornstarch, which is illustrated in Example 1.1.

Example 1.1 Glucose Production from Corn Starch

A process flow diagram for the production of glucose is shown in Figure 3. Identify each process unit according to the process operations listed in Table 3.

Although glucose could be obtained from many different natural sources, such as from various fruits, it is primarily obtained by hydrolysis of corn starch, which contains about 61% starch. Starch is a polymer consisting of glucose units combined to form either a linear polymer called amylose, containing 300 to 500 glucose units, or a branched polymer called amylopectin, containing about 10,000 glucose units. Glucose is a crystalline white solid, which exists in three isomeric forms: anhydrous α-D-glucose, α-D-glucose monohydrate and anhydrous β-D-glucose. Most of the glucose produced is used in baked goods and in confectionery as a sweetener. It is sold under the trivial name of dextrose, which has evolved to mean anhydrous α-D-glucose and α-D-glucose monohydrate.

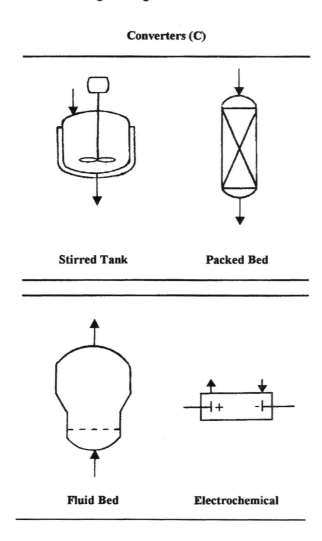

Figure 1.2 Flow-diagram symbols.

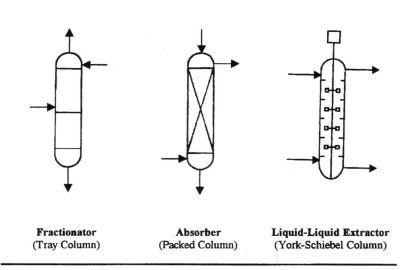

Figure 1.2 Continued.

Phase Separators (PS)

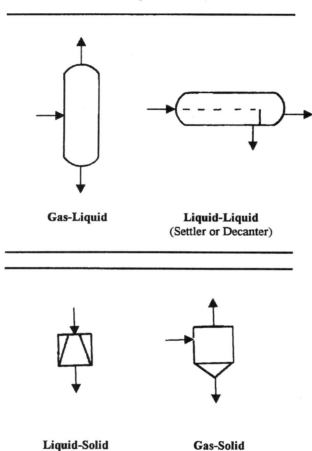

Gas-Liquid

Liquid-Liquid
(Settler or Decanter)

Liquid-Solid
(Centrifuge)

Gas-Solid
(Cyclone)

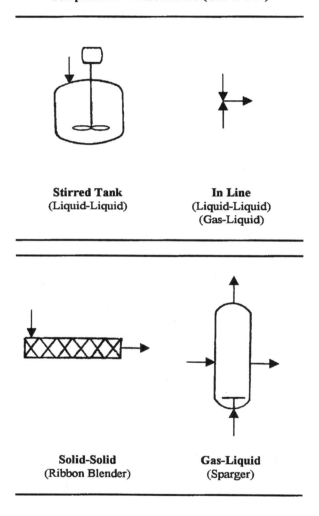

Component or Phase Mixers (CM or PM)

Stirred Tank
(Liquid-Liquid)

In Line
(Liquid-Liquid)
(Gas-Liquid)

Solid-Solid
(Ribbon Blender)

Gas-Liquid
(Sparger)

Figure 1.2 Continued.

Material and Energy Transfer

Pumps (P)

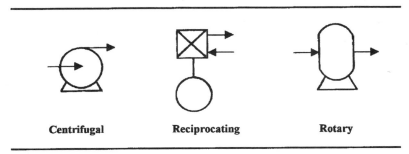

| Centrifugal | Reciprocating | Rotary |

Fans (F)

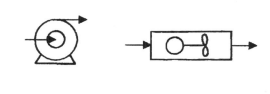

| Centrifugal | Axial |

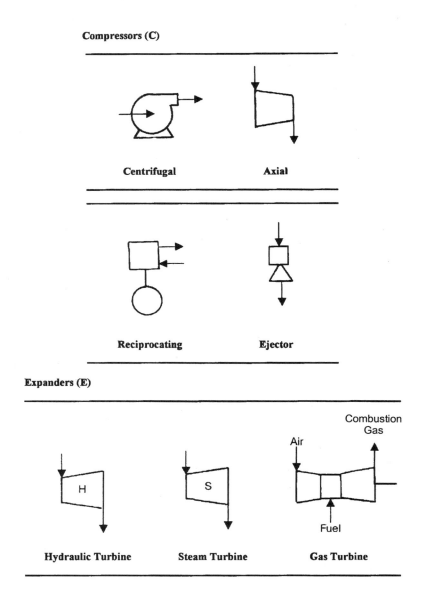

Figure 1.2 Continued.

Heat Exchangers (H)

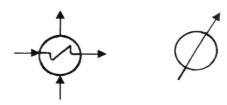

Interchanger   **Cooler**

                    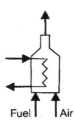

Fuel   Air

Heater   **Heater or Cooler**   **Furnace**

Storage (S)

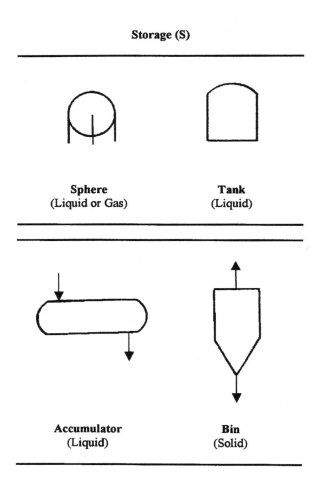

Sphere
(Liquid or Gas)

Tank
(Liquid)

Accumulator
(Liquid)

Bin
(Solid)

Figure 1.2 Continued.

Size Reducers (SR)

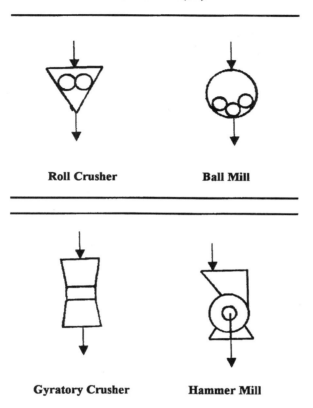

Roll Crusher   **Ball Mill**

Gyratory Crusher   **Hammer Mill**

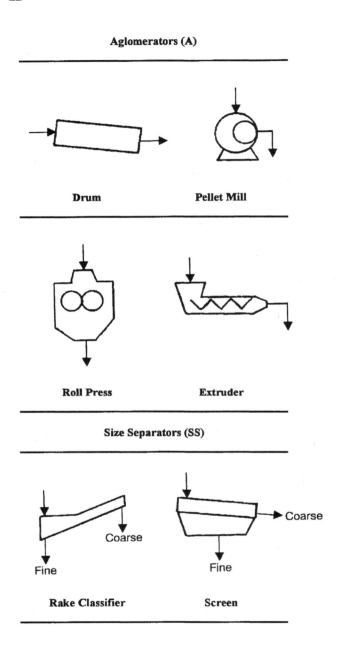

Figure 1.2 Continued.

Figure 1.3 shows the process flow diagram for converting starch into glucose. Table 1.4 identifies the basic process operations in the process, according to those given in Table 1.3. Sinclair [22] describes the process but it has been modified after discussion with Leiser [23]. Harness [24] describes the corn wet-milling process for producing a corn-starch slurry containing 30 to 40% solids, which flows to the first hydrolyzer, R-1. The first hydrolyzer converts 15 to 25% of the starch into glucose using alpha-amylase, an enzyme, which catalyzes the hydrolysis. Two process operations occur in the hydrolyzer – conversion and mixing – but the main purpose of the process unit is conversion. After hydrolysis the viscosity of the slurry is reduced. The centrifuge, PS-1, removes any residual oil and proteins, which were not removed in the corn wet-milling process. This is a phase-separation operation. The oil and protein will be processed to make animal feed.

The second hydrolyzer, R-2, completes the hydrolysis using glucoamylase, another enzyme. The reduction in viscosity of the starch slurry in R-1 aids in the mixing of glucoamylase and prevents the formation of a unhydrolyzable gelatinous material in R-2. Most of the remaining starch is hyrolyzed to glucose in 48 to 72 h in a batch operation. Aspergillus phoenicis, a mold, produces the glucoamylase enzyme in a fermentation process. The overall conversion of starch in this two-step hydrolysis is almost 100%. The effluent from R-1 is cooled by preheating the feed stream to R-1, which is an energy transfer operation. After the second stage of hydrolysis, the solution is decolorized in an adsorber, CS-1, packed with carbon. Because the hydrolysis is a batch operation, internal storage, S-1, of the solution is required to keep the next step of the process operating continuously.

After converting the starch into glucose, the rest of the process removes water from the glucose to obtain a dry product. The solution is pumped from storage to the first of three stages of evaporation (called effects) where some water is removed. To conserve steam and therefore energy, the first evaporator employs mechanical recompression of the water vapor evolved from the evaporation. Compressing the vapor elevates its temperature above the boiling point of the solution in CS-2 so that heat can be transferred to the boiling solution. Also, because the glucose is heat sensitive, the evaporation is carried out in a vacuum produced by the vacuum pump C-1. Each stage of evaporation is carried out in two steps. In the first step, a component-separation operation, energy is transferred to the solution in a boiler to evaporate some water, concentrating the glucose. Thus, the boiler is a component separator. In the second step, vapor and liquid are separated in a phase separator. After the first stage of evaporation, the solution is again decolorized in the adsorber, CS-3, and the small amounts of organic acids are removed in an ion exchanger. The ion exchanger, R-3, replaces anions with hydrogen ions and cations with hydroxyl ions, and thus the net effect is to replace the organic acids with water. Although the operation is a chemical reaction, the overall process is a separation because the ion exchanger is eventually regenerated and reused.

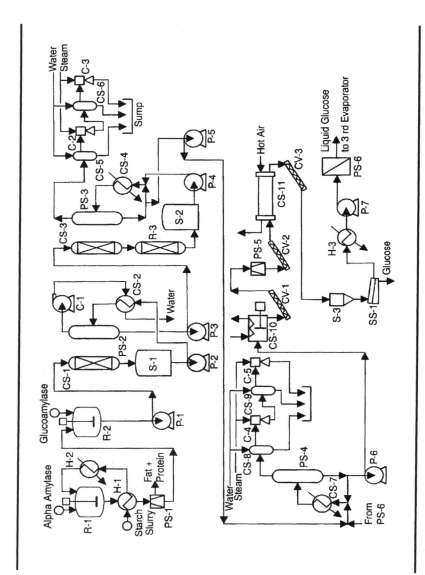

Figure 1.3 Glucose-process flow diagram.

Table 1.4 Glucose Production Process Operations

Process Unit	Process Conditions	Process Operations
Hydrolyzer, R-1	Time – 2 h Feed – 30 - 40% solids Temperature – 80 - 90 °C pH – 5.5 - 7.0 Hydrolysis – 15 - 25 %	Conversion Mixing
Interchanger, H-1		Energy Transfer
Heater, H-2		Energy Transfer
Centrifuge, PS-1	Solids Removed 0.3 - 0.4 % protein 0.5 - 0.6 % fat	Phase Separation
Hydrolyzer, R-2	Time – 48 - 72 h Temperature – 55 - 60 °C pH – 4.0 - 4.5 Dissolved Solids – 97.0 - 98.5 % glucose	Conversion Mixing
Pump, P-1		Material Transfer
Adsorber, CS-1		Component Separation
Tank, S-1		Storage
Pump, P-2		Material Transfer
Evaporators (contains three effects or stages)		
1st Effect 2nd Effect 3rd Effect	Product – 40 - 58% solids Product – 58 - 70% solids Product – 70 - 78% solids	
First Effect Evaporator, CS-2 Flash Drum, PS-2 Compressor, C-1 Pump, P-3 Adsorber, CS-3 Ion Exchanger, R-3 Tank, S-2		Component Separation Phase Separation Material Transfer Material Transfer Component Separation Conversion Storage

Second Effect Pump, P-4 Evaporator, CS-2 Flash Drum, PS-3 Barometric Condensers 2 stages, CS-5, CS-6 Steam Jet Ejectors 2 stages, C-2, C-3		Material Transfer Component Separation Phase Separation Component Separation & Phase Separation Material Transfer & Mixing
Third Effect		Same as Second Effect
Crystallizer, CS-10	Seed crystals – 20 - 25% of the batch Temperature – from 43 - 46 °C to 20 - 39 °C Time – 2 days Yield – 60% crystals	Component Separation
Conveyor, CV-1		Material Transfer
Centrifuge, PS-5	Product – 14% H_2O	Phase Separation
Conveyor, CV-2		Material Transfer
Rotary Dryer, CS-11		Component Separation
Conveyor, CV-3		Material Transfer
Bin, S-3		Storage
Melter, H-3		Energy Transfer
Pump, P-7		Material Transfer
Pressure Filter, PS-6		Phase Separation

The next two stages of evaporation are carried out in a vacuum produced by a two-stage steam ejector. The water vapor from the phase separator is first condensed by direct contact with cold water in the barometric condensers, C-5 and C-6. Each condenser contains a long pipe, where the condensate accumulates until the static pressure becomes great enough for the water to flow out of the condenser. Effectively, the barometric condenser is a pump. The remaining water vapor and non-condensable gases – from the gases dissolved in the feed solution, in the cooling water, and the air leaking into the system – are compressed to the pressure of the next stage by a steam-jet ejector before being condensed and compressed again. This operation is material transfer because the main purpose is to transfer the non-condensable gases and the remaining vapor to the atmosphere.

After the evaporation is complete, the glucose solution could be sold as a syrup or processed further to obtain powdered α-D-glucose monohydrate. To obtain the powder, the glucose is separated from the solution in horizontal cylindrical crystallizers by cooling and slowly mixing at 1.5 rpm. The concentrated solution is seeded with glucose crystals to promote crystallization. Approximately, 60% of the dextrose in the solution crystallizes as the monohydrate. After two days, the slurry is transferred by a screw conveyor, MT-1, to a perforated-screen centrifuge where the solution is partially separated from the crystals. The wet crystals, containing 14% water, are then conveyed to a rotary dryer to remove the remaining water. In this particular case, component separation occurs because water is being removed from the sugar solution that adheres to the crystals. As the water evaporates further crystallization of the glucose dissolved in the solution occurs. If water were removed from a insoluble solid by drying, such as from wet sand, then the operation is a phase separation.

The powdered glucose from the drier contains some oversized crystals, which must be removed to obtain a more marketable product of fine crystals. The oversized crystals are separated by the screen, SS-1, a size-separator. When removing a small amount of oversized crystals (less than 5%) from a feed, which consists predominately of fines, the operation is called "scalping". The oversized crystals are recovered by first melting and then pumping the liquid through a leaf filter to remove any insoluble material that has been carried through the process. After filtering, the liquid is recycled back to the evaporators for reprocessing.

REFERENCES

1. Wei, J., Russel, T.W.F., Swartzlander, T.W., The Structure of the Chemical Processing Industries, McGraw-Hill, New York, NY, 1979.
2. Greek, B.F., Petrochemicals Inch Toward Recovery, Chem. & Eng. News, p. 18, Nov. 22, 1982.
3. Matz, S.A., Modern Baking Technology, Sci. Am., 251, 5, 122, 1984.
4. Reisch, M., Aspirin is 100 Years Old, Chemical & Eng. News, p.12, Aug. 18, 1997.
5. Stinson, S.C., Bulk Drug Output Moves Outside U.S., Chem. & Eng. News, p. 25, Sept. 16, 1985.
6. Thayer, A.M., Use of Specialty Food Additives to Continue to Grow, Chem. & Eng. News, p. 25, June 3, 1991.
7. Stinson, S.C., Custom Synthesis Expanding for Drugs and Intermediates, Chem. & Eng. News, p. 25, Aug. 20, 1984.

8. Jaffe, G.M., Ascorbic Acid, Kirk-Othmer Encyclopedia of Chemical Technology, 3rd ed., H.F. Mark, D.F. Othmer, C.G. Overberger, G.T. Seaborg, eds., Vol. 24, p.8., John Wiley & Sons, New York, NY, 1998.
9. Layman, P.L., Cosmetics, Chem. & Eng. News, p.19, Apr. 29, 1985
10. Anonymous, Facts and Figures for the Chemical Industry, Chem. & Eng. News, p.40, June 29, 1998.
11. Albright, L.F., Polymerization of Ethylene, Chem. Eng., 73, 24, 127, 1966.
12. Mark, H.F., The Development of Plastics, Am. Sci., 72, 2, 156, 1984.
13. Sterzloff, S., Technology and Manufacture of Ammonia, John Wiley & Sons, New York, NY, 1981.
14. Haber, F., Le Rossignol, R., Production of Ammonia, US Patent 971, 501, Sept. 27, 1910.
15. Anonymous, Chementator, Chem. Eng., 89, 18, 17, 1982.
16. Anonymous, Worldwide Nitrogen Fixation Estimated, Chem. & Eng. News, p.34, 1976.
17. Fulks, B.D., Planning and Organizing for Less Troublesome Plant Start-ups, Chem. Eng., 89, 18, 96, 1982.
18. Trescott, M.M., Unit Operations in the Chemical industry: An American Innovation in Modern Chemical Engineering, A Century of Chemical Engineering, W.F. Furter, Ed., Plenum Press, New York, NY, 1982.
19. Ulrich, G.D., A Guide to Chemical Engineering Process Design and Economics, John Wiley & Sons, New York, NY, 1984.
20. Hill, R.G., Drawing Effective Flowsheet Symbols, Chem. Eng., 75, 1, 84, 1968.
21. Austin, D.G., Chemical Engineering Drawing Symbols, John Wiley & Sons, New York, NY, 1979.
22. Sinclair, P.M., Enzymes Convert Starch to Dextrose, Chem. Eng., 72, 18, 90, 1965.
23. Leiser, R., Personal Communication, A.E. Staley Manufacturing Co., Decatur, Il, Mar. 25, 1986.
24. Harness, J., Corn Wet Milling Industry in 1978, Product of the Corn Refining Industry in Food, Corn Refiners Association, Washington, DC, 1978.

2

Production and Capital Cost Estimation

Before initiating the development of a process, at various stages in its development, and before attempting the design of a process and plant, process engineers must make economic evaluations. The evaluation determines whether they should undertake a project, abandon it, continue with it (but with further research), or take it to the pilot plant stage. If they decide to proceed with process development, an economic evaluation will pinpoint those parts of the process requiring additional study. Winter [1] has stated that the economic evaluation of a project is a continuous procedure. As the process engineer gathers new information, he can make a more accurate evaluation followed by a reexamination of the project to determine if it should continue.

Even if insufficient technical information is available to design a plant completely, we must still make an economical evaluation to determine if it is economically and financially feasible. A project is economically feasible when it is more profitable than other competing projects, and financially feasible when management can raise the capital for its implementation. Although calculations may show that a given project could be extremely profitable, the capital requirements may strain the financial capabilities of the organization. In this case, the project may be abandoned unless partners can be found to share the risk. The economic evaluation of a process proceeds in several steps [1]. These are:

1. preparing a process flow diagram
2. calculating mass and energy flows
3. sizing major equipment
4. estimating the capital cost

5. estimating the production cost
6. forecasting the product sales price
7. estimating the return on investment

The main objective here is to determine the production cost of a chemical. Estimating the product-sales price and the return on investment is beyond the scope of this discussion. There are several texts, such as Valle-Riestra [20], Peters and Timmerhaus [4], and Holland and Wilkinson [38], that discuss methods of evaluating profitability and other aspects of process economics.

The difficulty in a process evaluation is not the computations, but the variability in the terminology that appears in the literature, which is a result of differences in company practice. Another difficulty is that in many cases the basis of the economic data reported in the literature is not clear as to what is included in the data. When economic data are not clearly defined, our only recourse is to compare data from several sources or to assume the worst case. Baasel [37] discusses the pitfalls of economic data.

CORPORATE CASH FLOW

The management of an organization needs estimates of the production cost and the capital required for a proposed process. Their responsibility is to raise the capital to construct the plant and to evaluate the process to maximize its profitability. Figure 2.1 depicts schematically the cash flow in an organization where the management of a firm is considered a bank, acquiring and dispensing funds. Corporate management acquires capital for various projects from profits earned by several existing divisions of the company, sale of bonds and stock, borrowed funds from banks and other organizations, income from licensing processes to other firms, various services to other firms, and return on investments obtained from other organizations. On the other hand, they dispense funds for payments of loans, purchase of stock, dividend payments, investments in other organizations, funds for a new plant, plant expansion, and improvements made on existing operations.

Corporate management provides funds, obtained from sales of products and return on investments for existing operations, such as a division of the corporation. Working capital is the funds required to keep a plant in operation. It flows in and out of an existing operation, as shown in Figure 2.2, and it is usually assumed to be completely recoverable at the end of a project without loss. Figure 2.2 shows that working capital is divided into two main categories, current liabilities and current assets. Current liabilities consist of bank loans and accounts payable (money owed to vendors for various purchases).

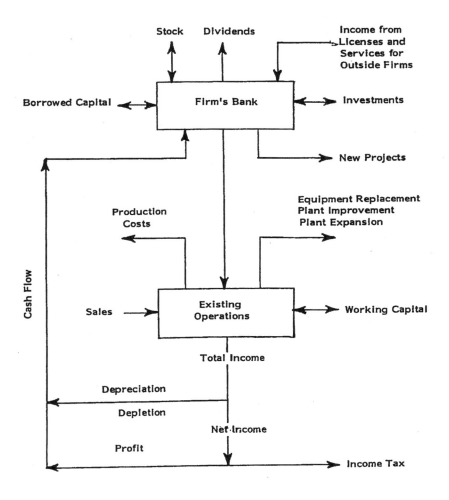

Figure 2.1 Cash flow in a corporation. Source: adapted from Ref. 2.

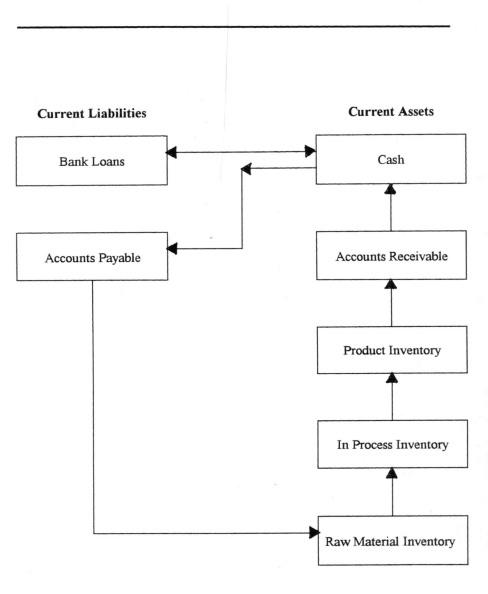

Figure 2.2 Flow of working capital. Source: adapted from Ref. 27.

Current assets consist of:

1. available cash – for salaries, raw material purchases, maintenance supplies, and taxes
2. accounts receivable – extended credit to customers
3. product inventory – material in storage tanks and bins
4. in-process inventory – material contained in pipe lines and vessels
5. raw material inventory – material in storage tanks and bins

Funds are continually required for equipment replacement, land improvement, and plant expansion, when economic conditions are favorable. Because funds for a project were originally provided by management, the division must return them as depreciation or depletion. Also, use of their capital management requires a profit. The sum of profit and depreciation or depletion constitutes cash flow.

PRODUCTION COSTS

To determine the financial attractiveness of a process, management requires both the total capital requirements and the production cost of a product. Operating cost and manufacturing cost have also been used synonymously with production cost. Figure 2.3 lists the various costs that contribute to the production cost. Peters and Timmerhaus [4] lists some of these costs. Perry and Chilton [3] give a more extensive list. Figure 2.3 groups costs under various categories. The important point is not under what category to include each cost, which is determined by the accounting practice of a firm, but more importantly not to omit any cost that influences the production cost.

Figure 2.3 divides the total production cost into three main categories direct costs, indirect costs, and general costs. Direct costs, also called variable costs, tend to be proportional to the production rate, whereas the indirect cost, composed of fixed cost and plant overhead cost, tend to remain constant regardless of the production rate. General costs include the costs of managing the firm, marketing the product, research and development on new and old products, and financing the operation.

Table 2.1, which corresponds to Figure 2.3, outlines a rapid method of estimating the production cost of a chemical using numerical factors given by Winter [1] and Humphreys [5]. These factors are only approximate, and they will vary with the type of process considered. They are useful, however, for preliminary estimates. Most companies will have their own factors that are specific for their processes.

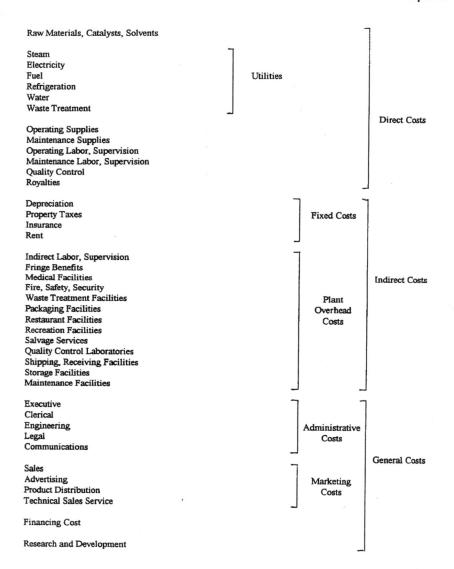

Figure 2.3 Components of the total production cost.

Table 2.1 Calculation Procedure for Production Cost

Direct Cost[a]	
Raw Materials	Amount of Incoming Stream x Cost
Catalysts and Solvents	Amount x Cost
Utilities	
Electricity	Power Consumed x Cost
Fuel	Power Consumed x Cost
Steam	Steam Consumed x Cost
Water	Water Consumed x Cost
Refrigeration	Heat Removed x Cost
Operating Labor	L x Cost
Operating Supervision	0.20 x Operating Labor Cost
Quality Control	0.20 x Operating Labor Cost
Maintenance Labor	0.027 x Fixed Capital Cost[b]
Maintenance Material	0.018 x Fixed Capital Cost
Operating Supplies	0.0075 x Fixed Capital Cost
Indirect Cost[a]	
Fixed Costs	
Depreciation[c]	$(1 - f_S)$ x (Depreciable Capital Cost) / (Plant Life)
Property taxes	0.02 x Fixed Capital Cost
Insurance	0.01 x Fixed Capital Cost
Plant Overhead Cost	
Fringe Benefits	0.22 x (Direct Labor + Supervision)
Overhead (less fringe benefits)	0.50 x (Direct Labor + Supervision)
General Costs[a]	
Administrative	0.045 x Production Cost
Marketing[d]	0.135 x Production Cost
Financing (interest)[f]	i x (Fixed Capital Cost + Working Capital[e])
Research and Development	0.0575 x Production Cost
Production Cost	Total of the Above Items

a. Numerical factors are obtained from Reference 5 except where indicated.
b. Fixed Capital Cost = Depreciable Capital Cost + Land Cost + Land Development Cost
c. Salvage fraction, f_S, is the fraction of the original depreciable capital cost.
d. Numerical factor is from Reference 1.
e. Working Capital = 0.20 x (Fixed Capital Cost)
f. Interest is at the current rate

DIRECT COSTS

Raw Materials

Sometimes raw material cost will dominate the production cost. A chemical company will attempt to protect its source of supply by arranging long term contracts, which also benefits the supplier. Raw material prices for preliminary estimates may be obtained from the sources listed in Table 2.2. Prices of chemicals depend on the quantity purchased. Published prices tend to be high, particularly, for Aldrich, Alfa Inorganic, and Fisher who sell small quantities of many chemicals for research. The most accurate source is the Chemical Marketing Reporter, which publishes prices for chemicals sold in bulk.

Catalysts

Catalysts are lost because of abrasion during use and regeneration. Also, some catalysts are eventually spent and must be replaced. Thus, the cost of catalysts must be included in the production cost. There are several corporations that specialize in manufacturing catalysts where the cost of catalysts may be obtained.

Solvents

Solvents are used in separation processes, such as in solvent extraction and gas absorption, and in liquid-phase reactions. The solvents are usually recovered within the process and reused, but losses occur because of leaks, incomplete recovery, and degradation. Leaks, however, are strictly regulated by the Environmental Protection Agency (EPA).

Utilities

Utilities include steam, electricity, fuel, cooling water, process water, compressed air, refrigeration, and waste treatment. Utility equipment is usually located outside of the process area and may supply several processes. We may consider each utility as a product, and estimate its cost according to the procedure outlined in Table 2.1. The cost of steam, electricity, and refrigeration depends mainly on fuel costs. Local utilities may give electric power costs, and the Federal Power Commission publishes rates for all public utilities in the United States. Table 2.3 lists approximate utility rates.

Table 2.2 Sources of Chemical Raw-Material Prices

Aldrich Chemical Catalog, Aldrich Chemical Co., Milwaukee, WI.
Alfa Inorganic Ltd., Beverly, MA.
The Chemical Marketing Reporter, New York, NY.
Fisher Chemical Index, Fisher Scientific Co., New York, NY.

Water, which is an increasingly important utility, is used both as a coolant and a process fluid. Its cost, as shown in Table 2.3, depends on the source or grade. Cooling water is obtained from reservoirs, rivers, and lakes and in many cases a cooling tower will recool the water. Process water quality depends on the needs of the process and may be city water, filtered, softened, demineralized cooling-tower water, condensate, distilled, and boiler feed water. The lowest grade of water is obtained from a well or river, which is filtered to remove suspended solids. The electronics industry needs an even purer grade called ultrapure water. Processing raw water to improve its grade increases its cost. A local water supplier or the Water Works Association can give the cost of city water.

Compressed air is mainly used to operate pneumatic instruments and control valves. Air is also used in aerobic fermentations for the production of chemicals and drugs and in biological waste treatment.

Refrigeration is needed when the required temperature is below the cooling-water temperature, such as in the production of liquid nitrogen and oxygen. Refrigeration is also used when the material being processed is sensitive to high temperatures, such as in food and pharmaceutical processes.

Fuel costs have a major impact on utility costs and will have an even greater impact in the future. When the price of oil rose in the 1970s, the chemical industry responded by increasing their efforts to improve the energy efficiency of their processes. Presently, the price of oil is low, but in the future the price of oil will rise again. Also, the consumption of oil and other fuels have an adverse effect on the environment so that efforts to conserve energy will continue.

Labor

Chemical plants require several types of labor. There is direct labor, consisting of operating labor to produce a chemical, and maintenance labor to maintain the process. There is also indirect labor, needed to operate and maintain facilities and services. Happel and Jordan [6] have pointed out that the contribution of labor costs to the product cost is small. But labor cost contributes to the cost of several other items, as shown in Table 2.1. When developing a new process, we can estimate the number of operators by visualizing the operations for the various

Table 2.3 Summary of Utility Costs

Utility	Condition	Quantity	Cost, $	Year
Cooling Water				
Well[a]		1000 gal	1.00	1993
River[a]		1000 gal	0.60	1993
Cooling Tower[b]	98 °F	1000 gal	0.40	1998
Process Water				
City [c]		1000 gal	0.80-1.80	1990
Filtered[c]		1000 gal	0.10-0.35	1990
Softened[c]		1000 gal	0.15-0.60	1990
Demineralized[c]		1000 gal	0.95-2.00	1990
Condensate[c]	212 °F	1000 gal	1.45-4.60	1990
Distilled[d]		1000 gal	2.25-4.00	1990
Boiler Feed[c]		1000 gal	1.95-5.60	1990
Saturated Steam				
High Pressure[b]	610 psig	1000 lb	6.00	1998
Medium Pressure[b]	160 psig	1000 lb	4.00	1998
Low Pressure[b]	30 psig	1000 lb	2.75	1998
Electricity	3 φ, 13.2 KV	MW-h	60.0	1993
Refrigeration[h,i]	−60 °F	t/d	1.50	1988
Air[c]	90 psig	1000 ft^3	1.00	1995
Fuel Oil[f]		1x10^6 Btu	2.28	1996
Fuel Gas[f]		1x10^6 Btu	2.00	1996
Nitrogen[g]	689 kPa	1000 m^3	47.0	1997

a. Source: Reference 21
b. Source: Reference 24
c. Source: Reference 5
d. Source: Reference 25
e. Source: Reference 22
f. Source: Reference 23
g. Source: Reference 24
h. Source: Reference 26
i. One ton of refrigeration per day, t/d, is defined as 12,000 Btu/h of heat absorbed.

process units based on previous experience. If experience is lacking, Cevidalli and Zaidman [7] propose using Equation 2.1.

$$L = \frac{K}{(1+p)^n} \frac{N}{m^b} \qquad (2.1)$$

This formula is a modification of a formula originally proposed by Wessel [8]. Cevidalli and Zaidman [7] examined several processes to determine the effect of production rate, process complexity, and degree of automation on the operating labor cost. In Equation 2.1, L is the number of hours required to produce one kilogram of product.

The process-productivity factor, K, is given in Table 2.4, which lists three process types: batch, continuous (normally automated), and continuous (highly automated). According to Table 2.4, a continuous, highly-automated process is the most efficient. We expect that the operating efficiency of the process will improve as engineers and technicians become more experienced in operating the plant. The improvement in operating efficiency is the yearly fractional increase in productivity, p. The base year for computing the operating labor is 1952. Thus, n is the number of years since 1952. By assuming that the fractional increase in labor productivity is 0.02, Cevidalli and Zaidman [7] found that the calculated operating labor using Equation 2.1 agrees with the actual labor requirement for several processes by 40%. This error is not unreasonable for an economic estimate.

Operating labor also depends on the the plant capacity, m, in kg/h. Table 2.4 shows that the exponent, b, in Equation 2.1 depends on the plant capacity. The exponent is 0.76 if the plant capacity is less than 5670 kg/h (12500 lb/h) and 0.84 if it is greater than 5670 kg/h. The economy of scale is evident in Equation 2.1, because the operating labor required to produce a kilogram of product decreases as the plant capacity increases. As shown in Table 2.1, once we calculate the operating labor we can calculate the operating supervision and maintenance labor.

The complexity of a process, as determined by the number of process units, N, also affects the operating labor required. The greater the number of process units the more complex the process is and the greater the operating labor. The number of process units is the most difficult term to evaluate in Equation 2.1. Bridgewater [9] defines a significant process unit as a unit that achieves a chemical or physical transformation of major process streams or any substantial and necessary side streams. Examples of process units are fractionation and filtration. Use the following guidelines for determining the number of process units:

1. Ignore the size of a process unit and multiple process units of the same type in series, such as the number of evaporators for multi-effect evaporation or the number of Continuously Stirred Tank Reactors (CSTRs).

2. Ignore pumps and heat exchangers unless substantial loads or unusual circumstances are involved, such as in a waste-heat boiler or quench tower.
3. Ignore storage unless it involves mechanical handling.
4. Ignore phase separators, such as gravity settlers. These are not significant process units, but a phase separator containing moving parts, such as a centrifuge, is considered a process unit.
5. Count mechanical operations, such as crushing, as a process unit.
6. Count utilities if they are specific to the process considered.

Estimates of the number of process units using these guidelines may vary, depending on the judgment of the process engineer.

Plant Maintenance

Maintenance costs consist of materials, labor, and supervision. Although maintenance cost increases as a plant ages, for economical estimates assume an average value for the life of the plant. The maintenance cost will vary from 3 to 6% of the fixed capital cost per year [5]. Use an average value of 4.5%, which consists of 60% labor and 40% materials [5].

Table 2.4   Process-Productivity Factor and Capacity Exponents for Equation 2.1

	Capacity Factor, b		Process-Productivity Factor, K	
Process Type	<5670 kg/h	>5670 kg/h	b = 0.76	b = 0.84
Batch	0.76	0.84	0.401	0.536
Continuous (normally automated)	0.76	0.84	0.296	0.396
Continuous (highly automated)	0.76	0.84	0.174	0.233

Operating Supplies

Supplies, which are not raw materials or maintenance supplies, are considered as operating supplies. Examples are custodial supplies, safety items, tools, column packing, and uniforms. The cost of operating supplies will vary from 0.5 to 1% of the fixed capital cost per year [5]. Use an average value of 0.75%.

Quality Control

Chemicals must meet certain specifications to be salable. Thus, analysis of process steams must be regularly made to determine their quality. Although there is a trend toward on-line analysis, samples of process streams must still be taken to check instrument performance. Also, there are still many analyses that cannot be made on-line. According to Peters and Timmerhaus [4] and Humphreys [5], the cost of quality control varies from 10 to 20% of operating labor. Use a value of 20% in Table 2.1.

INDIRECT COSTS

Indirect costs are those costs incurred that are not directly related to the production rate and consist of fixed and plant overhead costs, as shown in Figure 2.3.

Fixed Costs

During the life of a plant the production rate will vary, according to economic conditions, but depreciation, property taxes, insurance, and rent are independent of the production rate and will remain fixed. Instead of rent, land, which is not part of the fixed capital cost, is assumed to be purchased by borrowed capital and the interest paid yearly in the procedure outlined in Table 2.1.

Depreciation

Holland [11] has pointed out that depreciation has a number of different meanings of which the following are the most common:

1. a cost of operation
2. a tax allowance
3. a means of building up a fund to finance plant replacement
4. a measure of falling value

The value of a plant will decrease with time because of ware and technical obsolescence. In a sense, a plant will be consumed to manufacture a product. Depreciation determines the contribution of equipment cost to the production cost.

There are several depreciation methods, which are discussed in many economic texts. Since we want to develop a rapid method of estimating the production cost, we will use the simple linear depreciation method. For this method, divide the difference of the depreciable capital cost and its salvage value by the life of the plant, as shown in Table 2.1. An entire plant or individual equipment has three lives: an economic life, a physical life, and a tax life. The economic life occurs when a plant becomes obsolete, a physical life when a plant becomes too costly to maintain, and a tax life, which is fixed by the government. The plant life is usually ten to twenty years. The depreciable capital cost includes all the costs incurred in building a plant up to the point where the plant is ready to produce, except land and site-development costs. Care must be taken not to include costs that are not depreciable.

Plant Overhead

Plant overhead is the cost of operating the services and facilities required by the productive unit, as listed in Figure 2.3. Also included in this category are all the fringe benefits for direct as well as for indirect labor. It is common practice to include the fringe benefits of direct labor in the overhead rather than in direct costs.

GENERAL COSTS

General costs are associated with management of a plant. Included within general costs are administrative, marketing, financing, and research and development costs. Figure 2.3 divides general costs into their various components. Administrative costs vary from 3 to 6% of the production cost [1]. Use an average value of 4.5% in Table 2.1. Marketing costs include technical service, sales, advertising, and product distribution, consisting of packaging and shipping. If a plant sells a small quantity of a product to many customers, the plant will incur a higher cost than if it sells larger volumes to a few customers. Marketing costs vary from 5 to 22% of the production cost. Table 2.1 contains an average value of 13.5%.

In the past, the interest rate on borrowed capital has increased considerably. Usually, corporations and individuals will borrow capital when interest rates become favorable. Because the interest rate may change rapidly over short time intervals, Table 2.1 does not include a numerical value. The current interest rate can be obtained from the financial section of newspapers or from banks.

Finally, process and product improvements are continuously being sought. Thus, we must add the cost of research and development to the production cost, which varies from 3.6 to 8% of the production cost. Use an average value of 5.8% in Table 2.1.

CAPITAL COSTS

To calculate several of the cost items listed in Table 2.1, requires the depreciable and fixed capital costs. The depreciable capital cost is the capital required for equipment and its installation or modification in the process, and all the facilities required to operate the process. There is some variation in the definition of fixed capital cost. References [1-5], define the fixed capital as consisting of the depreciable capital cost, land cost, and site or land development cost. Woods [10], however, omits land cost and land development cost so that that the fixed capital cost equals the depreciable capital cost. We will adopt the first definition here. For now, assume that we know the depreciable capital cost. We will develop a procedure for its evaluation later. In Example 2.1 estimate the production cost using Table 2.1.

Example 2.1 Estimating the Production Cost of Ethylenediamine

Ethylenediamine is used to produce chelating agents and carbamate fungicides. Monoethanolamine (MEA), reacts with ammonia and hydrogen to produce ethylenediamine. The reaction occurs in the gas phase over a catalyst at temperatures < 300 °C (572 °F) and pressures > 250 bar (246.7 atm) [12]. Other details of the process are proprietary. The products are:

ethylenediamine   (EDA)   – 74%
diethyltriamine   (DETA) – 8%
piperazine   (PIP)   – 4%
aminoethylpiperazine (AEP)   – 10%
hydroxyethylpiperazine (HEP) – 4%

Data (March 1978)

Fixed Capital Cost	10.3×10^6
Interest Rate	10 % /yr
Utility Costs (kg total products)	
Electricity	3.56×10^6 J/kg total products @ 4.17×10^{-7} c/J
Steam (1.03 kPa)	18.1 kg st./kg total products @ 0.0113 c/kg
Water	0.1528 m^3/kg total products @ 3.17 c/m^3
Operating Labor Rate	8.61 $/h
Land Cost	0.015 x depreciable capital cost
Land Development Cost	0.021 x depreciable capital Cost
Total Raw Material Cost	79.8 c/kg

If the production rate is 10,000 t/yr (metric tons per year) (11030 tons/yr) of total product, find the production cost of one kilogram of total product?

The solution of the problem reduces to calculating the operating labor and fixed capital cost for one kilogram of total product. Following the procedure outlined in Table 2.1. Table 2.1.1 summarizes the results.

To allow time for maintenance during the year, the plant will operate only for 8000 h instead of 8760 h for a plant operating without interruption. Therefore, the production rate is

$$m = \frac{1 \times 10^4 \text{ t}}{1 \text{ yr}} \frac{1 \times 10^3 \text{ kg}}{1 \text{ t}} \frac{1 \text{ y}}{8000 \text{ h}} = 1.25 \times 10^3 \frac{\text{kg}}{\text{h}}$$

Table 2.1.1 Calculation of Ethylenediamine Production Cost

Direct Costs		Cost, c/kg
Raw Materials		79.80
Utilities		
Electricity	3.56×10^6 J/kg x 4.17×10^{-7} c/J =	1.485
Steam	18.1 kg st./kg x 0.01113 c/kg st. =	0.201
Water	0.1528 m^3/kg x 3.17 c/m^3 =	0.484
Operating Labor	3.224×10^{-3} h/kg x 861 c/h =	2.776
Operating Supervision	0.20 x 2.776 =	0.555
Quality Control	0.20 x 2.776 =	0.555
Maintenance Labor	0.027 x 10.3 =	0.278
Maintenance Material	0.018 x 10.3 =	0.185
Operating Supplies	0.0075 x 10.3 =	0.077
		86.396
Indirect Cost		
Fixed Costs		
Depreciation	(1 − 0.10) x 9.94 =	8.946
Property Taxes	0.02 x 10.3 =	0.206
Insurance	0.01 x 10.3 =	0.103
Plant Overhead Cost		
Fringe Benefits	0.22 x (2.776 + 0.555 + 0.278) =	0.794
Overhead	0.50 x (2.776 + 0.555 + 0.278) =	1.890
(less fringe benefits)		
		11.939
General Costs		
Administrative	0.045 x 131 =	5.895
Marketing	0.135 x 131 =	17.685
Financing (interest)	0.1 [10.3 + 0.2 (10.3) =	1.236
Research and Development	0.0575 x 131 =	7.533
		32.349
Production Cost		131.0

Operating Labor

To determine the operating labor calculate L, the number of hours to produce a kilogram of product, from Equation 2.1. First, determine the number of process units, N, using the guide lines discussed previously. To determine N examine the flow diagram for the process shown in Figure 2.1.1.

MEA, hydrogen, ammonia, and recyled gases mix before flowing into the packed-bed catalytic reactor, shown in Figure 2.1.1. After reaction the gas stream cools, condensing the condensable components. The gas-liquid stream leaving the condenser separates in a phase separator into a gas stream, consisting mainly of unreacted ammonia and hydrogen, and a liquid stream. The compressor then compresses the gases and recyles them back into the inlet of the reactor. Next, a series of distillation columns separates the liquid product stream. The distillation columns remove the more volatile components first. The first column removes ammonia, the second column water, and the third column separates EDA and PIP from HEP, AEP, DETA, and MEA. The MEA recycles backed to the reactor inlet. Because the process is proprietary, Figure 2.1.1 does not show the purification and the polyamine separation sections in any detail. To obtain an approximate labor cost, we will assume that one column in the Purifcation Section separates the EDA and PIP and two columns in the Polyamine Separation Section separates the more complex solution. According to the guidelines for determining N, the heat exchangers, compressors, and phase separators are not process units. Thus, there are six columns and one reactor for a total of seven process units.

Assuming that the process is highly automated, we find from Table 4 that b = 0.76 and that the process productivity, K = 0.174. Assume that p, the labor productivity increases at an annual rate of 2% since 1952. Twenty-six years have elapsed since 1952, therefore n = 26. Substituting into Equation 2.1, the number of hours of operating labor for a kilogram of the total product,

$$L = \frac{0.174}{(1 + 0.02)^{26}} \frac{7}{(1250)^{0.76}} = 3.224 \times 10^{-3} \text{ h/kg } (1.46 \times 10^{-3} \text{ h/lb})$$

Fixed Capital Cost

A typical plant life is ten years. Thus, the fixed capital cost, C_F, in dollars per kilogram of total product is,

$$C_F = \frac{10.3 \times 10^6 \text{ \$}}{10 \text{ yr}} \frac{1 \text{ yr}}{1 \times 10^4 \text{ t}} \frac{1 \text{ t}}{1 \times 10^3 \text{ kg}} = 0.103 \text{ \$/kg } (0.0467 \text{ \$/lb})$$

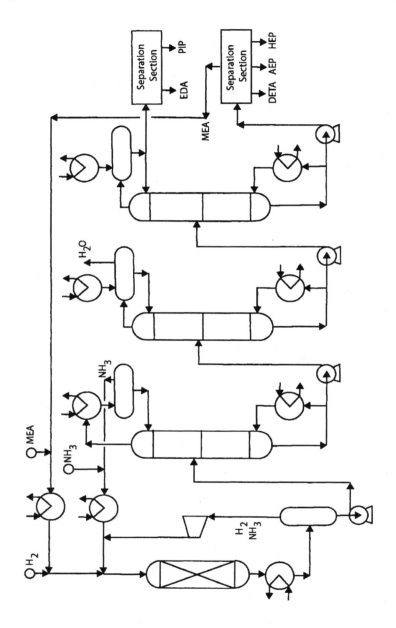

Figure 2.1.1 Ethylenediamine-synthesis process.

According to Table 2.1, the fixed capital cost equals the sum of the depreciable capital cost, land cost, and land development cost. Land cost is 0.015 times the depreciable capital cost and land development is 0.0211 times the depreciable capital cost for a fluid processing plant. Thus,

$$C_F = C_D + 0.015\ C_D + 0.0211\ C_D = 0.103\ \$/kg\ (0.0467\ c/kg)$$

Solving for C_D, we obtain

$$C_D = 0.0994\ \$/kg\ (0.0451\ \$/lb)$$

After, calculating the operating labor cost and depreciable capital cost, use the procedure outlined in Table 2.1 to calculate all other costs, except for the administrative, marketing, and research and development costs. First, calculate the production cost by summing up all the costs given in Table 2.1.1. These costs are the direct cost, indirect cost, administrative cost, marketing cost, financing cost, and the research and development cost. Thus, the production or manufacturing cost is,

$$C_M = 86.4 + 11.9 + 0.045\ C_M + 0.135\ C_M + 1.24 + 0.0575\ C_M$$

Solving for C_M, we obtain

$$C_M = 131\ c/kg\ (59.4\ c/lb)$$

We can now complete Table 2.1.1 for those items that depend on the production cost. The production cost for this process, reported by Kohn [12], is 119 c/kg (54.0 c/lb). Because the estimation of N in the operating labor cost requires judgment, we should expect that process engineers will differ in their estimates. If we estimate N to be 8 instead of 7, the production cost is 132 c/kg (59.9 c/lb), which is not significant.

CAPITAL COST ESTIMATION

Calculating the production cost requires estimating the depreciable capital cost and fixed capital cost. Before estimating the depreciable capital cost, the process engineer must first calculate mass and energy flow rates to size process equipment. He can then estimate the cost of all equipment and finally the depreciable and fixed capital costs. Besides sizing equipment he must also calculate utility requirements from the mass and energy flow rates. Two methods for estimating capital costs will be discussed: one is the average factor method and the second is the individual factor method. At the early stages of developing a process, you can use these simple methods. As the process development advances, then you should use more accurate methods.

DEPRECIABLE CAPITAL COST

Factor Methods

Figure 2.4 divides the depreciable capital costs into several categories. The two major categories are direct and indirect costs. Peters and Timmerhaus [4] and Humphreys [5] list these costs. Reference [3] gives a more detailed breakdown. As Figure 2.4 shows,

depreciable capital cost = the cost of:

delivered equipment
+ equipment placing
+ piping connections between equipment and to utilities
+ electrical equipment and wiring
+ instrumentation and controls
+ buildings
+ auxiliary facilities (offsites)
+ engineering
+ construction contractor's fee
+ contingency   (2.2)

In the factor methods for cost estimating, first calculate the purchased or delivered cost of all major equipment, for example, distillation columns, reactors, pumps, heat exchangers, etc. Then multiply the total equipment cost by factors to estimate the various other components of the depreciable capital cost given in Equation 2.2, such as piping and electrical wiring. Thus, we arrive at the cost of installing all the equipment and supplying all the services needed to produce an operational process.

It helps to visualize the process of constructing a plant to understand the calculation of depreciable capital cost. First, a purchasing agent orders equipment from various manufacturers from all over the world. The manufacturers then deliver the equipment to the plant site. Shipping charges, insurance, and taxes add to the cost of equipment, resulting in the delivered equipment costs.

After arriving at the plant site, construction workers set the equipment in place. This entails placing the equipment on concrete or steel structural supports, prepared in advanced. Because some equipment could weigh tons, a crane will lift the equipment onto supports. Then, construction workers secure the equipment in place. A factor will account for this cost.

Next, pipe fitters connect the equipment to other equipment and to steam and cooling water distribution systems. Piping and valves, which could weigh

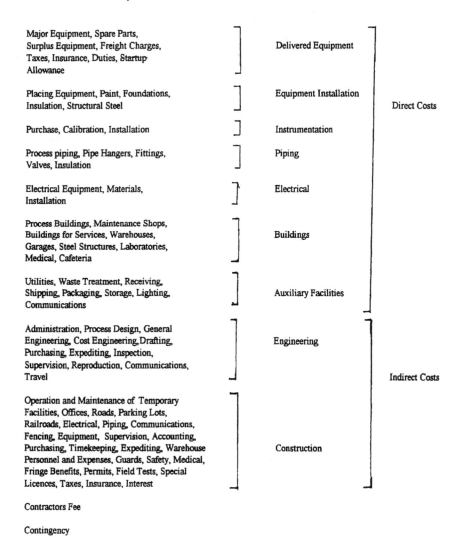

Figure 2.4 Components of depreciable capital cost for a chemical plant.

hundreds of pounds, must be supported by steel structures and pipe hangers, further adding to the cost of the plant. Another factor will account for this cost.

Because a typical plant will contain many items of machinery, such as pumps, compressors, and mixers, a plant will require an electric power distribution system. Electricians need wiring, switches, and other electrical equipment to connect the machinery to the electrical system.

To maintain the production rate, product quality, and plant safety requires a data acquisition and control system. This system consists of temperature, pressure, liquid level, flow rate, and composition sensors. Computers record data and may control the process. Modern chemical plants use program logic controllers (PLC) extensively. According to Valle-Riestra [20], instrumentation cost is about 15% of purchased equipment cost for little automatic control, 30% for full automatic control, and 40% for computer control.

Another factor is needed to estimate the cost of buildings to house the process and the various services required to operate the plant such as offices, maintenance shops, and laboratories. Process buildings, as described by Valle-Riestra [20], are mostly open structures rather than enclosed structures and are preferred for safety as well as economic reasons. Toxic or flammable gases or liquids released accidentally will dissipate more quickly in an open structure. A frequent arrangement is an open process tower five decks in height and constructed with I-beams [20]. Besides reducing the floor space occupied by equipment, the structure allows for gravity flow.

Auxiliary facilities provide services that are necessary for the operation of the process. Examples of these facilities are steam, electrical power, air, cooling water, refrigeration, and waste treatment. To account for this cost requires determining whether the facility will be dedicated or shared. If the facility is dedicated solely for the use of a single process, then its cost is assigned to the process. On the other hand, if other processes share the facility, then its cost is divided according to usage.

A plant is divided into four areas: the process area, storage, utilities, and services, as illustrated in Figure 2.5. The process area is called "battery limits" and the other areas auxiliary facilities. Battery limits derives from the time when oil refineries contained several stills in a row, resembling a gun battery. The battery limit contains all the equipment assigned to the process, but Valle-Riestra [20] pointed out that a process unit is not always physically located in one area of a plant.

Because the factor methods for calculating the depreciable capital cost are rapid methods and not based on a detailed design, many small items of equipment are knowingly omitted. Also, there are uncertainties in design and economic procedures, and bad weather, strikes, and other unforeseen events may cause delays. To correct for uncertainties and unforeseen events requires using a contingency factor or safety factor.

Engineers design, implement the design, and monitor the progress of construction. They organize the total construction effort. Besides chemical engineers,

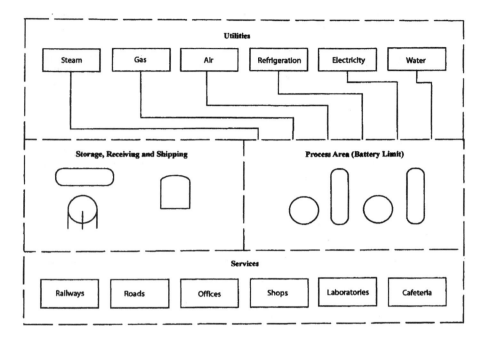

Figure 2.5 The process area and auxiliary facilities of a chemical plant.

plant design involves all the common branches of engineering – mechanical, civil and electrical engineering. Engineering and construction costs are indirect costs and are part of the depreciable capital cost.

Next, two methods for calculating the costs listed in Figure 2.4 are discussed. One method is the average factor method, and the other method is the individual factor method. The accuracy of a cost estimate should be considered. Table 2.5 contains the accuracy of various methods and their cost. Although the costs are out-of-date, they do show that as a process becomes well-defined the estimates become more costly. In the early stages, when a project is ill-defined, an accurate cost estimate is not warranted. The factor methods are study estimates and are less accurate than the detailed estimate.

Table 2.5 Typical Average Costs of Cost Estimates

Cost of Project[a]	Less than $2,000,000	$2,000,000 to $10,000,000	$10,000,000 to $100,000,000
Type of Estimate			
Order of Magnitude (± 30%)[a]	$3, 000	$6,000	$13, 000
Study (± 30%)	20,000	40, 000	60,000
Preliminary (± 20%)	50,000	80, 000	130, 000
Definitive (± 10%)	80,000	160, 000	320,000
Detailed (± 5%)	200,000	520,000	1,000, 000

[a]Accuracy of estimate.
Source: Adapted from Reference 39.

Average Factor Method

The average factor method is summarized by Equation 2.3.

$$C_D = (\Sigma_k f_{I\,k})(\Sigma_i C_{S\,i}) \tag{2.3}$$

where $\Sigma_k f_{I\,k}$ is an average factor that accounts for the cost of each item in Equation 2.2 required to install equipment. The installation factor accounts for all costs required to make the equipment operable. The average factor is the average of the individual factors of many pieces-of-equipment. Lang [14] originally proposed the factor method, and it is frequently called the Lang factor method.

The factor for buildings depends on the plant location and the plant type. To estimate the cost of buildings, we will consider three plant locations. These are a "grass-roots" plant, a plant at an existing site, and a plant addition. A "grass-roots" plant is isolated from an industrial complex and must provide all auxiliary facilities for its sole use. On the other hand, if a plant is part of an industrial complex, utilities – such as steam generation and water treatment facilities – may be shared with other processes located at the site. Sharing facilities reduces capital and production costs. The third type of plant is a plant addition, where the auxiliary facilities are again available.

The three types of processes considered are a solids process – such as a process producing lime, a solid-fluid process – such as a powdered-coffee process, and a fluid process – such as a methanol-synthesis process. No sharp division exists among these process types so that you must use some judgment to classify a process. Table 2.6 contains average cost factors for these process types. The factors for process equipment depend on the material of construction. Thus, Table 2.7

contains factors for both carbon steel and alloy steel. Alloy steels contain varying amounts of nickel and chromium, such as the stainless steels. The other factors in Table 2.6, i.e., for buildings, auxiliary facilities, indirect costs, contractor's fee, and contingency do not depend on the material of construction for a process. As an example, for a carbon-steel, fluid-processing plant constructed at an existing site. From Table 2.7, $f_{DC} = 1.86$, and from Table 2.6, the average factor, $\sum_k f_{I\,k} = 3.27 + f_{DC} = 3.27 + 1.86 = 5.13$.

The factors in Tables 2.6 and 2.7 are for an average process containing many pieces-of-equipment and should not be used for single piece-of-equipment and a small installation containing only a few pieces-of-equipment. For these cases, we will use the individual factor method, which will be described next.

Table 2.6 Cost Factors for Estimating Depreciable Capital Cost – Average Factor Method (Adapted from Reference 4.)

	Cost Factor, Fraction of Delivered Equipment Cost[a]		
	Solids Process	Solids-Fluid Process	Fluid Process
Direct Costs			
Delivered Equipment	1.00	1.00	1.00
Equipment Installation See Table 2.7	f_{DC}	f_{DC}	f_{DC}
Buildings (with services) Grass-Roots Plant Plant at an Existing Site Plant Addition	0.75 0.28 0.17	0.52 0.32 0.08	0.50 0.20 0.07
Auxiliary Facilities Grass-Roots Plant Plant at an Existing Site[b] Plant Addition	0.40 0	0.52[c] 0.55 0	0.70 0
Indirect Costs			
Engineering Construction	0.33 0.39	0.32 0.34	0.33 0.41
Contractor's Fee[d]	0.17	0.18	0.21
Contingency[e]	0.34	0.36	0.42

a)   Source offactors is Reference 2.4 except where indicated
b)   Includes installation cost
c)   Source: Reference 2.9
d)   5% of direct and indirect costs
e)   10% of direct and indirect costs

Table 2.7 Direct-Cost Factors for Process Equipment Installation – Average Factor Method

| | Cost Factor, Fraction of Delivered Equipment Cost | | | | | |
| | Solids Process | | Solids-Fluid Process | | Fluid Process | |
	Primarily Carbon Steel Equipment	Primarily Alloy Equipment	Primarily Carbon Steel Equipment	Primarily Alloy Equipment	Primarily Carbon Steel Equipment	Primarily Alloy Equipment
Equipment Installation						
Equipment Placing	0.15	0.10	0.20	0.12	0.25	0.15
Painting	0.06	0.03	0.07	0.03	0.06	0.03
Foundations	0.20	0.12	0.18	0.12	0.15	0.10
Insulation	0.05	0.03	0.10	0.06	0.20	0.15
Structural Steel	0.12	0.07	0.15	0.10	0.20	0.10
Instrumentation	0.12	0.10	0.20	0.15	0.25	0.20
Piping	0.25	0.15	0.45	0.40	0.60	0.55
Electrical	0.20	0.15	0.15	0.12	0.15	0.12
Total = f_{DC}	1.15	0.75	1.50	1.10	1.86	1.40

Source: Ref. 32.

Individual Factor Method

If we need the installation cost of a piece-of-equipment or a few pieces-of-equipment, then we have to use the individual factor method that Hand [28] first proposed. We can also use this method for large plants containing many pieces of equipment as well. Also, this method is more accurate than the average factor method. In this case, the capital cost for installed equipment,

$$C_D = \sum_i f_{I\,i}\, C_{P\,i} \tag{2.4}$$

where the subscript i refers to a piece-of-equipment. The installation factor for a piece-of-equipment, $f_{I\,i}$, contains the same cost items for installing equipment as those listed for the average factor method in Tables 2.6 and 2.7. For the individual factor method, the basis for the installation factor is the Free-On-Board (FOB) cost of the equipment, i.e., the cost at the manufacturer's doorstep instead of the delivered equipment cost.

The installation factor consists of direct costs, indirect costs, contingency cost, and a contractor's fee. The installation factor for a piece-of-equipment is given by

$$f_I = f_{DC} \, f_{IC} \, f_{CF} \qquad\qquad (2.4)$$

The direct-cost factor for equipment, f_{DC}, contained in Table 2.8, does not include buildings and auxiliary facilities. It includes the labor and materials needed to install equipment. The buildings and auxiliary costs will be accounted for after we calculate the depreciable capital cost for equipment.

EQUIPMENT COST ESTIMATION

Whether a process engineer uses the average factor method or the individual factor method, the major effort in estimating the depreciable capital cost is estimating the cost of equipment. After developing the process flow diagram and calculating mass and energy flow rates, he can then estimate the size of the equipment and the equipment cost. There are three sources of equipment cost data. These are: current vendor quotations, past vendor quotations, and literature estimates, in order of decreasing accuracy. Woods [10] has stated that correlation of equipment costs in the literature can have large errors − by as much as 100%. A correlation with a large error is not completely useless, but it will limit the conclusions that one can draw. Vendor quotations are the most accurate, but the effort required to prepare detailed specifications and quotations are not usually warranted in the early stages of a project. Thus, we rely on literature estimates and past quotations for quick estimates − in spite of their lower accuracy.

Equipment costs reported in the literature are either FOB, delivered, or installed cost. Usually, these costs are given at some time in the past. When reporting equipment costs, the date and shipping point should be specified, but the latter is frequently not given. Shipping cost (consisting of freight, taxes, and insurance) will vary from 10 to 25% of the purchased cost [10]. We will use 10%, a value recommended by Valle-Riestra [20]. Then, the delivered cost is

$$C_{Si} = 1.10 \, C_{Pi} \qquad\qquad (2.5)$$

Before adding equipment costs they must all be on the same basis − either FOB, delivered, or installed. For example, Table 2.6 requires that all equipment costs be on the delivered basis. If some equipment is reported on the installed basis, then add this cost after all other equipment costs are on the installed basis.

Table 2.8 Direct Cost Factors for Equipment Installation–Individual Factor Method

Equipment[a]	Factor	Equipment	Factor
Agitators (CS)[b]	1.3	Heat exchangers (shell/ tube)[d]	
Agitators (SS)[c]	1.2	SS/SS	1.9
Air Heaters, all types	1.5	CS/SS	2.1
		CS/Al	2.2
Beaters	1.4	CS/Cu	2.0
Blenders	1.3	CS/Monel	1.8
Blowers	1.4	Monel/Monel	1.6
		CS/Hastalloy	1.4
Boilers	1.5		
Centrifuges (CS)	1.3	Instruments, all types	2.5
Centrifuges (SS)	1.2	Miscellaneous (CS)	2.0
		Miscellaneous (SS)	1.5
Chimneys and stacks	1.2		
Columns, distillation (CS)	3.0	Pumps	
Columns, distillation (SS)	2.1	Centrifugal (CS)	2.8
		Centrifugal (SS)	2.0
Compressors, motor drive	1.2	Centrifugal, Hastalloy trim	1.4
Compressors, steam or	1.5	Centrifugal, nickel trim	1.7
gas drive		Centrifugal, Monel trim	1.7
Conveyors and elevators	1.4	Centrifugal, titanium trim	1.4
Cooling tower, concrete	1.2	All others (SS)	1.4
Crushers, classifiers, mills	1.3	All others (CS)	1.6
Crystallizers	1.9		
		Reactors	
Cyclones	1.4	Kettles (CS)	1.9
Dryers, spray and air	1.6	Kettles, glass lined	2.1
Dryers, other	1.4	Multitubular (SS)	1.6
		Multitubular (Cu)	1.8
Ejectors	1.7	Multitubular (CS)	2.2
Evaporators, calandria	1.5		
Evaporators, thin film (CS)	2.5	Refrigeration Plant	1.5
		Steam Drum	2.0
Evaporators, thin film (SS)	1.9	Sum of equipment costs (SS)	1.8
Extruders, compounding	1.5	Sum of equipment costs (CS)	2.0
Fans	1.4		

		Tanks	
Filters, all types	1.4	Process (SS)	1.8
Gas holders	1.3	Process (Al)	2.0
Granulators for plastics	1.5	Storage (SS)	1.5
		Storage (Al)	1.7
		Storage (CS)	2.3
		Field erected (SS)	1.2
		Field erected (CS)	1.4
Heat exchangers			
		Turbines	1.5
Air cooled (CS)	2.5	Vessels, pressure (SS)	1.7
Coil in Shell (SS)	1.7	Vessels, pressure (CS)	2.8
Glass	2.2		
Graphite	2.0		
Plate (SS)	1.5		
Plate (CS)	1.7		

a. Direct cost = materials + labor = indirect factor x equipment cost
b. Carbon steel (CS)
c. Stainless steel (SS)
d. Shell material/tube material
Source: Adapted from Reference 35 with permission.

Correcting Equipment Cost for Size

Usually, the cost literature contains equipment costs for capacities other than what is required. To scale the equipment cost to the required capacity, we usually assume that its cost varies to some power, usually fractional, of its capacity. Thus, the scaled cost will be

$$C_2 = C_1 \left(\frac{Q_2}{Q_1} \right)^n \tag{2.6}$$

If we know the cost of a piece-of-equipment at one capacity and the capacity exponent, n, then we can calculate its cost at another capacity. We can find cost data in References [10], [13], [15], [16], and [36]. More recent cost data are contained in References [4], [30], [31], and [37]. Table 2.9 contains costs and capacity exponents of some common equipment. The correlation range given in Table 2.9 gives the size limits for each piece-of-equipment. You should not extrapolate Equation 2.6 too far beyond the limits specified. For example, from Table 2.9, the cost of a propeller agitator for 3 hp in January 1990 is $2800, and the correlation range is 1 to 7 hp. Bringing the equipment cost up-to-date will be discuss later.

Equation 2.6 will be linear when plotted on log-log coordinates. The slope of the line is the capacity exponent, n. In most cases, the equipment size, cost, and capacity exponents in Table 2.9 were taken from Peters and Timmerhaus's log-log plots [4]. If the log-log plot was not linear, it was approximated by a straight line to maintain the simple relationship given by Equation 2.6. If you cannot find a capacity exponent for a piece-of-equipment, Lang [14] suggested using six tenths. This is called the six tenths rule. Drew and Ginder [33], however, found that six tenths is appropriate for pilot-scale equipment and seven tenths for large equipment. Because most exponents are less than one, doubling the equipment capacity will not double the equipment cost, which is an example of the economy of scale.

Correcting Equipment Cost for Design, Material of Construction, Temperature, and Pressure

Sometimes, the cost literature contains equipment cost at base conditions, $C_{B\,i}$ in Equation 2.7. The base conditions are a low temperature and pressure, carbon steel construction, and a specific design. If you need the actual cost of equipment, $C_{A\,i}$, at other conditions, multiply the base cost by correction factors. Thus,

$$C_{A\,i} = f_T\, f_P\, f_M\, f_D\, C_{B\,I} \tag{2.7}$$

where f_T corrects for temperature, f_P for pressure, f_M for material of construction, and f_D for a specific design. Table 2.10 contains values of f_T, f_P, and f_M for some equipment. For the case where the equipment is only available in one design, $f_D = 1$. The factors in Equation 2.7 depend on the type of equipment, and thus using the same correction factors for all equipment is an approximation. Also, if the equipment operates at extreme conditions of temperature, pressure, or with a corrosive fluid, the correction factors in Table 2.10 will be too low.

For shell-and-tube heat exchangers, the correction factors are defined differently. The shell material may be different than the tube material. If the process fluid is corrosive, for example, then the tube material could be stainless steel. Also, it is good practice to place the high-pressure fluid on the tube side to reduce the cost of metal. Table 2.11 contains material factors obtained from Guthrie [13] for combinations of shell-and-tube materials. Also, use the pressure and design correction factors given in Table 2.11 instead of Table 2.10. Because Guthrie [13] does not give any temperature correction factors use the factors given in Table 2.10, which will increase the heat-exchanger cost. To underestimate is worse than to overestimate, up to a point. Using Table 2.11, then, for heat exchangers the cost equation is

$$C_{A\,i} = f_T\, (f_P + f_D)\, f_M\, C_{B\,i} \tag{2.8}$$

Table 2.9 Equipment Cost Data Carb on Steel Construction

Equipment	Size	Capacity Units	FOB Cost[d], k$ January 1990	Correlation Range[d]	Capacity Exponent[d],n	Direct-Cost Factor[a], f_{DC}
Agitators						
Propeller	3	hp	2.8	1.0-7.0	0.50, e	Table 2.8
Turbine, single impeller	20.0	hp	12.0	3.0-100.0	0.30, e	Table 2.8
Air Coolers	1.0	ft²	0.137, j	–	0.8, j	Table 2.8
Blowers						
Centrifugal	4,000	ft³/min	60.0	800.0-1.8x10⁴	0.59, c	Table 2.8
Compressors & Drives						
Centrifugal, electric motor	600	hp	190	2.0x10²- 1.8x10⁴	0.32, c	Table 2.8
Centrifugal, steam or gas turbine	600	hp	210	2.0x10³- 2.1x10³	0.32, c	Table 2.8
Electric Motors						
Open Drip Proof	60	kW	3.0, g	0.20-5.0x10³	1.10, f	2.0 or 1.5, h
Totally Enclosed	60	kW	4.0, g	0.25-6.0x10³	1.10, f	2.0 or 1.5, h
Explosion Proof	100	kW	9.5, g	0.30x8.0x10³	1.10, f	2.0 or 1.5, h
Evaporators (installed)						
Forced Circulation	1,000	ft²	2500, i	1.0x10²- 7.0x10³	0.70, e	Table 2.8
Horizontal Tube	1,000	ft²	120, i	1.0x10²- 8.0x10³	0.53, e	Table 2.8
Vertical Tube	1,000	ft²	180, i	1.0x10²- 8.0x10³	0.53, e	Table 2.8
Fans						
Centrifugal, radial, low range	4,000	ft³/min	2.5	1.0x10³- 1.0x10⁴	0.44	Table 2.8
Centrifugal, radial, high range	10,000	ft³/min	40.0	1.0x10⁴- 1.0x10⁵	1.17	Table 2.8
Heat Exchangers (shell/tube)[b]						
Floating Head, CS/CS, 150 psia	1,000	ft²	14.0	1.0x10²- 5.0x10³	0.65, e	Table 2.8
Process Furnace	20,000	kJ/s	750, g	3.0x10³- 1.6x10⁵	0.85, g	Table 2.8
Pumps						
Centrifugal, high range	20	hp	9.0	2.68-335	0.42, g	Table 2.8
Centrifugal, low range	0.29	hp	2.3	0.10-2.0	0.29, g	Table 2.8
Gear, 100 psi	80	gpm	1.3	16-400	0.36, f	Table 2.8
Reactors [k]						
Stirred Tank, jacketed, CS, 50 psi	600	gal	17.0	30-6.0x10³	0.57, f	Table 2.8
Stirred Tank, glass lined, 100 psi	400	gal	33.0	30-4.0x10³,	0.54, c	Table 2.8
Rotary Vacuum Filters (SS)	30	ft²	60.0	4.0-600	0.67, c	Table 2.8
Tanks						
Storage, cone roof, low range	12.0x10⁶	gal	170	2.0x10⁵- 1.2x10⁶	0.32, f	Table 2.8
Storage, cone roof, high range	12.0x10⁶	gal	170	1.2x10⁶- 1.1x10⁷	0.32, f	Table 2.8

a. Source: Reference 2.35 except where indicated.
b. The shell-and-tube materials can differ. CS/SS means carbon steel shell and the stainless steel tubes.
c. Source: Reference 2.29.
d. Source: Reference 2.4 except where indicated. – January 1990 cost except where indicated.
e. Source: Reference 2.13.
f. Source: Reference 2.10.
g. Source: Reference 2.31 – mid-1982 cost.
h. Source: Reference: 2.31 – Use 2.0 for conveyors, crushers, grinders, gas-solid contactors, and mixers.
   – Use 1.5 for fans, compressors, and pumps.
i. Source: Reference 2.4 – Installed cost.
j. Source: Reference 2.37 – mid-1987 cost

k. no agitator

instead of Equation 2.7. Finally, the factors in Table 2.11 do not distinguish between shell and tube side conditions. Again, we will err on the high side by using the highest values of temperature and pressure, which, most likely, will be in the tubes.

Table 2.10 Equipment-Cost Correction Factors for Material of Construction, Temperature, and Pressure

Design Pressure		Correction Factor
psia	atm	
0.08	0.005	1.3
0.2	0.014	1.2
0.7	0.048	1.1
8 to 100	0.54 to 6.8	1.0
700	48	1.1
3000	204	1.2
6000	408	1.3
Design Temperature, °C		Correction Factor
− 80		1.3
0		1.0
100		1.05
600		1.1
5,000		1.2
10,000		1.4
Material of Construction		Correction Factor
Carbon steel (mild)		1.0
Bronze		1.05
Carbon/molybdenum steel		1.065
Aluminum		1.075
Cast steel		1.11
Stainless steel		1.28 to 1.5
Worthite alloy		1.41
Hastelloy C Alloy		1.54
Monel alloy		1.65
Titanium		2.0

Source: Adapted from Ref. 40.

Table 2.11 Cost Correction Factors for Shell-and-Tube Heat Exchangers
—Design, Materials, and Pressure (Source: Reference 13.)

Material Correction Factors

Material Shell/Tube	Surface Area, ft^2				
	up to 100	100 - 500	500 - 1,000	1,000 - 5,000	5,000 - 10,000
CS/CS[a]	1.00	1.00	1.00	1.00	1.00
SS/SS	2.50	3.10	3.26	3.75	4.50
CS/SS	1.54	1.78	2.25	2.81	3.52

a. Carbon steel shell/carbon steel tubes

Pressure Correction Factors

Pressure, psig	up to 150	up to 300	up to 400	up to 800	up to 1,000
Pressure Factor	0	0.10	0.25	0.52	0.55

Design Correction Factors

Heat-Exchanger Type	Design Factor
Floating Head	1.00
Fixed Tube Sheet	0.80
U-Tube	0.85
Kettle Reboiler	1.35

Correcting Equipment Cost for Inflation

Because the cost literature reports equipment costs for some time in the past, we must correct the costs for inflation. We can calculate the present value of cost of equipment, C_2, using an inflation index, I, as given by Equation 2.9.

$$C_{2i} = C_{1i} \frac{I_{2i}}{I_{1i}} \tag{2.9}$$

There are several inflation or cost indexes in use. Examples are the Chemical Engineering Cost Index (CE Index), and the Nelson Refinery Cost Index. *Chemical Engineering* magazine publishes the CE Index regularly, whereas the *Oil and Gas Journal* reports the Nelson Refinery Index. We will use the CE Index. Cost indexes are relative to some time in the past. *Chemical Engineering* magazine defined the CE Index as equal to 100 during 1957-1959 when plant costs were relatively stable.

Chemical Engineering magazine established their index in 1963, and revised it in 1982. To revise their index they surveyed the process industry, equipment manufacturers, contractors, and consultants, as described by Chilton and Arnold [17]. The magazine determined the fractional contribution to the CE index of the many components of the average chemical plant. Determining the fractional contribution is necessary because the components inflate at different rates. The types of plant studied were fluid, fluid-solid, and solids-processing plants, built as a new plant at a new site, a new plant at an existing site, and an expansion of an existing plant. Chilton and Arnold [17] discussed other details of the CE index. The major changes in the revised index were a reduction in the number of components comprising the index from 110 to 66, the replacement of many components with more suitable ones, and the lowering of the construction productivity factor from 2.50 to 1.75. Figure 2.6 shows the fractional contribution of the many components to the revised index. BLS in the first column in Figure 2.6 is an abbreviation for the Bureau of Labor Statistics.

Table 2.12 gives the CE Indexes since 1969. As shown in Figure 2.6 and Table 2.12, the CE Index is composed of four major parts – Equipment, Construction Labor, Buildings, and Engineering and Supervision. The equipment component, in turn, is subdivided into several components. Table 2.12 lists cost indexes for all major components of the CE Index. If we sum up the fractional contribution, given in Figure 2.6, of each component of the cost index, we obtain the plant cost index. Example 2.2 illustrates the calculation of the plant cost index from the component cost indexes. The cost indexes in Table 2.12 for a given year are time averaged for the year, and thus they are more representative of mid-year values as illustrated in Example 2.3.

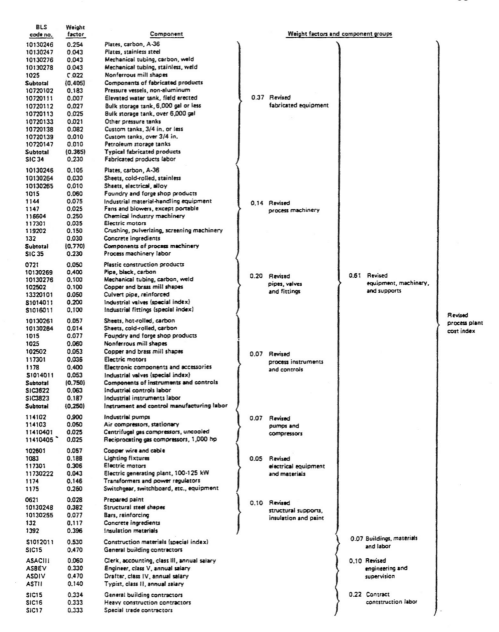

BLS code no.	Weight factor	Component	Weight factors and component groups			
10130246	0.254	Plates, carbon, A-36				
10130247	0.043	Plates, stainless steel				
10130276	0.043	Mechanical tubing, carbon, weld				
10130278	0.043	Mechanical tubing, stainless, weld				
1025	C.022	Nonferrous mill shapes				
Subtotal	(0.405)	Components of fabricated products				
10720102	0.183	Pressure vessels, non-aluminum				
10720111	0.007	Elevated water tank, field erected	0.37 Revised			
10720112	0.027	Bulk storage tank, 6,000 gal or less	fabricated equipment			
10720113	0.025	Bulk storage tank, over 6,000 gal				
10720133	0.021	Other pressure tanks				
10720138	0.082	Custom tanks, 3/4 in. or less				
10720139	0.010	Custom tanks, over 3/4 in.				
10720147	0.010	Petroleum storage tanks				
Subtotal	(0.365)	Typical fabricated products				
SIC 34	0.230	Fabricated products labor				
10130246	0.105	Plates, carbon, A-36				
10130264	0.030	Sheets, cold-rolled, stainless				
10130265	0.010	Sheets, electrical, alloy				
1015	0.060	Foundry and forge shop products				
1144	0.075	Industrial material-handling equipment	0.14 Revised			
1147	0.025	Fans and blowers, except portable	process machinery			
116604	0.250	Chemical industry machinery				
117301	0.035	Electric motors				
119202	0.150	Crushing, pulverizing, screening machinery				
132	0.030	Concrete ingredients				
Subtotal	(0.770)	Components of process machinery				
SIC 35	0.230	Process machinery labor				
0721	0.050	Plastic construction products				
10130269	0.400	Pipe, black, carbon	0.20 Revised	0.61 Revised		
10130276	0.100	Mechanical tubing, carbon, weld	pipes, valves	equipment, machinery,		
102502	0.100	Copper and brass mill shapes	and fittings	and supports		
13320101	0.050	Culvert pipe, reinforced				
S1014011	0.200	Industrial valves (special index)				
S1016011	0.100	Industrial fittings (special index)				
10130261	0.057	Sheets, hot-rolled, carbon			Revised process plant cost index	
10130284	0.014	Sheets, cold-rolled, carbon				
1015	0.077	Foundry and forge shop products				
1025	0.060	Nonferrous mill shapes				
102502	0.053	Copper and brass mill shapes	0.07 Revised			
117301	0.036	Electric motors	process instruments			
1178	0.400	Electronic components and accessories	and controls			
S1014011	0.053	Industrial valves (special index)				
Subtotal	(0.750)	Components of instruments and controls				
SIC3622	0.063	Industrial controls labor				
SIC3823	0.187	Industrial instruments labor				
Subtotal	(0.250)	Instrument and control manufacturing labor				
114102	0.900	Industrial pumps	0.07 Revised			
114103	0.050	Air compressors, stationary	pumps and			
11410401	0.025	Centrifugal gas compressors, uncooled	compressors			
11410405	0.025	Reciprocating gas compressors, 1,000 hp				
102601	0.057	Copper wire and cable				
1083	0.188	Lighting fixtures	0.05 Revised			
117301	0.306	Electric motors	electrical equipment			
11730222	0.043	Electric generating plant, 100-125 kW	and materials			
1174	0.146	Transformers and power regulators				
1175	0.260	Switchgear, switchboard, etc., equipment				
0621	0.028	Prepared paint	0.10 Revised			
10130248	0.382	Structural steel shapes	structural supports,			
10130255	0.077	Bars, reinforcing	insulation and paint			
132	0.117	Concrete ingredients				
1392	0.396	Insulation materials				
S1012011	0.530	Construction materials (special index)			0.07 Buildings, materials and labor	
SIC15	0.470	General building contractors				
ASACIII	0.060	Clerk, accounting, class III, annual salary			0.10 Revised	
ASBEV	0.330	Engineer, class V, annual salary			engineering and	
ASDIV	0.470	Drafter, class IV, annual salary			supervision	
ASTII	0.140	Typist, class II, annual salary				
SIC15	0.334	General building contractors			0.22 Contract	
SIC16	0.333	Heavy construction contractors			construction labor	
SIC17	0.333	Special trade contractors				

Figure 2.6 Components of the chemical engineering cost index. From Ref. 18 with permisson.

Table 2.12 CE Cost Indexes from 1969 to 2000 (Source: Ref. 34 with permission)

	1969	1970	1971	1972	1973	1974	1975	1976	1977	1978	2000
CE Index	119.0	125.7	132.3	137.2	144.1	165.4	182.4	192.1	204.1	218.8	394.1
Equipment	116.6	123.8	130.4	135.4	141.8	171.2	194.7	205.8	220.9	240.3	438.0
Heat Exchangers &Tanks	115.1	122.7	130.3	136.3	142.5	170.1	192.2	200.8	216.6	238.6	370.6
Process Machinery	116.8	122.9	127.9	132.1	137.8	160.0	184.7	197.5	211.6	228.3	439.4
Pipe, Valves, & Fittings	123.1	132.0	137.3	142.9	151.5	192.3	217.0	232.5	247.7	269.4	545.9
Process Instruments	126.1	132.1	139.9	143.9	147.1	164.7	181.4	193.1	203.3	216.0	368.5
Pumps & Compressors	119.6	125.6	133.2	135.9	139.8	175.7	208.3	220.9	240.2	257.5	665.3
Electrical Equipment	092.9	099.8	098.7	099.1	104.2	126.4	142.1	148.9	159.0	167.8	339.4
Structural Supports & Misc.	112.6	117.9	127.5	133.6	140.8	171.6	198.6	209.7	226.0	248.9	408.7
Construction Labor	128.3	137.3	146.2	152.2	157.9	163.3	168.6	174.2	178.2	185.9	279.2
Buildings	122.5	127.2	135.5	142.0	150.9	165.8	177.0	187.3	199.1	213.7	385.6
Engineering & Supervision	109.9	110.6	111.4	111.9	122.8	134.4	141.8	150.8	162.1	161.9	340.6

	1979	1980	1981	1982	1983	1984	1985	1986	1987	1988
CE Index	238.7	261.2	297.0	314.0	317.0	322.7	325.3	318.4	323.8	342.5
Equipment	264.7	292.6	323.9	336.2	336.0	344.0	347.2	336.3	343.9	372.7
Heat Exchangers & Tanks	261.7	291.6	321.8	326.0	327.4	334.1	336.3	314.6	321.6	357.2
Process Machinery	250.0	271.8	301.5	312.0	322.2	329.1	332.2	327.8	330.0	345.6
Pipe, Valves, & Fittings	301.2	330.0	360.1	383.2	366.6	381.2	385.0	374.5	388.3	431.1
Process Instruments	231.5	249.5	287.9	297.6	308.4	319.1	322.8	324.5	330.0	341.9
Pumps & Compressors	280.4	330.3	388.5	412.2	412.2	413.1	419.3	422.5	430.2	450.7
Electrical Equipment	183.2	206.1	222.4	235.4	242.7	248.0	251.8	251.9	256.2	269.6
Structural Supports & Misc.	273.6	297.7	322.0	338.2	338.5	343.1	346.9	342.3	344.6	370.4
Construction Labor	194.9	204.3	242.4	263.9	267.6	264.5	265.3	263.0	262.6	265.6
Buildings	228.4	238.3	274.9	290.1	295.6	300.3	304.4	303.9	309.1	319.2
Engineering & Supervision	185.9	214.0	268.5	304.9	323.3	336.3	338.9	341.2	346.0	343.3

Table 2.12 Continued

	1989	1990	1991	1992	1993	1994	1995	1996	1997	1998	1999
CE Index	355.4	357.6	361.3	358.2	359.2	368.1	381.1	381.7	386.5	389.5	390.6
Equipment	391.0	392.2	396.9	392.2	391.3	406.9	427.3	427.4	433.2	436.0	435.5
Heat Exchangers & Tanks	373.4	370.9	369.1	361.3	359.5	367.5	391.2	387.1	385.3	382.8	371.2
Process Machinery	359.2	366.3	375.4	378.2	392.2	393.5	408.6	415.5	424.8	430.8	433.6
Pipe, Valves, and Fittings	463.3	469.8	481.1	468.0	457.9	494.7	520.7	513.7	532.8	534.8	539.1
Process Instruments	352.3	353.3	353.8	356.7	376.1	365.4	377.7	372.1	371.5	365.3	363.5
Pumps & Compressors	482.6	502.9	531.0	597.4	550.9	584.2	600.8	614.5	632.2	648.5	658.5
Electrical Equipment	286.5	297.1	304.2	307.8	313.0	315.3	326.9	332.1	331.9	333.6	335.8
Structural Supports & Misc.	371.4	349.4	344.7	329.6	344.1	346.1	363.7	376.0	377.6	394.3	413.1
Construction Labor	270.4	271.4	274.8	273.0	270.9	272.9	274.3	277.5	281.9	287.4	292.5
Buildings	327.6	329.5	332.9	334.6	341.6	353.8	362.4	365.1	371.4	374.2	380.2
Engineering & Supervision	344.8	355.9	354.5	354.1	352.3	351.1	347.6	344.2	342.5	341.2	339.9

Source Ref.

Example 2.2 Calculation of the Plant Cost Index from Component Indexes

Calculate the plant cost index using the component cost indexes in 1998 from Table 2.12 and the fractional contribution of each component from Figure 2.6.

$I = 0.61\,[\,0.37\,(382.8) + 0.14\,(430.8) + 0.20\,(534.8) + 0.07\,(365.8) + 0.07\,(648.5)$

$+\,0.05\,(333.6) + 0.10\,(394.3)\,] + 0.22\,(287.4) + 0.07\,(374.2) + 0.10\,(341.2)$

$= 389.5$

The plant cost from Table 2.12 is 389.5, which agrees with the calculated value.

Example 2.3 Calculation of the Yearly-Average Cost Index

The monthly Chemical Engineering Cost Indexes for equipment are given below for 1980. Calculate the equipment cost index for the year.

January	277.2
February	281.2
March	284.6
April	290.5
May	291.3
June	292.2
July	295.3
August	296.0
September	296.9
October	300.0
November	301.7
December	304.1
Total	3511.0

The cost index for equipment for 1980 is the time averaged for the year. Thus, $I = (1/12)(3511) = 292.6$, which agrees with the equipment cost index in Table 2.12.

DEPRECIABLE CAPITAL COST

Calculation Procedures Using the Average Factor Method

There is now enough information to set up a calculating procedure using the average factor method for calculating the depreciable capital cost. Table 2.13 lists the equations and Table 2.14 outlines the calculation procedure. First, correct the equipment cost for size and inflation. Because equipment costs are sometimes correlated at an ordinary temperature, pressure, material of construction, and for a common design, the next step is to correct the base cost for the actual conditions. To use the cost factors in Tables 2.6 and 2.7 requires that we calculate the delivered equipment cost. After making these corrections, convert the FOB cost to the delivered cost by adding 10% to the FOB cost as recommended by Valle-Riestra [20]. The 10% accounts for freight, taxes, and insurance. Next, calculate the cost of installing the equipment. The installation cost includes direct, indirect, contractor's fee, and contingency costs. Use Table 2.7 for average direct-cost factors for equipment and Table 2.6 for the average indirect cost, contractor's fee, and contingency cost. After obtaining all equipment costs and cost factors at actual process conditions, calculate the installed equipment cost, C_{SAI}, using Equation 2.13.7 in Table 2.13.

Table 2.13 Summary of Equations for Depriciable Capital Cost – Average Factor Method (Based on the FOB Cost)

The subscript i refers to the FOB cost of a major piece-of-equipment.
The subscript k refers to a component of the installation cost listed in Table 2.6.

$$C_{P2i} = C_{P1i} \left(\frac{Q_{2i}}{Q_{1i}} \right)^n \qquad (2.13.1)$$

$$C_{PBi} = C_{P2i} \frac{I_{2i}}{I_{1i}} \qquad (2.13.2)$$

$C_{PAi} = f_{Ti} f_{Pi} f_{Mi} f_{Di} C_{PBi}$ —— for some equipment and (2.13.3)
$C_{PAi} = f_{Ti} (f_{Pi} + f_{Di}) f_{Mi} C_{PBi}$ —— for heat exchangers

$C_{SAi} = 1.10 C_{PAi}$ —— for actual conditions or (2.13.4)
$C_{SBi} = 1.10 C_{PBi}$ for base conditions

$f_I = \sum_k f_{Ik} + f_{DC}$ —— f_{Ik} from Table 2.6 (2.13.5)

$f_{DC} = f(\text{process type, material ....})$ —— from Table 2.7 (2.13.6)

$C_{SAI} = f_I (\sum_i C_{SAi})$ (2.13.7)

$C_{SBI} = f_I (\sum_i C_{SBi})$ (2.13.8)

f_{AB} = auxiliary-facilities factor + buildings factor, from Table 2.6 (2.13.9)

$C_D = C_{SAI} + f_{AB} C_{SBI}$ (2.13.10)

To calculate the depreciable capital cost we need to calculate the cost of buildings and auxiliary facilities. Table 2.6 contains factors for calculating these costs. Ulrich [31] pointed out that these costs are not affected by process-equipment operating temperature and pressure, materials of construction, or equipment design. Thus, we calculate the base installed cost, which is the installed cost of carbon-steel equipment at ordinary operating conditions and equipment design. To obtain the cost of auxiliary facilities and buildings, multiply C_{SBI} by f_{AB}. Now, we can now complete the calculation of the depreciable capital cost as outlined in Table 2.14.

Table 2.14 Calculation Procedure for Depreciable Capital Cost—
Average Factor Method

1. Obtain purchased-equipment costs for carbon steel at the tabulated size and capacity exponent from Table 2.9.

2. Calculate the equipment cost for the required size from Equation 2.13.1 in Table 2.13.

3. Obtain cost indexes from Table 2.12.

4. Calculate the equipment base cost at the present time, $C_{PB\,i}$, from Equation 2.13.2.

5. Calculate the actual equipment cost, C_{PAi}, at the design pressure and temperature, material-of-construction, and the required equipment design from Equation 2.13.3.

6. Calculate the delivered equipment cost from Equation 2.13.4.

7. Calculate the average installation factor for all equipment, f_I, from Equations 2.13.5 and 2.13.6.

8. Calculate the installed equipment cost at actual design conditions, C_{SAI}, from Equation 2.13.7.

9. Calculate the costs at the base conditions, C_{SBI}, from Equation 2.13.8.

10. Specify the process type.

11. Obtain the cost factors for buildings and auxiliary facilities from Table 2.6.

12. Calculate the combined factors for buildings and auxiliaries, f_{AB}, from Equation 2.13.9.

13. Calculate the depreciable capital cost from Equation 2.13.10.

Calculation Procedures Using the Individual Factor Method

Table 2.15 contains equations for calculating the depreciable capital cost using the individual factor method. The equations are similar to the average factor method. Table 2.16 outlines the calculation procedure. Table 2.8 contains direct-cost factors for several pieces-of-equipment, which depends on the material-of-construction. For indirect costs Guthrie [13] uses 1.34 for fluid processes and 1.29 for solids processes. He also uses 15% and 3% of the installation factor for contingency and the contractor's fee. Again, because process operating conditions and materials of construction do not affect the cost of buildings and auxiliary facilities, we use the base installed costs to calculate these costs. For quick estimates Guthrie [36], uses 2 to 6% of the installed costs for buildings and 17 to 25% for auxiliary facilities. Use averages of 4% and 21% respectively for both costs. A calculation procedure for the depreciable capital costs is outlined in Table 2.16.

Table 2.15 Summary of Equations for Depreciable Capital Cost – Individual Factor Method (Based on Purchased Equipment Cost (FOB))

The subscript i refers to a major piece-of-equipment.

$$C_{PB2i} = C_{PB1i} \left(\frac{Q_{2i}}{Q_{1i}} \right)^n \tag{2.15.1}$$

$$C_{PBi} = C_{PB2i} \ \frac{I_{2i}}{I_{1i}} \tag{2.15.2}$$

$C_{PAi} = f_{Ti} \, f_{Pi} \, f_{Mi} \, f_{Di} \, C_{PBi}$ —— for some equipment (2.15.3)
$C_{PAi} = f_{Ti} (f_{Pi} + f_{Di}) \, f_{Mi} \, C_{PBi}$ —— for heat exchangers

$f_{Ii} = f_{DCi} \, f_{ICi} \, f_{CFi}$ —— f_{DCi} from Table 2.8 (2.15.4)
$f_{ICi} = 1.34$ for a fluid process or 1.29 for a solids process
$f_{CFi} = 1.18$

$$C_{AI} = \sum_i f_{Ii} \, C_{PAi} \tag{2.15.5}$$

$$C_{BI} = \sum_i f_{Ii} \, C_{PBi} \tag{2.15.6}$$

$$C_D = C_{AI} + f_{AB} \, C_{BI} \tag{2.15.7}$$

f_{AB} = auxiliary-facilities factor + buildings factor = $0.04 + 0.21 = 0.25$ (2.15.8)
$f_{AB} = 0$ for a plant addition

Table 2.16   Calculation Procedure for Depreciable Capital Cost—Individual Factor Method

1. Obtain purchased equipment costs for carbon steel at the tabulated size and also the capacity exponent from Table 2.9.

2. Calculate the equipment cost at the required size from Equation 2.15.1 in Table 2.15.

3. Obtain cost indexes from Table 2.12.

4. Calculate the equipment base cost, C_{PB2i}, at the present time from Equation 2.15.2.

5. Calculate the equipment cost, C_{PAi}, at the design pressure and temperature, material of construction, and the required equipment design from Equation 2.15.3.

6. Obtain the direct-cost factors for each piece-of-equipment from Table 2.8.

7. Calculate the equipment installation factor, f_{Ii}, from Equation 2.15.4.

8. Repeat steps 1 to 7 for all major equipment.

9. Calculate the installed equipment cost at actual process conditions, C_{AI}, from Equation 2.15.5.

10. Calculate the installed equipment cost at the base conditions, C_{BI}, from Equation 2.15.6.

11. Obtain f_{AB}, the combined factors for buildings and auxiliaries, from Equation 2.15.8.

12. Calculate the depreciable capital cost from Equation 2.15.7.

TOTAL CAPITAL COST

The total capital cost consists of the depreciable capital cost, land cost, land or site development cost, startup cost, and working capital. In theory, land cost is completely recoverable when a plant shuts down, and □herefore is not depreciable. Land cost varies from 0.01 to 0.02 times the depreciable capital cost. Use an average value of 0.015.

Land development cost, which is not depreciable, consists of such items as site clearing, construction of roads, walkways, railroads, fences, parking lots, wharves, piers, recreational areas, and landscaping. Presumably, these items improve the value of the land, and their costs, to a certain extent, are recoverable. Table 2.17 lists land development cost for three process types as a fraction of the depreciable capital cost.

To startup a plant requires additional capital. It is expected that some equipment will not work after it is installed. Each process unit requires testing because of possible leaks, incorrect wiring of electrical equipment, and many mechanical problems. Humphreys [5] divides startup costs into two parts, those costs resulting from technical difficulties and those costs associated with personnel, which we will call operations startup. The technical costs are associated with process modifications and consists of equipment alterations, modifications, and adjustments to make the process operable. The operations startup consists of such items as operator training, extra operators and supervisors, raw materials that result in off-grade products, and several other items. Humphreys [5] gives a detailed list of the many items contained in startup costs. Startup occurs continuously over time as equipment is installed, and it overlaps the constructional and production phases of a plant. It is usually not clear at what time construction ends and production begins. Startup is usually defined as the period between the completion of plant construction and when steady-state operation begins. According to Humph-

Table 2.17 Factors for Estimating Land-Development Cost

Plant at an Existing Site
Fraction of Depreciable Capital Cost, f_L

Solids Processing Plant	Solids-Fluid Processing Plant	Fluid Processing Plant
0.0285	0.0249	0.0211

reys [5], the cost of a plant should include all costs required to make the plant operational. Startup cost seldom exceeds 10% of the fixed capital cost. Peters and Timmerhaus [4] recommend 8 to 10%. According to Peters and Timmerhaus, the startup cost can be accounted for in the first year of plant operation or in the total capital investment. We will assume that it will be accounted for in the first year of operation.

Working capital is the money required to finance the daily operations of a plant. As stated earlier, it consists of the money required to buy raw materials and store products, accounts receivable, and storage of various supplies, which are necessary to keep the plant operating. About one month's supply of raw materials and products, and one month of accounts receivable would suffice, or 20% of the fixed capital cost. Thus, the total capital cost,

$$C_T = C_D + 0.015\ C_D + f_L\ C_D + 0.20\ C_F \qquad\qquad [28]$$

Example 2.4 Capital-Cost Estimation of an Allyl-Choride-Synthesis Process

Allyl alcohol and glycerin can be synthesized from allyl chloride (3-chloro-1-propene) [19]. Gas-phase thermal chlorination of propene has been proposed as a route to allyll alcohol. In this process, shown in Figure 2.4.1, a process furnace heats propylene, which then mixes with chlorine in a mixer. The intersecting streams in the eductor-mixer create turbulence and hence enhance mixing. The chlorine reacts with propylene inside the tubes of two parallel shell-and-tube reactors. Dowtherm A, a heat-transfer fluid used at high temperatures, removes the enthalpy of reaction. A pump circulates the Dowtherm A through the shell of the reactors and through a water cooler. The products flow through air-cooled heat exchangers, where fans blow cool air across the tubes of the cooler to remove heat. The cooled product stream then condenses to form a crude allyl chloride stream containing several by-products. Finally, the crude allyl chloride flows to a separation section of the process.

The process design for the production of allyl chloride has been completed. Table 2.4.1 lists the specifications for the major pieces-of-equipment. Estimate the depreciable capital cost and the total capital cost as of mid-1998. The process is a plant addition at an existing site, i.e., buildings and auxiliary facilities are available. The cost of the eductor-mixer as of mid-1998 is $ 1,000.

First, convert all equipment costs to a common basis of FOB costs as of mid-1998. Table 2.9 contains the costs of some common equipment as of January 1990, except where indicated. Since the allyl chloride section of the process is a small installation, use cost indexes for specific equipment rather than the plant cost index, which is an average of all equipment. Follow the calculation procedure outlined in Table 2.16, which uses the equations listed in Table 2.15.

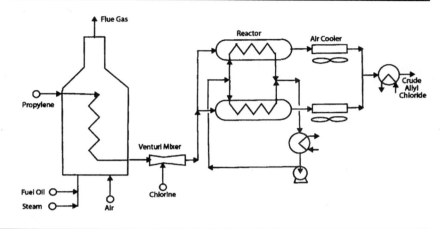

Figure 2.4.1 Allyl-chloride-synthesis process.

Table 2.4.1 Equipment Specificatons for the Allyl Chloride Process

Equipment	No	Size	Units	Design Temperature °F	Design Pressure psia	Material
Propylene Heater	1	5.5x10^6	Btu/h	2,000	60	CS
Chlorinators (Fixed-Tube Sheet)	2	330	ft²	2,000	60	CS/CS[a]
Dowtherm Pump (Centrifugal Pump)	1	7.5 290 65	hp gpm ft(head)	550	assume 30 psig	CS
Air Coolers	2	145	ft²	1,040	50	CS/CS
Dowtherm Cooler (Fixed-Tube Sheet)	1	63	ft²	550	50	CS/CS
Condenser (Fixed-Tube Sheet)	1	364	ft²	200	50	CS/CS
Eductor-Mixer	1	7560	lb/h	1,000	60	CS

a. CS/CS means carbon steel shell and carbon steel tubes.

Propylene Heater

Table 2.9 contains the cost of process furnaces, also called process heaters. The cost of a furnace with a heating rate of 20,000 kJ/s is $ 750,000 in mid 1982. Converting the heating rate, 5.5×10^6 Btu/h, given in Table 2.4.1, to kJ/s we obtain 1.612×10^3 kJ/s. This heating rate is below the lower limit of the correlation range given in Table 2.9. Because we have no other data, use the data in Table 2.9 to estimate the heater cost. From Equation 2.15.1 in Table 2.15, we find that the base cost,

$$C_{PB2i} = 750,000 \left(\frac{1,612}{20,000} \right)^{0.85} = \$88,200$$

Next, correct for inflation. Adjust the base cost from January 1990 to mid-1998 using Equation 2.15.2. The cost indexes for equipment are listed in Table 2.12,

$$C_{PBi} = 88,200 \ \frac{436.0}{336.2} = \$ 114,400$$

To obtain the cost at design conditions, correct the base cost for temperature, pressure, material of construction, and equipment design. In Table 2.4.1, the operating temperature is specified as 2,000 °F. From Table 2.10, the temperature is between 600 and 5000 °C. Taking the high value, the temperature correction factor is 1.2. The pressure is at base conditions, and therefore the pressure correction factor is 1.0. Because the furnace is constructed of carbon steel the material correction factor is also 1.0. In this case, the design factor is assume to be 1.0. Thus, from Equation 2.15.3, the furnace cost at design conditions is

$$C_{PAi} = 1.2 \ (1.0) \ (1.0) \ (1.0) \ (1.144 \times 10^5) = \$ 1.373 \times 10^5$$

From Table 2.8, the direct-cost factor for a furnace is 1.3, and from Equation 2.15.4 in Table 2.15, the indirect-cost factor is 1.34 for a fluid process, and the factor for contingency and the contractor's fee is 1.18. Then, from Equation 2.15.4, the installation factor for the furnace,

$$f_{Ii} = 1.3(1.34)(1.18) = 2.056$$

Chlorinators

Because there appears to be no cost data for chlorinators, we will approximate the cost by using a fixed-tube, shell-and-tube heat exchanger. From Table 2.9, the cost of a 1,000 ft^2 floating-head heat exchanger in January 1990 was $14,000. As indicated in Table 2.4.1, each chlorinator requires 330 ft^2 of surface area.

The cost of a chlorinator in January 1990 was

$$C_{PB2i} = 14,000 \left(\frac{330}{1,000} \right)^{0.65} = \$6,810$$

To obtain the January 1990 cost index, interpolate between 1989 and 1990, the mid-year indexes for heat exchangers given in Table 2.12. The cost of the chlorinator in mid-1998,

$$C_{PBi} = 6810 \, \frac{382.8}{372.2} = \$7,004$$

Table 2.11 contains correction factors for pressure, material-of-construction, and design. Table 2.10 contains the correction factor for temperature. Thus, the cost of the chlorinator at design conditions from Equation 2.15.3 is,

$$C_{PAi} = 1.2 \, (0 + 0.8) \, (1.0) \, (7004) = \$6,724$$

Table 2.8 does not contain a direct-cost factor for a CS/CS heat exchanger. Use 2.0, which is close to other factors for shell-and-tube heat exchangers. Again, the indirect-cost factor is 1.34, and the factor for contingency and contractor's fee is 1.18. Thus, the installation factor for the chlorinator,

$$f_{Ii} = 2.0 \, (1.34) \, (1.18) = 3.162$$

Dowtherm Cooler

The installed-cost calculation for the Dowtherm cooler follows the same procedure as the calculation for the chlorinator.

Correct for size.

$$C_{PB2i} = 1.4 \times 10^4 \left(\frac{63}{1,000} \right)^{0.65} = \$2,321$$

Correct for inflation from January 1990 to mid-1998. Use the cost index for heat exchangers.

$$C_{PB\,i} = 2321 \, \frac{382.8}{372.2} = \$2,387$$

Correct for temperature ($f_T = 1.1$), pressure (0), heat exchanger design ($f_D = 0.8$), and material of construction ($f_M = 1.0$).

$$C_{PA\,i} = 1.1 \, (0 + 0.8) \, (1.0) \, (2387) = \$2,101$$

Calculate the installation factor.

$$f_{I\,i} = 2.0 \, (1.34) \, (1.18) = 3.162$$

Condenser

Repeat the above calculation for the condenser.

$$C_{PB2i} = 1.4 \times 10^4 \left(\frac{364}{1,000} \right)^{0.65} = \$7,258$$

$$C_{PB\,i} = 7258 \, \frac{382.8}{372.2} = \$7,465$$

$$C_{PA\,i} = 1.05 \, (0 + 0.8) \, (1.0) \, (7465) = \$6,271$$

$$f_{I\,i} = (2.0) \, (1.34) \, (1.18) = 3.162$$

Air Coolers

Correct for size.

$$C_{PB2i} = 137 \left(\frac{145}{1} \right)^{0.8} = \$7,342$$

Correct for inflation from mid-1987 to mid-1998. Use the cost index for heat exchangers.

$$C_{PB\,i} = 7,342 \; \frac{382.8}{321.6} = \$8,739$$

Correct for temperature ($f_T = 1.1$), pressure ($f_P = 1.0$), and materials-of-construction ($f_M = 1.0$). There is no information on a correction factor for design. Use $f_D = 1.0$.

$$C_{PA\,i} = 1.1 \, (1.0) \, (1.0) \, (1.0) \, (8,739) = \$9,613$$

Calculate the installation factor.

$$f_{I\,i} = (2.5) \, (1.34) \, (1.18) = 3.953$$

Dowtherm Pump

Correct for size.

$$C_{PB2i} = 9,000 \left(\frac{7.5}{20.0} \right)^{0.42} = \$5,961$$

Correct for inflation from January 1990 to mid-1998. Use the cost index for pumps.

$$C_{PB\,i} = 5,961 \; \frac{648.5}{492.8} = \$7,844$$

$$C_{PA\,i} = 1.1 \, (1.0) \, (1.0) \, (1.0) \, (7844) = \$8,628$$

$$f_{I\,i} = 2.8 \, (1.34) \, (1.18) = 4.427$$

Eductor-Mixer

The cost of the mixer is given as \$1000 FOB – mid-1998. No inflation or size correction is required.

$C_{PB\,i} = \$1,000$

$C_{PA\,i} = 1.1\,(1.0)\,(1.0)\,(1.0)\,(1,000) = \$1,100$

From Table 2.8, the direct-cost factor for miscellaneous equipment for carbon steel is 2.0.

$f_{I\,i} = 2.0\,(1.34)\,(1.18) = 3.162$

Installed Cost

Equation 2.15.5 gives the total installed cost for all major equipment at actual process conditions. As Figure 2.4.1 shows, there are two reactors and two air coolers. Add all installed costs for the equipment at actual conditions.

$C_{AI} = 2.056\,(137,200) + 2\,(3.162)\,(6,724) +\ 3.162\,(2101) + 3.162\,(6,271)\ + 2\ (3.953)\,(9,613) + 4.427\,(8,628) + 3.162\,(1,100) = \$468,800$

Equation 2.15.6 gives the total installed cost for all major equipment at the base conditions. Add all installed costs for the equipment at the base conditions.

$C_{BI} = 2.056\,(114,400) + 2\,(3.162)\,(7,004) + 3.162\,(2,387) + 3.162\,(7,456) + 2\ (3.953)\,(12,930) + 4.427\,(7,844) + 3.162\,(1,000) = \$417,600$

We see that the cost of the installation at ordinary process conditions, C_{BI}, is less than the cost at actual conditions, C_{AI}. If this process were a grass-roots plant, we would have to add the additional cost of buildings and auxiliary facilities. In this case, the process is a plant addition at an existing site where the buildings and auxiliary facilities are available. Therefore, $f_{AB} = 0$, as given in Equation 8.15.8. Thus, the depreciable capital cost is equal to \$468,800.

The fix capital cost equals the sum of the depreciable capital cost, land cost, and site development cost. Because this process will be built at an existing site, the land cost is not a consideration. Table 2.17 lists site-development factors for a plant at an existing site. For a fluid processing plant, the factor is 0.0211. Thus, the fixed capital cost,

$C_F = 468,800 + 0.0211\,(468,800) = \$478,700$

Finally, the total capital cost for the project equals the sum of the fixed capital cost and working capital.

$$C_T = 478,700 + 0.20\,(478,700) = \$574,400$$

NOMENCLATURE

b capacity exponent for operating labor estimation

C cost

C_A actual equipment cost

C_{AI} actual equipment cost, installed

C_B base equipment cost

C_{BI} base equipment cost, installed

C_D depreciable capital cost

C_F fixed capital cost

C_M production or manufacturing cost

C_S delivered equipment cost

C_P purchased or FOB cost

C_{PA} purchased equipment cost based on actual process conditions

C_{PB} purchased equipment cost based on base process conditions (ordinary pressures, temperatures, and materials)

C_T total capital cost

C_W working capital

f_{AB} auxiliary-facilities factor + buildings factor (fraction of base cost)

f_{CF} contingency and contractor's fee factor (fraction of actual or base cost)

f_D design factor (fraction of base equipment cost)

f_{DC} direct-cost factor (fraction actual or base cost)

f_{IC} indirect-cost factor (fraction of actual or base cost)

f_I installation factor

f_L site development factor (fraction of depreciable capital cost)

f_M material factor (fraction of base equipment cost)

f_P pressure factor (fraction of base equipment cost)

f_S fractional salvage value

f_T temperature factor (fraction of base equipment cost)

i   fractional interest rate

I   inflation index

K   plant productivity

L   operating-labor man hours, h/kg

m   plant capacity, kg/h

n   number of years since 1952 or equipment-cost capacity component

N   number of process units

p   fractional, yearly labor-productivity increase per year

Q   equipment capacity

Subscripts

i   refers to a piece-of-equipment

k   component of the installation factor listed in Table 7

REFERENCES

1.   Winter, O., Preliminary Economic Evaluation of Chemical Processes at the Research Level, Ind. Eng. Chem., 61, 4, 45, 1969.
2.   Uhl, VW., Hawkins, A.W., Technical Economics for Engineers, AIChE Continuing Education Series, 5, American Institute of Chemical Engineers, New York, NY, 1971.
3.   Perry, R.H., Chilton, C.H., eds., Chemical Engineer's Handbook, 5th ed., McGraw-Hill, New York, NY, 1973.
4.   Peters, M.S., Timmerhaus, K.D., Plant Design and Economics for Chemical Engineers, 4th ed., McGrawHill, New York, NY, 1991.
5.   Humphreys, K.K., ed., Jelen's Cost and Optimization Engineering, 3rd ed., McGraw-Hill, New York, NY, 1970.
6.   Happel, J., Jordan, D. G., Chemical Process Economics, Marcel Dekker, New York, NY, 1970.
7.   Cevidalli, G., Zaidman, B., Evaluate Research Projects, Chem. Eng., 87, 14, 145, 1980.
8.   Wessel, H.E., New Graph Correlates Operating Labor Data for Chemical Processes, Chem. Eng., 59, 7, 209, 1952.

9. Bridgewater, A.V., The Functional Approach to Rapid Cost Estimation, AACE Bulletin, 18, 5, 153, 1976.
10. Woods, DR., Financial Decision Making in the Process Industry, Prentice Hall, Englewood Cliffs, NJ, 1975.
11. Holland, F.A., Watson, F.A., Wilkinson, J.K., Introduction to Process Economics, John Wiley & Sons, New York, NY, 1974.
12. Kohn, P., Ethylenediamine Route Eases Pollution Worries, Chem. Eng., 85, 7, 90, 1969.
13. Guthrie, K.M., Data and Techniques for Preliminary Cost Estimating, Chem. Eng., 76, 7, 114, 1969.
14. Lang, H.J., Simplified Approach to Preliminary Cost Estimates, Chem. Eng., 55, 6, 112, June 1948.
15. Popper, H., ed., Modern Cost Engineering Techniques, McGraw-Hill, New York, NY, 1970.
16. Chilton, C., Popper, H., Norden, R.B., Modern Cost Engineering: Methods and Data, McGraw-Hill, New York, NY, 1979.
17. Chilton, C.H., Arnold, T.H., New Index Shows Plant Cost Trends, Chem. Eng., 70, 4, 143, 1963.
18. Matley J., CE Plant Cost Index - Revised, Chem. Eng., 89, 8, 153, 1982.
19. Student Contest Problem, American Institute of Chemical Engineers, New York, NY, 1973.
20. Valle-Riestra, J.F., Project Evaluation in the Chemical Process Industries, McGraw-Hill, New York, NY, 1983.
21. Student Contest Problem, American Institute of Chemical Engineers, New York, NY, 1993.
22. Student Contest Problem, American Institute of Chemical Engineers, New York, NY, 1995.
23. Student Contest Problem, American Institute of Chemical Engineers, New York, NY, 1996.
24. Student Contest Problem, American Institute of Chemical Engineers, New York, NY, 1997.
25. Student Contest Problem, American Institute of Chemical Engineers, New York, NY, 1998.
26. Student Contest Problem, American Institute of Chemical Engineers, New York, NY, 1985.
27. Holland, F. A., How to Evaluate Working Capital, Chem. Eng., 81, 7, 7 1, Aug. 5,1974.
28. Hand, W. E., From Flow Sheet to Cost Estimate, Petroleum Refiner, 3 7, 9, 13 3, 1958.
29. Remer, D. S., Chai, L.H., Design Factors for Scaling-up Engineering Equipment, Chem. Eng. Progr., 87, 8, 77, 1990.
30. Walas, S.M., Chemical Process Equipment, Butterworths, Boston, MA, 1988.

31. Ulrich, G.D., A Guide to Chemical Engineering Process Design and Economics, John Wiley & Sons, New York, NY, 1984.

32. Wilden, W., Personal Communication, Allied Chemical Co., Morristown, NJ, about 1975.

33. Drew, J. W., How to Estimate the Cost of Pilot-Plant Equipment, Chem. Eng., 77, 3, 100, 1970.

34. Gillis, M., Personal Communication, Chemical Engineering, McGraw-Hill, New York, NY, May 25, 1999.

35. Cran, J., Improved Factor Method Gives Preliminary Cost Estimate, Chem. Eng., 88, 7, 65, 1981.

36. Guthrie, K. W., Process Plant Estimating, Evaluation, and Control, Craftsman Book Company of America, Sloana Beach, CA, 1974.

37. Baasel, W.D., Preliminary Chemical Engineering Plant Design, 2nd ed., VanNostrand, New York, NY, 1990.

38. Holland, F.A., Wilkinson, J.K., Process Economics, Perry's Chemical Engineering Handbook, 7th ed., Perry, R.H., Green, D.W., eds., McGraw-Hill, New York, NY, 1997.

39. Pikulik, A., Diaz, H.E. Cost Estimating for Major Process Equipment, Chem, Eng., 84, 21, 106, 1977.

40. Allen, D.H., Page R.C., Revised Techniques for Predesign Cost Estimating, Chem. Eng., 82, 5, 142, 1975.

3

Process Circuit Analysis

We can view any process as a circuit analagous an electrical circuit. Instead of voltage differences between points in the circuit, there are pressure differences. Instead of current flow, there is mass flow. Before a process can be completely designed, all the mass flow rates, compositions, temperatures, pressures and energy requirements in all parts of the process must be known. Process engineers usually specify pressure drops and temperatures from experience. They calculate mass flow rates, which are traditionally treated in a course in mass and energy balances. However, mass and energy balances are only a partial set of equations that process engineers can write when analyzing a process circuit.

The objective of process circuit analysis is to determine specifications for the process. These include temperatures, pressures, composition, and flow rates of all streams. Also included is the energy transferred and the degree of separation or reaction required of heat exchangers, reactors, and separators. After specifying recoveries and conversions of components, the process engineer can calculate the mass and energy requirements for a process. The process engineer will generate specifications for all process units, which must be fulfilled by equipment design experts. In a sense, process engineers are conductors, controlling the design of the process. It is their responsibility to see that all the pieces fit.

STRATEGY OF PROBLEM SOLVING

Before proceeding, we will examine the structure of problem solving by considering the following procedure:

1. list the appropriate relations and unknown variables for the problem
2. calculate the degrees of freedom
3. specify or unspecify variables until the degrees of freedom are zero
4. determine a solution procedure
5. solve the equations
6. organize the results in tabular or graphical form
7. check the solution.

When implementing this procedure, proceed step-by-step. Do not carry out a step before completing the preceding step, particularly when executing step one. Also, do not combine steps, e.g., attempting to carry out steps four and five before completing steps two and three. Formulate the problem first, i.e., complete steps one to three. Then, it will be certain that a solution exists. Frequently, steps one to four are executed simultaneously. The numerical solution to the problem is begun, and equations are introduced along the way as needed. Eventually a solution is obtained. With experience the process engineer can recognize that certain problems have solutions, however, in most cases, it is not initially evident that there is enough information or what the most efficient solution procedure should be.

Polya [1], who has examined the nature of problem solving, has devised a similar procedure. He states, "First, we have to understand the problem; we have to see clearly what is required. Second, we have to see how the various items are connected, how the unknown is linked to the data, in order to obtain the idea of the solution, to make a plan. Third, we carry out our plan. Fourth, we look back at the completed solution, we review and discuss it."

Executing the steps systematically uncovers what information is missing and results in better insight into the structure of the problem. We learn continuously. Polya [1] again states that, "Our conception of the problem is likely to be rather incomplete when we start the work; our outlook is different, when we have made some progress; it is again different when we have almost obtained the solution."

PROCESS-CIRCUIT RELATIONSHIPS

Executing steps one to three in the procedure is the process of defining a problem. Before solving a set of equations, you must clearly show that the number of equations equals the number of unknowns. Circumventing this step will result in considerable wasted effort. The relationship between the number of equations and unknowns is expressed by

$$F = V - R \tag{3.1}$$

where F is the degrees of freedom, V the number of variables, and R the number of independent relations. If F is positive, the number of variables is in excess and the problem is under-specified. If F is negative, the number of equations is in excess, and the problem is over-specified. Only if F is zero can you calculate values for all variables. Usually, when formulating the problem, the number of variables is in excess and we must specify additional variables. First, however, you must be certain that you have not omitted any relations. The excess variables are called degrees of freedom, supposedly because we are "free" to designate numerical values for any of the variables in the equation set to obtain zero degrees of freedom.

To execute step one requires knowing what relations are available for analyzing process circuits. Mass and energy balances h`ave already been mentioned. Below is a list of relations.

1. mass balance
2. energy balance
3. momentum balance
4. rate equations
    a) heat transfer
    b) mass transfer
    c) chemical reaction
5. equilibrium relations
    a) phase
    b) chemical
6. economic relations
7. system property relationships
    a) thermodynamic
    b) transport
    c) transfer
    d) reaction
    e) economic data

Generally, when analyzing process circuits our only interest is in the macroscopic behavior of each process unit, i.e., the relationship between the inlet and outlet streams. We will not consider the microscopic behavior of the components within the unit. At this point, our interest is in what the process unit does, not how it accomplishes its task. To do otherwise will greatly increase the complexity of the analysis. The problem usually is: given the flow rates, compositions, temperatures, and pressures of all inlet streams, determine these properties for all the outlet streams. One way to avoid considering the detailed behavior of a process unit is to obtain a relationship between the exit streams. For example, for a partial condenser, the exit streams are the vapor and liquid streams. To predict accurately the composition of the exit streams will require considering simultaneous heat and mass transfer rates in the condenser and integrating a set of differential equations. Integration requires knowing the length of the condenser, which is the objective of the analysis. A quicker approach is to specify recoveries, compositions or an approach to equilibrium of the components, whatever we know from experience or pilot-plant studies. If we expect from experience that the exiting vapor and liquid streams will approach equilibrium for a reasonable condenser length, then we can calculate the compositions of the exit streams. Later, the heat exchanger designer, the expert, will satisfy the equilibrium condition by designing a condenser of sufficient length to approach equilibrium. Then, he will have to consider the rates of mass and heat transfer because rate processes determines the size of all equipment.

Mass Balances

In general, for unsteady state, the component mass or mole balance for each process unit may be stated as

rate of flow in + rate of depletion + rate of formation by reaction =
rate of flow out + rate of accumulation + rate of disappearance by reaction   (3.2)

Because the system either gains or loses mass, drop either of the rate terms for depletion or accumulation. To apply Equation 3.2 to a specific situation, the first decision requires determining whether the process operation is steady or unsteady state. The unsteady-state operations are:

1. startup
2. change over to a new operating conditions
3. periodic
4. disturbances

An example of the application of Equation 3.2 can be seen in Figure 3.1. Consider the steady-state operation of the steam stripper. Steam stripping is a common operation in waste-water treatment for removing small amounts of organic compounds from water. Nathan [4] discusses processes for removing chlorinated hydrocarbons from wastewater. In this example, we will consider removing ethlyene dichloride. It is good practice to always analyze a problem by starting with a general relationship, like Equation 3.2, and drop those terms that do not apply or are too small to be of any significance. For steady state, drop both the rates of depletion and accumulation terms. Because there is no chemical reaction, drop the chemical reaction terms. Thus, Equation 3.2 reduces to

rate of flow in = rate of flow out   (3.3)

To apply Equation 3.3, first begin by numbering the process steams, as shown in Figure 3.1. We will always designate the flow rate as m regardless of the units employed: mass, molar, English or S.I., and we will frequently designate the concentration variable as y regardless of its units. Also, use numerical subscripts

Subscripts:
First Subscript – stream number
Second Subscript
Ethylene Dichloride – 1
Water – 2

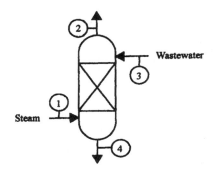

Figure 3.1 Steam stripping of ethylene dichloride from wastewater.

for the stream number and the component being considered. Write the stream number first, according to Figure 3.1, and the component number second, where 1 indicates ethylene dichloride and 2 water. Thus, $y_{4,1}$ means that in stream four the concentration of ethylene dichloride is $y_{4,1}$. If we use molar flow rates, y must be in mole fraction units. A component balance may be written for each component and for an n component system, n independent component balances may be written. In this case, we may write, according to Equation 3.3, for ethylene dichloride and water

$$y_{3,1}\ m_3 = y_{2,1}\ m_2 + y_{4,1}\ m_4 \tag{3.4}$$

$$y_{1,2}\ m_1 + y_{3,2}\ m_3 = y_{2,2}\ m_2 + y_{4,2}\ m_4 \tag{3.5}$$

Because for stream 1 contains no ethylene dichloride, we can also write

$$y_{2,1} + y_{2,2} = 1 \tag{3.6}$$

$$y_{2,1} + y_{2,2} = 1 \tag{3.7}$$

$$y_{4,1} + y_{4,2} = 1 \tag{3.8}$$

The total mole balance is not an independent equation because by adding the component balances and then substituting Equations 3.4 to 3.8 into the sum will yield the total mole balance,

$$m_1 + m_3 = m_2 + m_4 \qquad\qquad (3.9)$$

If you decide to use the total balance, then you must eliminate one of the equations from 3.4 to 3.8, given above. You may eliminate any one of the equations. The equation eliminated will depend on the particular problem. Even if the total balance is not an independent equation, it still must be satisfied and could be use as a check on your computations.

Energy Balances

The macroscopic energy balance is used whenever energy changes occur, particularly energy exchange with the surroundings. Energy exchange occurs frequently because of the need to cool or heat process streams and to transfer liquids, gases or solids from one process unit to another. Energy exchange usually occurs more frequently than separation and chemical reaction. The energy balance is given by

$$\Delta h + \frac{v^2}{2g} + g\,\Delta z = Q - W \qquad\qquad (3.10)$$

which states that the change in enthalpy in the process unit must be compensated for by a change in kinetic energy, potential energy, heat transferred into the system, and work done by the system. In many processes, the kinetic and potential energy changes are small when compared to the magnitude of the other terms and may be neglected.

Rate Equations

All physical and chemical transformations take time. Some physical phenomena, such as the vaporization at a boiling liquid surface, occurs very rapidly and for all practical purposes are instantaneous. Also, some chemical reactions, such as combustion reactions, are very rapid, but mass transfer and many chemical reactions are very slow by comparison. For such phenomena to occur to the extent desired requires allowing sufficient time, which is achieved by allowing sufficient equipment volume or surface area. Rate equations, then, are necessary to determine equipment sizes. For example, the well-known expression for the rate of heat transfer,

$$Q = U\,A\,(\Delta t)_{LM} \qquad\qquad (3.11)$$

is frequently used to determine the surface area required, A, for the required heat transferred, Q. In process circuit analysis, as discussed earlier, the stream properties of a process circuit can be determined by initially avoiding the complication of considering rate equations by specifying an approach to equilibrium. Later, to determine the size of the process units to achieve the required energy transfer, chemical conversion, and degree of separation, requires using rate equations.

Equilibrium Relations

From the previous discussion, equilibrium relations required for process circuit analysis are evidently important. To achieve equilibrium requires equipment infinite in size, which is a physical and economical impossibility. We must be satisfied with an economical approach to equilibrium conditions. In some cases, because of rapid mass transfer or chemical reaction, the difference between actual and equilibrium conditions is insignificant.

By assuming chemical equilibrium at the exit of a reactor, we can write a relationship between the composition of the components in the exit stream. For example, for the oxidation of SO_2 with O_2 to give SO_3

$$2\ SO_2 + O_2 \rightarrow 2\ SO_3 \tag{3.12}$$

At equilibrium,

$$K_P = \frac{(p_{SO3})^2}{p_{O2}\,(p_{SO2})^2} = \frac{(y_{SO3})^2}{y_{O2}\,(y_{SO2})^2\,P} \tag{3.13}$$

We can write an equilibrium relation for each independent reaction.

Similarly, for a single stage separator, if we assume equilibrium between phases leaving the separator, we may write a relationship between the composition of a component in each phase leaving the separator. Consider a solution of methane and propane being flashed across a valve. Downstream of the valve, we may write an equation to express the phase equilibrium of methane in a way that is similar to chemical equilibrium

$$CH_4\ (1) \rightarrow CH_4\ (g) \tag{3.14}$$

The relationship between the composition of methane in the vapor and liquid phases is

$$K_M = \frac{y_{VM}}{y_{LM}} \tag{3.15}$$

We can write a similar relationship for propane. One equilibrium relationship can be written for each component in a mixture.

System Properties

After writing mass balances, energy balances, and equilibrium relations, we need
system property data to complete the formulation of the problem. Here, we divide
the system property data into thermodynamic, transport, transfer, reaction properties, and economic data. Examples of thermodynamic properties are heat capacity,
vapor pressure, and latent heat of vaporization. Transport properties include viscosity, thermal conductivity, and diffusivity. Corresponding to transport properties are the transfer coefficients, which are friction factor and heat and mass transfer coefficients. Chemical reaction properties are the reaction rate constant and
activation energy. Finally, economic data are equipment costs, utility costs, inflation index, and other data, which were discussed in Chapter 2.

There frequently seems to be insufficient system property data. We may obtain accurate system property data from laboratory measurements, which are expensive. To avoid making measurements, we must rely on correlations or empirical equations for estimating these data. Reid et al. [2] have compiled many useful
methods for estimating thermodynamic as well as transport properties. In most
cases, these methods are empirical or at best semi-empirical with limited accuracy.
The accuracy of system property data may limit the accuracy of process calculations. Without experimental data, we can attempt to estimate the thermodynamic
property from a knowledge of the molecular structure of a molecule. For example,
if we know the molecular structure of a pure organic compound, its heat capacity
may be estimated by adding the contribution to the heat capacity made by various
functional groups, such as —CH$_3$, —OH, —O—, etc., as illustrated by Reid et al.
[2]. We can estimate other properties by these "group methods." An ultimate goal
of physical property research is to be able to calculate accurately any physical
property of a compound from its basic molecular properties. Thus, we can reduce
the need for costly property measurements.

Temperature and composition affect physical properties, but the effect of
pressure is generally small and we can neglect it. One exception is gas density. A
well known example of the effect of temperature is the variation of heat capacity
of a gas with temperature, which is generally curve fitted in the form of a polyno-
mial.

$$c_p = a + b\,T + c\,T^2 + d\,T^3 \qquad\qquad (3.16)$$

An equation of state describes the variation of molar density of a gas with
pressure and temperature. For a gas at high temperature and low pressure, the
ideal gas law,

$$\rho = P/RT \qquad\qquad (3.17)$$

is sufficiently accurate, but we may use it at a high pressure if we are willing to sacrifice some accuracy for simplicity. Accurate equations of state are more complicated than the ideal gas law. For example, the Redlich-Kwong equation,

$$P + \frac{a}{[\, T^{1/2} \, v \, (v + b) \,]} (v - b) = R \, T \tag{3.18}$$

a modification of Van der Waal's equation, is a more accurate equation of state than the ideal gas law. Engineers always face "tradeoffs" between accuracy and simplicity.

For mixtures, the problem is estimating a property of a mixture, given that property for the pure components. Estimating thermodynamic properties of mixtures requires a "mixing rule" to calculate a property for a mixture from the pure-component properties. If the solution is ideal, the mole fraction average,

$$P = \sum_{i}^{n} y_i \, P_i \tag{3.19}$$

of the property is sufficient. Reid et al. [2] shows that viscosity, a transport property, has a more complex mixing rule than the mole-fraction average.

Transfer properties, the heat and mass transfer coefficient and friction factor, depend not only on transport and thermodynamic properties but also on the hydrodynamic behavior of a fluid. The geometry of the system will influence the hydrodynamic behavior. By reducing the parameters by arranging them into dimensionless groups, we can reduce the number of parameters that have to be varied to correlate any of the transfer properties. For example, the friction factor equation,

$$f = 0.1 \, [\, (\varepsilon / d) + (68 / Re)]^{0.25} \tag{3.20}$$

one of many correlations reviewed by Olvjic [3], has been correlated in terms of the dimensionless roughness factor, ε/d, and the Reynolds group.

Rates of reaction require rate constants and activation energies. These parameters are obtain from experiments.

Economic Relations

Usually, there is more than one solution to an engineering problem that is technically feasible, and socially, environmentally, and even esthetically acceptable. Among these solutions, the engineer selects the solution that is the least costly and is financially feasible. Even though a project may appear profitable, there may be insufficient capital available to implement the project so that financial feasibility is also an important consideration. Assuming that a particular solution meets all the constraints, including financial feasibility, then we design the process for the minimum total cost,

$$C_T = C_D + C_I + C_G \tag{3.21}$$

which is the sum of direct, indirect, and general costs.

For a quicker solution to a design problem than that obtained by solving Equation 3.21, we could use a "rule-of-thumb." For example, for a heat exchanger using water to cool a process stream, we can assign an approach temperature difference between the exiting water stream and entering process stream. Thus, for this particular heat exchanger we may write the approach temperature difference, based on economic experience, as

$$t_P - t_W = 5 \text{ K} \tag{3.22}$$

Equation 3.22 means that as the exit temperature of the water, t_W, approaches its maximum value, t_P, the heat-exchanger surface area will become larger and larger. When $t_P = t_W$, the area will be infinite.

If we use Equation 3.22 in place of Equation 3.21 to find the optimum, cooling-water temperature, we assume that the calculated heat-exchanger area will approximate an optimum value. The approach-temperature difference is not a constant, but it will vary with time and location, reflecting equipment, and local energy, labor, and other costs. Because of the oil embargo in the 1970s and the subsequent rise in oil prices, and its effect on all energy costs, many of the old rules-of-thumb appearing in early publications required revision. Now, oil prices are again high, so rules-of-thumb must reflect the change.

There are other rules-of-thumb based on economic experience, which the reader will recognize, such as the optimum reflux ratio in distillation and the optimum liquid to gas ratio in gas absorption. You may also specify recoveries of key components or their concentrations in an exit stream for separators. When we use any of these rules, the assumption is that the calculated separator size will be of reasonable cost, approximating the optimum-size separator. Similarly, for chemical reactors we may specify conversion of a desirable compound, its exit composition or an approach temperature difference. For chemical reactors, the approach temperature difference is the difference between the actual temperature and the chemical-equilibrium temperature. Again, we assume that a reactor that approximates the optimum-size reactor will result when using this rule.

PROCESS ANALYSIS EXAMPLES

To illustrate the foregoing discussion, we will begin first by analyzing single process units. Later, we will assemble the individual process units into a process. After writing the appropriate relations for a process unit, we calculate the degrees of freedom or the number of variables that we must specify before attempting to solve the relations. Following this, to determine which variables to specify, thus completing the formulation of the problem. The problem then reduces to a mathematical problem of determining a solution procedure and "grinding" out an answer, which are not trivial steps.

Example 3.1 Purging Air from a Tank

For the first example, consider the operation of purging a storage tank of air before filling it with a flammable liquid. Purging has two meanings. One meaning is purging a process unit by displacing the air with an inert gas to conduct safe plant operations and maintenance. Another meaning is withdrawing a stream to limit the concentration of contaminants within a process. Later, we will examine the latter application of the purge. When plants are shut down for routine maintenance, workers must frequently enter vessels – used to process or store flammable or toxic chemicals – for cleaning or repairs. In many cases vessels require welding. For safety reasons, it is essential to remove all traces of a chemical before allowing workers to enter a vessel. Explosions triggered by a welder's torch occur frequently. The New York Times [5] reported that an explosion killed a welder who entered an "empty" compartment of a barge used to transport oil. Apparently, his welding torch ignited residual fumes left by the oil.

Organic vapors, and some inorganic gases, have flammability limits when mixed with air. To burn these gases requires a mixture composition between a minimum and maximum fuel concentration. The fuel concentration from the minimum to the maximum is the flammability range for the gas. Figure 3.1.1 shows the range for a number of gases. Outside the flammability range, the mixture is too diluted with either air or fuel to sustain combustion. For example, when a car "floods" and will not start, the gasoline is in excess, and the air-fuel ratio is outside the flammability range. Figure 3.1.1 shows that the flammability range is very narrow for gasoline. We can obtain the flammability limits for many more chemicals from the Chemical Engineering Handbook [7] and chemical manufacturers. Usually, the wider the flammability range, the more unsafe the gas is. Other factors influence the manageability of a gas, such as the minimum energy required to ignite an air-gas mixture. Thus, for a gas to burn, the air-gas mixture must be within the flammable range and must have an ignition source. The source, such as a flame, a spark or a hot metal, must be capable of supplying sufficient energy for ignition. A good rule to follow is that if a gas is within the flamability range, ignition is inevitable, and if repairing a tank requires welding, a welder's torch is certainly sufficient.

If a vessel contained a flammable gas or if it will contain a flammable gas, then we must purge the vessel with a gas. Purging is dilution of a flammable gas

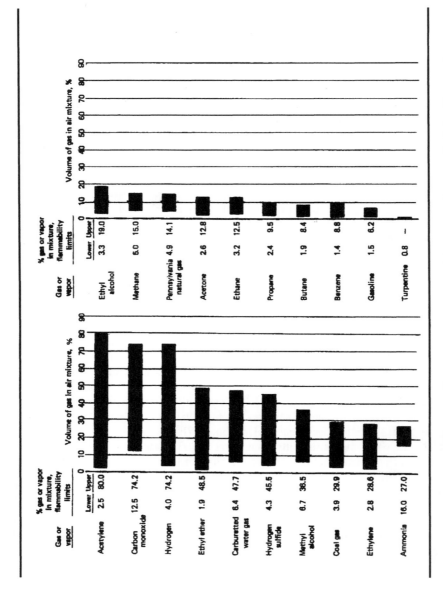

Figure 3.1.1 Flammability range for selected combustible gases. Source: Ref. 6 with permission.

with an inert gas – such as combustion gases, nitrogen or carbon dioxide – until the mixture composition is below its lower flammability limit. Thus, there is no possibility of ignition. Carbon dioxide cannot be produced on site at a low cost, and it is not inert in some applications. Nitrogen is usually the preferred purge gas.

Process Analysis

To illustrate the purging operation, consider the operation of filling a storage tank with liquefied natural gas (LNG). In 1965, Exxon contracted to build two storage tanks, each with a capacity of 40,000 m^3, in Barcelona, Spain [9]. A liquefaction plant built at Marsa el Brega in Libya supplied the storage facility with LNG by ship. Before filling with LNG, the oxygen concentration in the tanks must be at a safe level. The tanks were purged of oxygen using nitrogen, delivered from a liquid-nitrogen storage tank, at 180 l/s (at 20 °C, 1 atm). The liquid nitrogen is vaporized before flowing into the LNG tank. Samples of the gas taken during purging at various heights in the LNG storage tank showed that the oxygen content in the tank was essentially the same. Calculate how long it takes to purge the LNG storage tank.

As stated earlier, formulate or define the problem first before attempting to obtain a numerical solution. At this point there may not be enough information. After defining the problem, the information required will be evident. If we refer to the list of available relationships, the first step is to make a mole balance. Since there are two components, we can make two component balances or the total balance and one of the component balances. Also, the oxygen analysis shows that the gas in the tank is well mixed. Thus, the gas composition in the tank and in the exit stream are equal. Figure 3.1.2 is the flow diagram for the process. The first subscript, either 1 and 2, identifies the stream number. The second subscript, either oxygen or nitrogen, identifies the component.

This problem isan unsteady state problem because the oxygen concemtration will change with time. On the left side of Equation 3.2 – discussed at the beginning of the chapter – the rates of oxygen flow into the tank and formation of oxygen by chemical reaction are zero. On the right side of Equation 3.2, the rates of accumulation and disappearance of oxygen by chemical reaction are also zero. Thus, Equation 3.2 reduces to

The rate of depletion is expressed by

$$\text{rate of depletion} = -\frac{d\,(y_{2,1}\,N)}{dt} \tag{3.1.2}$$

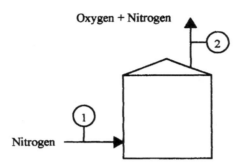

Figure 3.1.2 Purging a liquefied-natural-gas storage tank.

where the moles of oxygen in the tank at any time is $y_{2,1}$ N, and the negative sign is needed because the derivative is negative.

The moles of oxygen flowing out of the tank is

$$\text{rate of flow out} = y_{2,1}\, m_2 \qquad\qquad\qquad\qquad (3.1.3)$$

After substituting Equations 3.1.2 and 3.1.3 into Equation 3.1.1, the oxygen mole balance reduces to Equation 3.1.4 in Table 3.1.1. Because Equation 3.1.4 is an unsteady-state, first-order differential equation, we need an initial condition to calculate the constant of integration. Initially, the tank contains air, which has an oxygen concentration of approximately 21 % by volume. We could also write the mole balance for nitrogen, but in this case it is more convenient to write the total mole balance, which results in Equation 3.1.5. Once we write Equations 3.1.4 to 3.1.6, the nitrogen mole balance is not an independent equation. Equation 3.1.7 states that the molar flow rate is equal to the product of the molar density and the volumetric flow rate.

Assume that the storage tank is well insulated, and the nitrogen flowing into the tank is at the same temperature as the gas mixture in the tank as given by Equation 3.1.12. Thus, the purging operation is isothermal, eliminating the energy equation. Also, experience shows that the pressure drop across the tank will be very small, eliminating the momentum balance. The pressure at the storage tank exit, p_2, will be known because it is fixed by the design of the system. None of the rate processes and phase or chemical equilibrium occur. Equation 3.1.8 states that

Table 3.1.1 Summary of Equations for Calculating the Purging Time of a Storage Tank

---

Subscripts: $O_2 = 1$, $N_2 = 2$

Mole Balances

$$-\frac{d\,(y_{2,1}\,N)}{dt} = y_{2,1}\,m_2 \quad \text{—} \quad \text{at } t = 0,\, y_{2,1} = 0.2 \tag{3.1.4}$$

$$m_1 = m_2 \tag{3.1.5}$$

$$y_{2,1} + y_{2,1} \tag{3.1.6}$$

$$m_1 = \rho_1\,Q_1' \tag{3.17}$$

Thermodynamic Properties

$$\rho_2 = N/V' \tag{3.1.8}$$

$$p_1 = \rho_1\,R'\,T_1' \tag{3.1.9}$$

$$p_2' = \rho_2\,R'\,T_1' \tag{3.1.10}$$

Design Specifications

$$p_1 \approx p_2 \tag{3.1.11}$$

$$T_1' = T_2 \tag{3.1.12}$$

Variables

$$y_{2,1} - y_{2,2} - t - N - m_1 - m_2 - \rho_1 - \rho_2 - p_1 - T_2$$

Degrees of Freedom

$$F = V - R = 10 - 9 = 1$$

---

molar density equals the number of moles in the storage tank divided by the tank volume. The only system property data needed is a relationship for the gas molar density, given by Equations 3.1.9 and 3.1.10. At ambient conditions the ideal gas law is adequate.

At this point, we have used all the relationships available. Now, determine if we have completely defined the problem by calculating the degrees of freedom.

First, prime all the known variables, as shown in Table 3.1.1. Then list all the unknowns and calculate the degrees of freedom as shown. Because there is one degree of freedom, no solution is possible. We must specify another variable.

To calculate the purging time from Equation 3.1.4, we must specify the final oxygen concentration. When filling the tank with methane, it must be certain that the methane concentration will never be within the flammability limits. The triangular diagram in Figure 3.1.3 shows the flammability or ignition limits for mixtures of oxygen, nitrogen, and methane. Ignition could occur for any mixture of the three gases within the flamability curve shown in Figure 3.1.3. Before filling the tank with methane, reduce the oxygen content in the tank to avoid creating a flammable mixture. In Figure 3.1.3, the sides and base of the triangle represent two component mixtures. The base represents mixtures of oxygen and nitrogen, the left side, mixtures of oxygen and methane, and the right side, mixtures of nitrogen and methane. If we do not purge the tank with nitrogen before filling with methane, the concentration of the three component mixture will pass through the flammability range. The mixing line in Figure 3.1.3 shows the mixing of methane with air. The mixing line begins at the base of the triangle at 21% oxygen and ends at the apex of the triangle, which represents 100% methane and 0% nitrogen. By reducing the oxygen concentration to about 12% by adding nitrogen, the mixing line will be tangent to the flammability curve when adding methane, as shown in Figure 3.1.3. To be safe, however, reduce the oxygen concentration to 1% in the nitrogen–oxygen mixture. The base of the triangle represents the mixing of nitrogen with air. After the oxygen concentration reaches 1%, then stop the nitrogen flow. When filling the storage tank with methane initially, the methane will contain an excessive amount of nitrogen. The storage facility will have to be designed to dispose of the gases until the concentration of methane in the storage tank reaches an acceptable level of purity. Essentially, the nitrogen-oxygen mixture is now being purged with methane.

Now that we have specified the final oxygen concentration, the degrees of freedom are zero, and we can solve the set of equations in Table 3.1.1. The next step is to outline a solution procedure, i.e., to determine the order we will solve the equations. In this case, the procedure is simple, and we can arrive at a suitable order by inspection. When the number of equations increases, a greater effort will be required to set up an efficient solution procedure.

After integrating Equation 3.1.4, the oxygen concentration in the tank at any time becomes

$$y_{2,1} = K \exp(-m_2 t / N) \tag{3.1.13}$$

where K is a constant of integration. Using the initial condition that at $t = 0$, $y_{2,1} = 0.21$ in Equation 3.1.13, we obtain $K = 0.21$.

Thus, Equation 3.1.13 becomes

$$y_{2,1} = 0.21 \exp(-m_2 t / N) \tag{3.1.14}$$

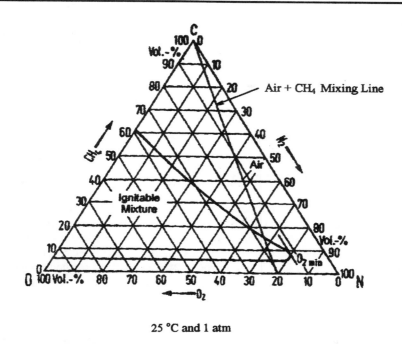

Figure 3.1.3 Flammability limits for oxygen-nitrogen-methane mixtures. From Ref 10 with permission.

Next relate m_2 in Equation 3.1.14 to the tank volume and the volumetric flow rate of nitrogen into the storage tank. From Equations 3.1.9 to 3.1.11, we find that the gas density at the inlet and outlet of the tank is the same.

Solving Equations 3.1.5, 3.1.7, and 3.1.8 we find that

$$m_2 = N\, Q_1/V \qquad (3.1.15)$$

Substitute Equation 3.1.15 into Equation 3.1.14. After solving for the purging time, we obtain

$$t = \frac{V}{Q_1} \ln \frac{y_{2,1}}{0.21} \qquad (3.1.16)$$

For a final oxygen concentration of 1.0 % at a nitrogen flow rate of 180 l/s or 0.180 m^3/s,

$$t = - \frac{40000}{0.180} \ln \frac{0.01}{0.21} = 6.766 \times 10^5 \tag{3.1.17}$$

The purging time is 6.766×10^5 s (188 h).

Example 3.2 Cooling-Tower Analysis

Water, from lakes, rivers, and the sea, is a common coolant. Because of water shortages or the environmental effects of discharging heated water, air may also be use as a coolant, either directly or indirectly. In the direct method, called the dry system, a fan blows air directly over a heat exchanger surface. Because of the low heat capacity of air, a large quantity is required. In the indirect method, called the wet system, water is the primary coolant. Air cools the water by evaporating a small fraction of the water in a tower. The cooled water is then returned to the process. A process engineer will have to choose either the dry or wet method. Cooling water is not a main part of the process but an "offsite" operation, i.e., it is generally located off to one side of the process area. We may consider cooling and treating the water to remove dissolved salts as a sub-process.

Figure 3.2.1 shows the mechanical-draft crossflow tower, which is the most commonly used cooling tower [11]. Water enters the top of the tower and flows downward over packing, called fill. The fill increases the surface area for mass transfer by breaking up the water into droplets or spreading it into a thin film. A cooling tower, like a packed bed absorber or stripper, must provide good contact between air and water to promote rapid evaporation. Good contact reduces the size of the tower and also the pressure drop, called "draft" by cooling-tower design engineers. A fan, located at the top of the tower and shown in Figure 3.2.1, draws air into the tower. Louvers distribute incoming air, which then flows across the tower, removing evaporated water.

During the operation of the tower, water is lost by evaporation, water drop-lets entrained in the outgoing air, and in a water purge, called blowdown. To re-duce carry-over of water droplets the air flows across drift eliminators. The water droplets impinge on the drift eliminators and then flows down to the bottom of the tower. The droplet water loss is about 0.2% of the incoming water [11]. After leaving the drift eliminators, air flows up and out of the tower. Evaporation of water into air transfers heat from the water to the air. Cooling the water requires about 1.0 % evaporation for every 5.56 °C (10.0 °F) drop in the water tempera-ture[11]. To reduce scale formation in the tower because of dissolved calcium or

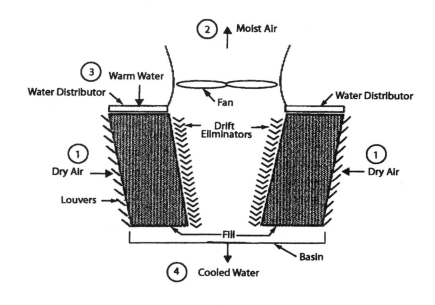

Figure 3.2.1 Induced-draft cooling tower.

magnesium salts in the water, requires periodic "blowdown" or purging of the water. About 0.3% of the water is lost for every 10 °F of cooling because of blow-down [11]. These water losses must be made up, which adds to the operating cost of the tower.

Process Analysis

As a further illustration of the general method of analysis, we will consider the design of a cooling tower, shown in Figure 3.2.1. It is required to cool 40 m³/min (10,500 gal/min) of water at 43.0 °C (109.0 °F) using air having a dry-bulb temperature of 34.4 °C (94.0 °F) and a wet-bulb temperature of 26.7 °C (80.0 °F). Besides the water flow rate, we will need the air flow rate to size the tower and the fan. For now, however, we will only devise a procedure to calculate the air flow rate but not to size the tower or fan. Again, formulate the problem first by listing the appropriate relations, and then determine if sufficient information is available to obtain a solution. After completing the analysis, outline a solution procedure.

Again, start with Equation 3.2, the generalized mole balance relation, and drop those terms that do not apply or are too small to be considered. Clearly, there is no chemical reaction, and the tower operates at steady state. Thus, Equation 3.2 reduces to the flow rate into the tower equal the flow rate out of the tower. Table 3.2.1 lists the appropriate relations (the subscript three means Chapter 3, two, example two, and one, table one). As established before, the first subscript in the composition variable, y, indicates the stream number, shown in Figure 3.2.1, and the second subscript the component, one for water and two for air. The primed variables indicate specified variables. Thus, in Table 3.2.1, Equation 3.2.1 is the water mole balance and Equation 3.2.2 the air mole balance (three means Chapter 3, two, Example two, and two Equation 1). Nitrogen and oxygen are only slightly soluble in water and, therefore, we treat air as a single, unabsorbed component. The water and air mole balances together with the mole fraction summations, given by Equations 3.2.3 and 3.2.4, are all the mole balance relations that we can write. If it is more convenient to use the total mole balance instead of a component balance, then drop one of the equations in the set from Equations 3.2.1 to 3.2.4.

Because cooling water is not an isothermal process, we must use the energy equation. The general energy balance, Equation 3.10, is modified to fit the cooling tower. We define the system by a boundary that cuts across all the streams and encloses the tower, but not the fan, which is located in the upper part of the tower. The kinetic and potential energy changes of the streams across this boundary are small compared to the enthalpy change. Although the fan does work on the air stream to overcome the resistance to air flow in the tower, no work crosses the boundary selected. At a later stage in the design, we will need a mechanical energy balance to calculate the fan power. Finally, because no heat flows across the boundary, the heat-transfer term will be zero. Therefore, enthalpy is conserved, and the cooling-tower energy equation reduces to Equation 3.2.5 in Table 3.2.1.

Equation 3.2.6 gives the concentration of water vapor in the inlet air as function of t_{1w}, y_{1w}, and Δh_{vw}, where the subscript, w, means wet bulb. The equations are in functional notation to indicate that these data may be available in tables, graphs or equations. The wet-bulb temperature, t_{1w}, will be discussed later. Equation 3.2.7 expresses the mole fraction of water vapor in the exit air in terms of the vapor pressure at saturation. The air leaving the tower is assumed to be 90% saturated, a value recommended by Walas [12].

Before solving the equations, we need system property data, which, in this case, are thermodynamic properties. Equations 3.2.9 and 3.2.11 states that we may obtain vapor pressures for water from steam tables, such as those compiled by Chaar et al. [13]. Equation 3.2.10 also states that we can find the enthalpy of vaporization in the steam tables. We assume that the air-water mixture is ideal to calculate the enthalpy of air, so we can use the mole-fraction average of the pure-component enthalpies. Equations 3.2.12 and 3.2.13 in Table 3.2.1 give the mole fraction average of the inlet and outlet enthalpy. Table 3.2.1 also lists pure component enthalpies for water vapor (Equations 3.2.14 and 3.2.16) and for air (Equa-

Table 3.2.1 Summary of Equations for Calculating Cooling-Tower Air Flow Rate

Subscripts: $H_2O = 1$, Air $= 2$

Mole Balances

$$y_{1,1} \, m_1 + m_3' = y_{2,1} \, m_2 + m_4 \tag{3.2.1}$$

$$y_{1,2} \, m_1 = y_{2,2} \, m_2 \tag{3.2.2}$$

$$y_{1,1} + y_{1,2} = 1 \tag{3.2.3}$$

$$y_{2,1} + y_{2,2} = 1 \tag{3.2.4}$$

Energy Balance

$$h_1 \, m_1 + h_3 \, m_3' = h_2 \, m_2 + h_4 \, m_4 \tag{3.2.5}$$

Transport Relation

$$y_{1,1} = f(t_{1W}', y_{1W}, \Delta h_{VW}) \tag{3.2.6}$$

Equilibrium Relations

$$y_{2,1} = 0.9 \, p_{2,1\,S} / P' \tag{3.2.7}$$

$$y_{1W} = p_{1W} / P' \tag{3.2.8}$$

Thermodynamic Properties

$$p_{1W} = f(t_{1W}') \;—\; \text{steam tables} \tag{3.2.9}$$

$$\Delta h_{VW} = f(t_{1W}') \;—\; \text{steam tables} \tag{3.2.10}$$

$$p_{2,1S} = f(t_2) \;—\; \text{steam tables} \tag{3.2.11}$$

$$h_1 = y_{1,1} \, h_{1,1} + y_{1,2} \, h_{1,2} \tag{3.2.12}$$

$$h_2 = y_{2,1} \, h_{2,1} + y_{2,2} \, h_{2,2} \tag{3.2.13}$$

$$h_{1,1} = f(t_1') \tag{3.2.14}$$

$$h_{1,2} = f(t_1') \tag{3.2.15}$$

$$h_{2,1} = f(t_2) \tag{3.2.16}$$

$$h_{2,2} = f(t_2) \tag{3.2.17}$$

$$h_3 = f(t_3') \tag{3.2.18}$$

$$h_4 = f(t_4) \tag{3.2.19}$$

Economic Relations

$$t_4 - t_{w1}' = 9.0\ ^\circ F \tag{3.2.20}$$

$$m_3' / m_1 = 2.09\ \text{lbmol water/lbmol air} \tag{3.2.21}$$

Variables

$y_{1,1} - y_{1,2} - y_{2,1} - y_{2,2} - y_{1w} - m_1 - m_2 - m_4 - p_{1w} - p_{2,1S} - \Delta h_{vw} - h_1 - h_2 - h_3 - h_4 - h_{1,1} - h_{1,2} - h_{2,1} - h_{2,2} - t_2 - t_4$

Degrees of Freedom

$$F = 21 - 21 = 0$$

---

tions 3.2.15 and 3.2.17). Equations 3.2.18 and 3.2.19 give the enthalpies for pure water.

Finally, Table 3.2.1 contains two economic relations or rules-of-thumb. Equation 3.2.20 states that the approach temperature differences for the water, which is the difference between the exit water temperature and the wet-bulb temperature of the inlet air, is 5.0 °C (9 °F). The wet-bulb temperature of the surrounding air is the lowest water temperature achievable by evaporation. Usually, the approach temperature difference is between 4.0 and 8.0 °C. The smaller the approach temperature difference, the larger the cooling tower, and hence the more it will cost. This increased tower cost must be balanced against the economic benefits of colder water. These are: a reduction in the water flow rate for process cooling and in the size of heat exchangers for the plant because of an increase in the log-mean-temperature driving force. The other rule-of-thumb, Equation 3.2.21, states that the optimum mass ratio of the water-to-air flow rates is usually between 0.75 to 1.5 for mechanical-draft towers [14].

As before, prime all the known variables in the equations listed in Table 3.2.1. The table shows that there are twenty-one unknowns and equations, resulting in zero degrees of freedom. Thus, we have completely defined the problem.

Before obtaining numerical answers, we must derive equations for the functional relationships expressed in Table 3.2.1, which are given in Table 3.2.2. Equation 3.2.27 is the psychrometric relation, derived by Bird et al. (3.15). This relation

Table 3.2.2 Revised Summary of Equations for Calculating Cooling-Tower Air Flow Rate

Subscripts: $H_2O = 1$, $Air = 2$

Mole Balances

$$y_{1,1} \ m_1 + m_3' = y_{2,1} \ m_2 + m_4 \tag{3.2.22}$$

$$y_{1,2} \ m_1 = y_{2,2} \ m_2 \tag{3.2.23}$$

$$y_{1,1} + y_{1,2} = 1 \tag{3.2.24}$$

$$y_{2,1} + y_{2,2} = 1 \tag{3.2.25}$$

Energy Balance

$$h_1 \ m_1 + h_3 \ m_3' = h_2 \ m_2 + h_4 \ m_4 \tag{3.2.26}$$

Transport Relation

$$f(t_{1W}', y_{1W}, \Delta h_{VW}) = y_{1W} - y_{1,1} = \left(\frac{h'}{k'} \right) \frac{1}{\Delta h_{VW}} (t_1' - t_{1W}')(1 - y_{1W}) \tag{3.2.27}$$

Equilibrium Relations

$$y_{2,1} = 0.9 \ p_{2,1S} / P' \tag{3.2.28}$$

$$y_{1W} = p_{1W} / P' \tag{3.2.29}$$

Thermodynamic Properties

$$p_{1W} = f(t_{1W}') \quad \text{— steam tables} \tag{3.2.30}$$

$$\Delta h_{VW} = f(t_{1W}') \quad \text{— steam tables} \tag{3.1.31}$$

$$\ln \frac{p_{2,1S}}{p_{1W}} = -\frac{\Delta h_{VW}}{R'} \left(\frac{1}{T_2} - \frac{1}{T_{1W}} \right) \tag{3.2.32}$$

$$T_2 = t_2 + 460.0 \tag{3.2.33}$$

$$T_{1W} = t_{1W}' + 460.0 \tag{3.2.34}$$

Table 3.2.2 continued

$$h_1 = y_{1,1}\, h_{1,1} + y_{1,2}\, h_{1,2} \tag{3.2.35}$$

$$h_2 = y_{2,1}\, h_{2,1} + y_{2,2}\, h_{2,2} \tag{3.2.36}$$

$$h_{1,1} = c_{P1,1}'\, (t_1' - t_R') + \Delta h_{VR}' \tag{3.2.37}$$

$$h_{1,2} = c_{P1,2}'\, (t_1' - t_R') \tag{3.2.38}$$

$$h_{2,1} = c_{P2,1}'\, (t_2 - t_R') + \Delta h_{VR}' \tag{3.2.39}$$

$$h_{2,2} = c_{P2,2}'\, (t_2 - t_R') \tag{3.2.40}$$

$$h_3 = c_{P3}'\, (t_3 - t_R') \tag{3.2.41}$$

$$h_4 = c_{P4}'\, (t_4 - t_R') \tag{3.2.42}$$

Economic Relations

$$t_4 - t_{1W}' = 9.0\ {}^\circ F \tag{3.2.43}$$

$$m_3' / m_1 = 2.09\ \text{lbmol water/lbmol air} \tag{3.2.44}$$

Variables

$y_{1,1}$ - $y_{1,2}$ - $y_{2,1}$ - $y_{2,2}$ - y_{1W} - m_1 - m_2 - m_4 - p_{1W} - $p_{2,1S}$ - Δh_{VW} - h_1 - h_2 - h_3 - h_4 - $h_{1,1}$ - $h_{1,2}$ - $h_{2,1}$ - $h_{2,2}$ - t_2 - t_4 - T_2 - T_{1W}

Degrees of Freedom

$$F = 23 - 23 = 0$$

gives us the mole fraction of water in air – in this case the water mole fraction in the incoming air. When the tip of a thermometer in a high-velocity air stream is covered with a wet wick, the wick will reach a steady-state temperature, called the wet-bulb temperature. At the wet-bulb temperature, the heat removed from the wick by the evaporating water just equals the heat transferred to the wick from the air. To calculate the water concentration at the wet-bulb temperature, y_{1W}, use Equations 3.2.29 and 3.2.30. As Equation 3.1.31 states, the heat of vaporization at the wet-bulb temperature, Δh_{VW}, is found in the steam tables at t_{1W}. The ratio of heat to mass transfer coefficients, (h/k), calculated by using data taken from Bird et al. [15], is 5.93 Btu/lbmol $^\circ$F (24.8 kJ/kg mol-K).

Over a small temperature range, the enthalpy of vaporization is essentially constant. Thus, we may use the Clausius-Clapyeron equation, Equation 3.2.32, to express the vapor pressure of water as a function of temperature. Next, calculate the mole fraction of water in the exit air using Equation 3.2.28, where $p_{2,1S}$, is the vapor pressure of water at the exit-air temperature. We assume that heat capacity of air and water vapor is constant over the temperature range of interest. Using data taken from Reid et al. [2], calculate the heat capacities at 100 °F (37.8 °C). Thus, $c_{p1} = 8.2$ Btu/lbmol-°F (34.3 kJ/kg mol-K) and $c_{p2} = 7.2$ Btu/lbmol-°F (30.1 kJ/kg mol-K). The heat capacity of water, 18.0 Btu/lbmol-°F (75.4 kJ/kg mol-K), is also assumed constant. We select 32.0 °F as the reference temperature, t_R, to correspond to the steam tables. Thus, Equations 37 to 42 in Table 3.2.2 are the pure component enthalpies of all the components.

The next step in the problem solving procedure is to outline a solution procedure for the Equations listed in Table 3.2.2. There are algorithms available for determining in what order to solve a set of algebraic equations, which is called the precedence order. See, for example, Rudd and Watson [17] and Myers and Seider [18] for a discussion of some of these algorithms. Sometimes, we can develop a procedure by inspection of an equation set, as in the procedure given in Table 3.2.3.

Table 3.2.3 Calculation Procedure – Cooling-Tower Analysis

1. Obtain p_{1w} from the steam tables at t_{1w} (Equation 3.2.30 in Table 3.2.2)

2. Calculate y_{1w} from Equation 3.2.29.

3. Obtain Δh_{1w} from the steam tables at t_{1w}, Equation 3.2.31.

4. Calculate $y_{1,1}$ from the psychrometric relation, Equation 3.2.27.

5. Assume an exit air temperature, t_2.

6. Calculate $p_{2,1S}$ from Equations 3.2.32 to 3.2.34.

7. Calculate $y_{2,1}$ from Equation 3.2.28.

8. Calculate m_1, m_2, m_4, $y_{1,2}$ and $y_{2,2}$ from Equations 3.2.22 to 3.2.25 and Equation 3.2.44.

9. Calculate h_1, h_2, h_3, and h_4 from Equations 3.2.35 to 3.2.43.

10. Substitute h_1, h_2, h_3, h_4, m_1, m_2, m_3, and m_4 into Equation 3.2.26 to check the assumed value of t_2.

11. Repeat steps 5 to 10 until Equation 3.2.26 is satisfied within a sufficient degree of accuracy.

Table 3.2.4 Specified Variables – Cooling Tower Analysis

Variable	Quantity	Units
m_3	291300	lbmol/h
t_1	94.0	°F
t_{1W}	80.0	°F
t_3	109.0	°F
t_R	32.0	°F
$\Delta h_{V,32\,F}$	19350	Btu/lbmol
P	14.7	psia
$c_{P1,1} \approx c_{P2,1}$	8.2	Btu/lbmol-°F
$c_{P1,2} \approx c_{P2,2}$	7.2	Btu/lbmol-°F
$c_{P3} \approx c_{P4}$	18.0	Btu/lbmol-°F
h/k	5.93	Btu/lbmol-°F
R	1.986	Btu/lbmol-°F

Table 3.2.4 lists the specified variables. The cooling tower is processing 40 m³/min (1410 ft³/min) of water at 109 °F (43.8 °C), which from the steam tables, has a specific volume of 0.01616 ft³/lb (1.01x10⁻³ m³/kg). Thus,

$$m_3 = \frac{40.0\ \text{m}^3}{1\ \text{min}}\ \frac{60\ \text{min}}{1\ \text{h}}\ \frac{35.31\ \text{ft}^3}{1\ \text{m}^3}\ \frac{1}{0.01616\ \text{ft}^3}\ \frac{\text{lb}}{18\ \text{lb}}\ \frac{1\ \text{lbmol}}{} = 291300\ \frac{\text{lbmol}}{\text{h}}$$

Finally, we can solve the equations listed in Table 3.2.2 simultaneously using POLYMATH [19] or some other suitable mathematical software. The solution procedure used in POLYMATH is the bounded Newton-Raphson method described by Shacham and Shacham [20]. Table 3.2.5 lists the stream properties, which include the solution to the equations and specified temperatures and pressures at each line. The difference in the water flow rates into and out of the cooling tower is the water evaporated. Thus, to cool 164,700 lbmol/h (74,700 kg mol/h) water requires evaporating 5,200 lbmol/h (2,360 kg mol/h) of water. The evaporated water, along with water lost because of leaks, blowdown, and drift are a cost of operation.

Table 3.2.5: Stream Properties – Cooling-Tower Analysis

Stream Number	Temperature °F	Pressure psia	Flow Rate lbmol/h[a]	Concentration, Mole Fraction	
				Water	Air
1	94.0	14.7	164700	0.0302	0.9698
2	100.7	–	169800	0.0525	0.9475
3	109.0	–	291300	0	1.0
4	89.0	–	286100	0	1.0

[a]Multiply by 0.4536 to obtain kg mol/h.

Example 3.3 Flash Valves, Partial Condensers, and Partial Vaporizers

Flashing, partial condensation, and partial vaporization are frequently occurring process operations. Because partial separation occurs during these operations, they are all separations. We will treat them together because the equations for calculating downstream conditions are almost identical, differing only in the heat-transfer term in the energy equation. The flash valve is essentially adiabatic, the condenser removes heat, and the vaporizer adds heat to a process stream. The pressure drops across these units also differ considerably. The pressure drop across flash valves is about 1 Mpa (145 psi), for condensers, 10 kPa (1.45 psi), and for the vaporizers, 1 kPa (0.145 psi). In all these units, we assume equilibrium between the vapor and liquid streams leaving each process unit. This implies that sufficient contact time will be allowed to reach equilibrium. The turbulence between the vapor and liquid streams in the flash valve and the vaporizer insures good contact and hence a rapid approach to equilibrium. In the condenser, equilibrium may not be completely attained. Nevertheless, we will assume equilibrium.

Frequently, vapor-liquid phase separators follow and are combined with the component separators, and equilibrium is assume between the exit streams of this combination. Here, the phase separators are omitted as shown in Figure 3.3.1 to keep the two kinds of separators divided according to their major function – one where essentially component separation occurs and the other where essentially phase separation occurs.

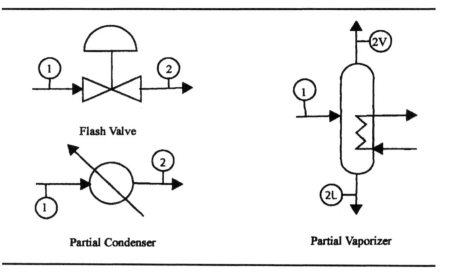

Figure 3.3.1 Single-stage component separators.

To illustrate the method of analysis, we will consider the separation of propane from methane, obtained from natural gas. Both methane and propane have fuel and non-fuel uses, but using these compounds as fuels dominates the market. Swearingen [21] describes a cryogenic process for recovering propane from a mixture of methane and propane involving several flashing steps. In one part of this process, a liquid mixture from a fractionator flashes across a valve to provide a cold liquid stream for use in a heat exchanger. When the pressure drops, the "hot liquid" converts into a vapor-liquid stream. The large enthalpy of vaporization is supplied by cooling the entire stream. The principle, cooling by evaporation, is the same as that employed to produce cooling water in a tower.

The objective in analyzing these units is to calculate the temperature, the composition, and the flow rates of the vapor and liquid exit streams, given the properties of the entering streams. First, write the mole balances. For two components, we write two component balances and a mole fraction summation for each unknown stream as given by Equations 3.3.1 to 3.3.4 in Table 3.3.1. There are two phases in equilibrium leaving the valve, condenser and vaporizer, although the phases have not, as yet, been separated. A phase separator will separate the phases. For a vaporizer, both component and phase separation occur in the same process unit. As stated before, the first numerical subscript is the line number and the second the component number. We also identify the phases by an additional subscript, V for vapor and L for liquid. Because we are assuming equilibrium between the vapor and liquid for each component downstream of the valve, we can eliminate the rate equations. Therefore, we can write two equilibrium relations, which are given by Equations 3.3.6 and 3.3.7. The energy balance for the three process units, which differ only in the heat transfer term, Q, is given by Equation 3.3.5. For the flash valve, Q = 0, and the first of the three equations applies.

Table 3.3.1 Summary of Equations for Calculating the Exit Temperature of Single-Stage Component Separators

Subscripts: methane = 1, propane = 2

Mole Balances

$$y_{1,1}' \, m_1' = y_{2V,1} \, m_{2V} + y_{2L,1} \, m_{2L} \tag{3.3.1}$$

$$y_{1,2}' \, m_1' = y_{2V,2} \, m_{2V} + y_{2L,2} \, m_{2L} \tag{3.3.2}$$

$$y_{2V,1} + y_{2V,2} = 1 \tag{3.3.3}$$

$$y_{2L,1} + y_{2L,2} = 1 \tag{3.3.4}$$

Energy Balance

$$h_1 \, m_1' = h_{2V} \, m_{2V} + h_{2L} \, m_{2L} \;\text{---}\; \text{flash valve (Q = 0)} \tag{3.3.5}$$

$$h_1 \, m_1' = h_{2V} \, m_{2V} + h_{2L} \, m_{2L} + Q \;\text{---}\; \text{a partial condenser or}$$

$$h_1 \, m_1' + Q = h_{2V} \, m_{2V} + h_{2L} \, m_{2L} \;\text{---}\; \text{a partial vaporizer}$$

Equilibrium Relations

$$K_{2,1} = y_{2V,1} / y_{2L,1} \tag{3.3.6}$$

$$K_{2,2} = y_{2V,2} / y_{2L,2} \tag{3.3.7}$$

Thermodynamic Properties

$$K_{2,1} = f(\, T_2, P_2'\,) \tag{3.3.8}$$

$$K_{2,2} = f(\, T_2, P_2'\,) \tag{3.3.9}$$

$$h_1 = y_{1,1}' \, h_{1,1} + y_{1,2}' \, h_{1,2} \tag{3.3.10}$$

$$h_{2V} = y_{2V,1} \, h_{2V,1} + y_{2V,2} \, h_{2V,2} \tag{3.3.11}$$

$$h_{2L} = y_{2L,1} \, h_{2L,1} + y_{2L,2} \, h_{2L,2} \tag{3.3.12}$$

Table 3.3.1 Continued

$$h_{1,1} = f(T_1') \tag{3.3.13}$$

$$h_{1,2} = f(T_1') \tag{3.3.14}$$

$$h_{2V,1} = f(T_2) \tag{3.3.15}$$

$$h_{2V,2} = f(T_2) \tag{3.3.16}$$

$$h_{2L,1} = f(T_2) \tag{3.3.17}$$

$$h_{2L,2} = f(T_2) \tag{3.3.18}$$

Variables

$y_{2V,1}$ - $y_{2V,2}$ - $y_{2L,1}$ - $y_{2L,2}$ - m_{2V} - m_{2L} - T_2 - $K_{2,1}$ - $K_{2,2}$ - h_1 - $h_{1,1}$ - $h_{1,2}$ - h_{2V} - $h_{2V,1}$ - $h_{2V,2}$ - h_{2L} - $h_{2L,1}$ - $h_{2L,2}$

Degrees of Freedom

$$F = 18 - 18 = 0$$

Next the equations that we can write are for calculating system properties. Because equilibrium is assumed, the rate equations and, therefore, the transport and transfer properties are of no concern. In general, the thermodynamic properties of mixtures will depend on temperature, pressure, and composition, we will assume that the mixture is an ideal solution to simplify the computation of thermodynamic properties. Thus, we can write the enthalpies of the mixtures as mole fraction averages of the pure component enthalpies, without an enthalpy of mixing term. We can also write the phase equilibrium relations as functions of temperature and pressure only and not composition. The pure component enthalpies of liquids generally do not depend strongly on pressure, but there may be some effect of pressure on the vapor-phase enthalpy. We will neglect this effect for simplicity.

The next step in the problem solving format is to prime the specified variables in the equations listed in Table 3.3.1. Next, list the unknown variables and calculate the degrees of freedom. The degrees of freedom are zero, and therefore, a solution is possible. Now that the problem is completely formulated, the next step is to outline a solution procedure.

The solution of the equations listed in Table 3.3.1 requires an iterative procedure. Thus, it is good strategy to examine the variables to determine if there are limits on their values. For example, the mole fractions of the components will vary from zero to one. This fact greatly simplifies the solution procedure. Also, the final flash temperature will lie somewhere between the bubble and dew-point temperatures. The bubble-point temperature is that temperature at which the first bubble of vapor forms. It is also the temperature at which the last bubble of vapor condenses. Similarly, the dew-point temperature is the temperature at which the first drop of liquid condenses or the last drop of liquid that vaporizes. Table 3.3.2 lists the equations for calculating the bubble-point temperature, and Table 3.3.3 lists the equations for calculating the dew-point temperature. These calculations do not require mass and energy balances. We could solve this set of equations simultaneously in its present form, after substituting appropriate expressions for the equations shown in functional notation.

Table 3.3.2 Summary of Equations for Calculating the Bubble-Point Temperature

Subscripts: Methane = 1, Propane = 2

Equilibrium Relations

$$y_{2V,1B} + y_{2V,2B} = 1 \qquad\qquad\qquad (3.3.19)$$

$$K_{2,1B} = y_{2V,1B} \, / \, y_{2L,1B}' \qquad\qquad\qquad (3.3.20)$$

$$K_{2,2B} = y_{2V,2B} \, / \, y_{2L,2B}' \qquad\qquad\qquad (3.3.21)$$

Thermodynamic Properties

$$K_{2,1B} = f\,(\,T_{2B},\,P_2'\,) \qquad\qquad\qquad (3.3.22)$$

$$K_{2,2B} = f\,(\,T_{2B},\,P_2'\,) \qquad\qquad\qquad (3.3\ 23)$$

Variables

$$y_{2V,1B} - y_{2V,2B} - T_B - K_{2,1B} - K_{2,2B}$$

Degrees of Freedom

$$F = 5 - 5 = 0$$

Table 3.3.3 Summary of Equations for Calculating the Dew-Point Temperature

Subscripts: Methane = 1, Propane = 2

Equilibrium Relations

$$y_{2L,1D} + y_{2L,2D} = 1 \tag{3.3.24}$$

$$K_{2,1D} = y_{2V,1D}' \, / \, y_{2L,1D} \tag{3.3.25}$$

$$K_{2,2D} = y_{2V,2D}' \, / \, y_{2L,2D} \tag{3.3.26}$$

Thermodynamic Properties

$$K_{2,1D} = f \, (T_{2D}, P_2') \tag{3.3.27}$$

$$K_{2,2D} = f (T_{2D}, P_2') \tag{3.3.28}$$

Variables

$$y_{2L,1D} - y_{2L,2D} - T_{2D} - K_{2,1D} - K_{2,2D}$$

Degrees of Freedom

$$F = 5 - 5 = 0$$

    To simplify the solution procedure, first, inspect the equations to determine if some rearrangement of them will simplify their solution. Although this problem requires solving equations for a two-component system, we will generalize the solution for multicomponent systems.

    Starting with Equations 3.3.1 and 3.3.2 in Table 3.3.1, the mole balance for the i th component is

$$y_{L,i} \, m_1 = y_{2V,i} \, m_{2V} + y_{2L,i} \, m_{2L} \tag{3.3.29}$$

    The equilibrium relation for the i th component is

$$K_{2,i} = y_{2V,i} \, / \, y_{2L,i} \tag{3.3.30}$$

Solving Equations 3.3.29 and 3.3.30 simultaneously for $y_{2L,i}$ and $y_{2V,I}$, the mole fraction for the i th component in the liquid,

$$y_{2L,i} = \frac{y_{1,i}}{K_{2,i} \dfrac{m_{2V}}{m_1} + \dfrac{m_{2L}}{m_1}} \qquad (3.3.31)$$

and the mole fraction in the vapor phase,

$$y_{2V,i} = \frac{K_{2,i} \; y_{1,i}}{K_{2,i} \dfrac{m_{2V}}{m_1} + \dfrac{m_{2L}}{m_1}} \qquad (3.3.32)$$

If Equation 3.3.29 is summed up for all components, the total mole balance is

$$m_1 = m_{2V} + m_{2L} \qquad (3.3.33)$$

Solving Equation 3.3.33 for m_{2L}/m_1 and after substituting the result into Equations 3.3.31 and 3.3.32, the equations become

$$y_{2L,i} = \frac{y_{1,i}}{(K_{2,i} - 1)(m_{2V}/m_1) + 1} \qquad (3.3.34)$$

and

$$y_{2V,i} = \frac{K_{2,i} \; y_{1,i}}{(K_{2,i} - 1)(m_{2V}/m_1) + 1} \qquad (3.3.35)$$

After summing up Equations 3.3.34 and 3.3.35,

$$\sum y_{2L,i} = \sum \frac{y_{1,i}}{(K_{2,i} - 1)(m_{2V}/m_1) + 1} = 1 \qquad (3.3.36)$$

and

$$\sum y_{2V,i} = \sum \frac{K_{2,i}\, y_{1,i}}{(K_{2,i}-1)(m_{2V}/m_1)+1} = 1 \tag{3.3.37}$$

When Equation 3.3.36 is subtracted from Equation 3.3.37, the final flash equation is

$$\sum \frac{(K_{2,i}-1)\, y_{1,i}}{(K_{2,i}-1)(m_{2V}/m_1)+1} = 0 \tag{3.3.38}$$

According to King [22], Equation 3.3.38 is mathematically well behaved. The equation has no spurious roots and maximum or minimum. Also, the fraction of liquid vaporized, $m_{2,V}/m_1$, varies between 0 to 1 and is linear.

Similarly, we can also reduce the energy equation for $Q = 0$, Equation 3.3.5, to a more usable form. First, divide Equation 3.3.5 by m_1 to obtain Equation 3.3.39.

$$h_{2V}\, \frac{m_{2V}}{m_1} + h_{2L}\, \frac{m_{2L}}{m_1} - h_1 = 0 \tag{3.3.39}$$

The enthaply of the vapor phase,

$$h_{2V} = \sum y_{2V,i}\, h_{2V,i} \tag{3.3.40}$$

and the enthalpy of the liquid phase,

$$h_{2L} = \sum y_{2L,i}\, h_{2L,i} \tag{3.3.41}$$

After subsituting Equation 3.3.35 into Equation 3.3.40 and Equation 3.3.34 into 3.3.41,

$$h_{2V} = \sum \frac{K_{2,i}\, y_{1,i}}{(K_{2,i}-1)(m_{2V}/m_1)+1}\, h_{2V,i} \tag{3.3.42}$$

and

$$h_{2L} = \sum \frac{y_{1,i}}{(K_{2,i}-1)(m_{2V}/m_1)+1}\, h_{2L,i} \tag{3.3.43}$$

After substituting Equations 3.3.42 and 3.3.43 into Equation 3.3.39, and with some algebraic manipulation, we obtain the final form of the energy equation, Equation 3.3.44.

$$\sum \frac{y_{1,i}}{(K_{2,i}-1)(m_{2V}/m_1)+1} \left[K_{2,i}\, h_{2V,i}\, \frac{m_{2V}}{m_1} + h_{2L,i}\left(1-\frac{m_{2V}}{m_1}\right) \right] - h_1 = 0 \quad (3.3.44)$$

The calculation procedure using Equations 3.3.38 and 3.3.44 is outlined in Table 3.3.4.

Table 3.3.4: Procedure for Calculating the Temperature of a Flashed Liquid

1. Calculate the bubble-point temperature. Assume a temperature and then calculate values for the equilibrium relations from Equations 3.3.22 and 3.3.23 in Table 3.2.2. Next, calculate the vapor-phase mole fractions from Equations 3.3.20 and 3.3.21. Check the results using Equation 3.3.19. Assume a new temperature and repeat the calculation until temperature converges to a desired degree of accuracy.

2. Similarly, calculate the dew-point temperature. Assume a temperature and then calculate values for the equilibrium relations from Equations 3.3.27 and 3.3.28 in Table 3.3.3. Next, calculate the liquid-phase mole fractions from 3.3.25 and 3.3.26. Check the results using Equation 3.3.24. Assume a new temperature and repeat the calculation until temperature converges to a desired degree of accuracy.

3. Assume a temperature, T_2, between the bubble and dew point temperatures.

4. Calculate values for the equilibrium relations at T_2 from Equations 3.3.8 and 3.3.9 in Table 3.3.1.

5. Solve for the mole fractions for the liquid and vapor from Equations 3.3.3, 3.3.4, 3.3.6, and 3.3.7.

6. Substitute these values into Equation 3.3.38 and solve for m_{2V}/m_1 by trial.

7. Calculate the pure-component enthalpies from Equations 3.3.13 to 3.3.18 and the enthalpy of the feed solution from Equation 3.3.10.

8. Check the guess of T_2 by substituting all calculated quantities into the energy balance, Equation 3.3.44.

9. Assume a new value of T_2, and repeat steps 3 to 7 until the energy equation is satisfied within a sufficient degree of accuracy.

Example 3.4 Packed-Bed, Catalytic Reactor

In this problem, we will analyze a packed-bed catalytic reactor. Heat may be either transferred into or out of a reactor, depending on whether the reaction is exothermic or endothermic. One design for transferring heat is to pack the catalyst into tubes, approximately 5.0 cm (2 in) in diameter, and arrange them in parallel inside a shell. A heat-transfer fluid flows into the shell surrounding the tubes, removing or adding heat. We will consider the production of formaldehyde synthesized by oxidizing methanol with air. Formaldehyde ranks 25th by volume among all chemicals produced. Its major end uses are 60% for adhesives and 15% for plastics [23].

Process Chemistry

Because formaldehyde synthesis is exothermic, the reactor requires a coolant to remove the excess enthalpy of reaction. Thermodynamically, we should run the reaction at as low a temperature as possible to increase conversion, but at low temperatures, however, the rate of reaction decreases.   At high reaction temperatures unwanted side reactions occur. Commercially, the reaction occurs from 600 °C (1110 °F) to 650 °C (1200 °F), which results in a methanol conversion of 77 to 87 % when using a silver catalyst [24]. Because formaldehyde and methanol can form flammable mixtures with oxygen, we should carry out the reaction with mixture compositions outside of its flammability range. The oxygen used is less than the stoichiometric amount.

Process Analysis

Methanol flows at the rate of 1000 kmol/h (2205 lb mol) into the reactor, shown in Figure 3.4.1, where methanol is oxidized catalytically to formaldehyde under non-adiabatic conditions. The reactants enter the reactor at 500 °C (932 °F), and the products exit at 600 °C (1110 °F). The methanol in stream 1 and air in stream 2 are both at 500 °C, and the methanol conversion is 80 %. To minimize possible combustion of methanol and formaldehyde, we set the molar flow rate of oxygen at 80% of the stoichiometric quantity. The reaction is

$$CH_3OH(g) + 1/2\ O_2(g) \rightarrow HCHO(g) + H_2O(g)\ (-37{,}280\ cal,\ 298\ K) \qquad (3.4.1)$$

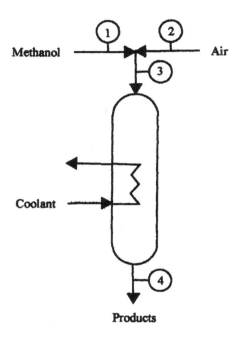

Figure 3.4.1 Packed-bed, catalytic reactor.

Table 3.4.1 lists the component balances, the energy balance, and thermo-dynamic property relations. Because moles are generally not conserved in a chemical reaction, we must include a source term in the component balances to account for the depletion or generation of moles. The balances are given in Table 3.4.1 by Equations 3.4.4 to 3.4.8. In this case, the conversion is an experimental value. If the conversion is unknown and the reaction is at equilibrium, then we can write an equilibrium relation for the reaction to calculate the conversion. Besides the general list of relationships, discussed earlier, there is a specification relationship. Equation 3.4.11 specifies that the moles of oxygen should be 80% of the stoichiometric amount to minimize the risk of the methanol and formaldehyde igniting and burning.

Table 3.4.1 Summary of Equations for Calculating Heat Transfer to a Reactor

Subscripts: $CH_3OH = 1$, $O_2 = 2$, $CH_2O = 3$, $H_2O = 4$, $N_2 = 5$

Mole Balances

$$y_{3,1} \, m_3 = y_{4,1} \, m_4 + x_1' \, y_{3,1} \, m_3 \tag{3.4.4}$$

$$y_{3,2} \, m_3 = y_{4,2} \, m_4 + (1/2) \, x_1' \, y_{3,1} \, m_3 \tag{3.4.5}$$

$$x_1' \, y_{3,1} \, m_3 = y_{4,3} \, m_4 \tag{3.4.6}$$

$$x_1' \, y_{3,1} \, m_3 = y_{4,4} \, m_4 \tag{3.4.7}$$

$$y_{3,5} \, m_3 = y_{4,5} \, m_4 \tag{3.4.8}$$

$$y_{3,1} + y_{3,2} + y_{3,5} = 1 \tag{3.4.9}$$

$$y_{4,1} + y_{4,2} + y_{4,3} + y_{4,4} + y_{4,5} = 1 \tag{3.4.10}$$

Reaction Specification

$$y_{3,2} / y_{3,1} = (0.80)(1/2) \tag{3.4.11}$$

Energy Balance

$$\Delta h_3 \, m_3 + \Delta h_R' \, x_1' \, y_{3,1} \, m_3 = Q - \Delta h_4 \, m_4 \tag{3.4.12}$$

Thermodynamic Properties

$$h_3 = y_{3,1} \, h_{3,1} + y_{3,2} \, h_{3,2} + y_{3,5} \, h_{3,5} \tag{3.4.13}$$

$$h_4 = h_{4,1} \, h_{4,1} + y_{4,2} \, h_{4,2} + y_{4,3} \, h_{4,3} + y_{4,4} \, h_{4,4} + y_{4,5} \, h_4 \tag{3.4.14}$$

$$h_{3,1} = f(T_3') \tag{3.4.15}$$

$$h_{3,2} = f(T_3') \tag{3.4.16}$$

$$h_{3,5} = f(T_4') \tag{3.4.17}$$

$$h_{4,1} = f(T_4')$$ (3.4.18)

$$h_{4,2} = f(T_4')$$ (3.4.19)

$$h_{4,3} = f(T_4')$$ (3.4.20)

$$h_{4,4} = f(T_4')$$ (3.4.21)

$$h_{4,5} = f(T_4')$$ (3.4.22)

Variables

$y_{3,1} - y_{3,2} - y_{3,5} - y_{4,1} - y_{4,2} - y_{4,3} - y_{4,4} - y_{4,5} - m_3 - m_4 - Q - h_3 - h_4 - h_{3,1} - h_{3,2} - h_{3,5} - h_{4,1} - h_{4,2} - h_{4,3} - h_{4,4} - h_{4,5}$

Degrees of Freedom

$$F = 21 - 19 = 2$$

Equation 3.4.12, the energy balance for the reactor, requires some explanation. We write the general energy equation, Equation 3.10 at the beginning of the chapter, for the boundary that encloses the process stream, but not the coolant. We can again neglect the kinetic and potential energy terms. Also, the reactor does no work on the reacting gases so that Equation 3.10 for the reactor becomes

$$\Delta h = Q$$ (3.4.2)

where Q is the heat transferred from the coolant to the process stream, and Δh is the enthalpy change of the process stream across the reactor. Since enthalpy is a state function, you can chose any path to evaluate Δh, starting from the state at the entrance and ending at the state at the exit of the reactor. Because enthalpies of reaction are given at 25 °C, select the path shown in Figure 3.4.2 for evaluating Δh. First, cool the reactants to 25 °C, then let them react isothermally at 25 °C, and finally heat the exit gases to the exit temperature. Thus, the enthalphy change across the reactor becomes

$$\Delta h = \Delta h_3 \, m_3 + \Delta h_R \, x_1 \, y_{3,1} \, m_3 + \Delta h_4 \, m_4$$ (3.4.3)

After substituting Equation 3.4.3 into Equation 3.4.2, we obtain Equation 3.4.12 in Table 3.4.1. Physically, Equation 3.4.12 means that the enthalpy flowing into the reactor with the reactant stream plus part of the enthalpy released in the reactor by chemical reaction will raise the temperature of the products to 600 °C. The coolant removes the remaining enthalpy of reaction as heat. For simplicity, we again assume that we can use the mole fraction average of the pure component enthalpies for the enthapy of gas mixtures as given by Equations 3.4.13 and 3.4.14. Equations 3.4.15 to 3.4.22 are the pure component enthalpies we need for Equations 3.4.13 and 3.4.14.

The reactor analysis given in Table 3.4.1 shows that there are two degrees of freedom, and thus we have not completely defined the problem. We must either write two additional equations or specify two additional variables. In this case, we see that in Figure 3.4.1 the methanol and air streams mix before entering the reactor. Mixing is a process step even though the mixer may only be two intersecting streams. Table 3.4.2 lists the equations for the mixer, which are three additional mole balances. The equations, however, contain an additional variable, m_2. We have already written the mole fraction summation for stream 3. The air and methanol streams are at the same temperature so that we do not need an energy balance for the mixer.

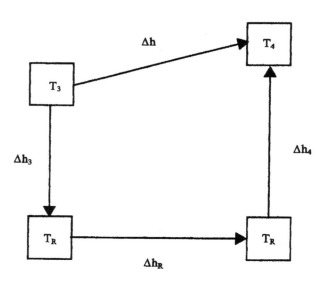

Figure 3.4.2 Thermodynamic path for a gas phase reaction.

After counting all the equations and variables in Tables 3.4.1 and 3.4.2, we find that we now have zero degrees of freedom. Thus, we have defined the problem, and we can now outline the solution procedure. The twenty-two equations are decoupled, i.e., it is not necessary to solve all them simultaneously. By inspection we find that we can solve the mole balance equations independently of the energy balance. This frequently occurs, usually when the temperatures in some of the lines are known. Furthermore, in this case, we do require an iterative calculation procedure. We again obtained a solution procedure by inspection, which is given in Table 3.4.3.

Frequently, we do not analyze simple process problems by the approach given in Tables 3.4.1 to 3.4.3. Instead, from the beginning, we assume that a solution is possible, and we carry out the calculations, introducing equations as needed. With experience one can recognize that certain problems have solutions, however, in most cases it is not evident that there is enough information to solve a problem, particularly when the solution contains many equations. In this problem, we will calculate the mole balance quickly without a formal analysis, once we know that the degrees of freedom are zero.

Because there is 1000 kmol/h (2204 lb mol/h) of methanol in line 3, there will be 200 kmol/h (440.8 lb mol/h) of methanol and 800 kmol/h of formaldehyde in line 4 because the conversion is 80 %. If 80 % of the stoichiometric quantity of oxygen is required, there will be 0.8 (1/2) (1000) = 400 kmol/h (881.6 lb mol/h) of oxygen at lines 2 and 3 and zero at line 4. The nitrogen flow rate in lines 2, 3 and 4 is (0.79/0.21) (400) = 1505 kmol/h (3317 lb mol/h) . It is good practice to tabulate

Table 3.4.2 Summary of Equations for a Mixer

Mole Balances

$$m_1' = y_{3,1} \, m_3 \tag{3.4.23}$$

$$y_{2,2}' \, m_2 = y_{3,2} \, m_3 \tag{3.4.24}$$

$$y_{2,5}' \, m_2 = y_{3,5} \, m_3 \tag{3.5.25}$$

Variables

Additional Variable is m_2.

Degrees of Freedom

$F = 22 - 22 = 0$

Table 3.4.3 Calculation Procedure for Calculating Heat Transfer to a Reactor

1. Solve Equations 3.4.9, 3.4.11 and 3.4.23 to 3.4.25 simultaneously to obtain m_2, m_3, $y_{3,1}$, $y_{3,2}$ and $y_{3,5}$.

2. Solve Equations 3.4.4 to 3.4.8 in terms of m_4. Substitute these derived equations into Equation 3.4.10 and solve for m_4.

3. Solve for $y_{4,1}$, $y_{4,2}$, $y_{4,3}$, $y_{4,4}$, and $y_{4,5}$ using Equations 3.4.4 to 3.4.8.

4. Calculate the pure component enthalpies from Equations 3.4.15 to 3.4.22.

5. Calculate the mixture enthalpies from Equations 3.4.13 and 3.4.14.

6. Calculate Q from Equation 3.4.12.

---

the results of a calculation for later reference and for checking the solution. Table 3.4.4 lists the steam properties – temperature, pressure, flow rate and composition. From experience we specify a 0.81 bar (0.8 atm) pressure drop across the reactor.

Now, calculate the enthalpy for each component by using an average heat capacity from the inlet temperature to the base temperature of 25 °C and from the base temperature to the outlet temperature. Thus, Equations 3.4.15 to 3.4.22 for each component reduce to

$$\Delta h = c_P \Delta T \tag{3.4.26}$$

Next, calculate the enthalpy change for each component and then add them to obtain the enthalpies of streams 3 and 4. Table 3.4.5 summarizes the results of these calculations.

From Equation 3.4.12 in Table 3.4.1

$$Q = -13.341 \times 10^6 + 800\,(-37420) + 16.96 \times 10^6 = -2.63 \times 10^7 \text{ kcal/h}$$
$$(-10.4 \times 10^7 \text{ Btu/h}). \tag{3.4.27}$$

Because heat added to the system is defined as positive, the minus sign means that we must remove heat.

Table 3.4.4: Stream Properties – Formaldehyde-Synthesis Reactor

Line No.	Temperature °C	Pressure atm	Flow Rate[a] kmol/h	Composition (mole fraction)				
				CH_3OH	O_2	H_2O	HCHO	N_2
1	500	1.8	1000	1.0	0	0	0	0
2	500	1.8	1950	0	0.210	0	0	0.790
3	500	1.8	2905	0.344	0.138	0	0	0.518
4	600	1.0	3305	0.061	0	0.242	0.242	0.455

[a]To convert to lb mol/h multiply by 2.205.

Table 3.4.5 Energy Balance Summary – Formaldehyde-Synthesis Reactor

$m_3 \Delta h_3 \ (10^6 \ kcal/h)^a$				
	m	c_P	Δt	Δh
HCHO	–	–	–	–
N_2	1505	(7.16)	(–475) =	–5.119
H_2O	–	–	–	–
O_2	400	(7.52)	(–475) =	–1.429
CH_3OH	1000	(14.3)	(–475) =	–6.793
				–
				13.34

[a] To convert to Btu/h multiply by 3.968.

Table 3.4.5 Continued

$m_4 \Delta h_4 (10^6$ kcal/h$)^a$				
	m	c_P	Δt	Δh
HCHO	800	(10.8)	(575) =	4.968
N_2	1505	(7.23)	(575) =	6.257
H_2O	800	(8.68)	(575) =	3.993
O_2	–	–	–	–
CH_3OH	200	(15.2)	(575) =	1.748
				17.00

a) To convert to Btu/h multiply by 3.968.

Example 3.5 Methanol-Synthesis Process

In this problem, we will determine the degrees of freedom of a process circuit composed of several process units by examining a methanol-synthesis process. Methanol was first synthesized from carbon monoxide and hydrogen on a commercial scale in 1923 by Badische Anilindund Soda-Fabrik (BASF) in Germany [25]. Methanol is an important basic bulk chemical used in the synthesis of formaldehyde and acetic acid [28] and it has been proposed as an automobile fuel and fuel additive [26]. Methanol has also been proposed as a substrate to produce a bacterium suitable as a protein source (single-cell protein). The bacterium would be a soy meal and fishmeal substitute for animal and poultry feeds [27]. If these applications should ever develop, the demand for methanol will increase considerably.

Process Chemistry

A two-step-reaction sequence describes the methanol synthesis. In the first step, steam reforming, a packed bed reactor (reformer) converts methane into a mixture of hydrogen and carbon monoxide (synthesis gas), according to Equation 3.5.1. Then, in the second step, a second packed-bed reactor (converter) converts the synthesis gas into methanol, as shown by Equation 3.5.2.

$$CH_4 + H_2O \rightarrow 3 H_2 + CO \ (+49{,}269 \text{ cal}, 298 \text{ K}) \tag{3.5.1}$$

$$2H_2 + CO \rightarrow CH_3OH \ (-21,685 \text{ cal}, 298 \text{ K}) \tag{3.5.2}$$

Methanol formation is exothermic, requiring removal of the enthalpy of reaction. Thermodynamically, the conversion to methanol increases by reacting at low temperatures. Also, there is a reduction in the number of moles during reaction, according to Equation 3.5.2, indicating that the converter should operate at a high pressure to increase conversion.

The Imperial Chemical Industries (ICI) has developed a reactive copper oxide catalyst [28], which allows operating the converter at low pressures, around 100 atm. Even though a high pressure increases conversion, a low pressure saves on gas compression and material of construction costs. The zinc-oxide, chromic-oxide catalyst, developed early in the history of the process, requires temperatures well above 300 °C for a reasonable rate of reaction, but conversions are low. To compensate for this lower catalytic activity, the converter pressure must be at 200 atm or higher. Because the reactivity of the new copper-oxide catalyst is high, the converter temperature can be lowered, favoring a high thermodynamic conversion. Sulfur containing compounds, however, easily poison the copper-oxide catalyst. Furthermore, iron pentacarbonyl forms by reaction of carbon monoxide with iron, but the reaction is less favored at low temperatures and pressures. Therefore, carbon steel instead of the more expensive stainless steel can be used for piping, reactors, and other process equipment.

Besides methanol formation, side reactions also occur, forming high molecular weight alcohols, dimethyl ether, carbonyl compounds, and methane. Because of the numerous side products formed, these compounds are divided into two groups, called the low-boiling and high-boiling compounds. No methane forms in the converter [31].

According to Equation 3.5.2, methanol synthesis requires a ratio of two moles of hydrogen to one mole of carbon monoxide, whereas Equation 3.5.1 shows that steam reforming produces a ratio of three to one. Thus, the excess hydrogen, as well as the inert gases (methane and nitrogen), will accumulate in the process and must be removed. One way of removing the excess hydrogen is to add carbon dioxide to the reformer feed gas to react with the hydrogen according to Equation 3.5.3.

$$2CO_2 + H_2 \rightarrow CO + H_2O \ (9,855 \text{ cal}, 298 \text{ K}) \tag{3.5.3}$$

Equation 3.5.3 is called the reverse-shift reaction because it occurs opposite to the normal direction. Carbon dioxide will react with hydrogen in the converter according to Equation 3.5.4 to form methanol.

$$CO_2 + 3 H_3 \rightarrow CH_3OH + H_2O \ (-11,830 \text{ cal}, 298K) \tag{3.5.4}$$

Another way of removing the excess hydrogen and inert gases is to use a purge stream. Unless carbon dioxide is available at low cost, purging is usually employed [28]. Because the purge stream is combustible, it may be used as a fuel to supply some of the enthalpy of reaction for the endothermic reforming reaction. If it is economical, the hydrogen in the purge stream could also be recovered.

Thermodynamically, the reforming reaction, Equation 3.5.1, shows that the reformer should be operated at the lowest pressure and highest temperature possible. The reforming reaction occurs on a nickel-oxide catalyst at 880 °C (1620 °F) and 20 bar, which results in a 25 °C approach to the equilibrium temperature [25,29]. Methane conversion increases by reducing the pressure, but natural gas is available at a high pressure. It would be costly to reduce the reformer pressure and then recompress the synthesis gas later to 100 bar (98.7 atm) for the converter. The steam to carbon monoxide ratio is normally in the range of 2.5 to 3.0 [30]. The ratio favors both the conversion of methane to carbon monoxide and the carbon monoxide to carbon dioxide as indicated by Equations 3.5.1 and 3.5.3. If the ratio is decreased, the methane concentration increases in the reformed gas, but if the ratio is set at three, the unreacted methane is small. The methane is a diluent in the synthesis reaction given by Equation 3.5.2.

Process Description

The process generates three hot gas streams: flue gas, reformer gas, and converter gas. We must recover the enthalpy of these streams to have an economically viable process. Thus, methanol synthesis plants are designed to generate 70% of their energy requirements internally [30]. The excess enthalpy generates high-pressure steam for steam-turbine drivers needed to compress the synthesis gas and the converter recycle gas. This is an example of a process where the process engineer must integrate several energy-transfer steps with reaction and separation steps for an energy-efficient process.

Figure 3.5.1 is the flow diagram for the Imperial Chemical Industries (ICI) process. The solid lines in the diagram are for the process streams, and dashed lines are for the steam system, which is really a subprocess of the main process – just as the cooling-water supply system is also a subprocess. Sulfur-containing compounds present in most natural gas streams will poison the reforming and synthesis catalysts. A hydrodesulphurization reaction removes these compounds by a using a catalyst in a packed bed. If there is no hydrogen present in the natural gas, purge gas from the synthesis loop, which is hydrogen rich, can be mixed with the natural-gas feed stream. Hydo-desulpurization forms hydrogen sulfide, which then reacts with zinc oxide in a packed bed to form zinc sulfide. Both the hydrogenation-catalyst and the zinc-oxide beds may be contained in the same vessel.

After removing hydrogen sulfide and mixing the stream with steam, the mixture flows to the reformer. Combustion gas heats the reformer to supply the enthalpy of reaction. To cool the hot reformed gas, steam is generated first and then vapor in the reboilers of the methanol-recovery section of the process. Cooling the reformed gas reduces the temperature and therefore the gas volume, which reduces the energy of compression. During cooling, water condenses and is removed in gas-liquid separators at various points in the process. After compressing the reformed gas in the first stage of compression, the gas then mixes with recyle gas to form feed gas. The feed gas is compressed and then preheated by the converter gas in an interchanger before entering the converter.

Because the reaction is exothermic, the synthesis gas is injected at several points in the converter to cool the reacting gases, which prevents overheating the catalyst. After leaving the converter, the gases are first cooled by preheating the feed to the converter and then cooled by water to condense out crude methanol. Then, a gas-liquid separator separates the crude methanol from the noncondensible gases. Purging part of the recycle stream from the separator removes excess hydrogen and inert gases from the process. Then, the purged gases mix with natural gas and air and finally burned to heat the reformer.

The crude methanol from the separator, containing methanol, water, low boiling compounds, and high boiling compounds, flows to the fractionation section. In the fractionation section, the crude methanol first flashes, and then the vapor-liquid stream flows to a "topping" column to remove the low-boiling compounds. Finally, the bottom stream from the "topping" column flows to a "refining" column to remove the high-boiling compounds, producing a purified methanol product and a wastewater stream.

Process Analysis

To analyze the process circuit, consider only a small segment of the methanol process – the synthesis loop – as indicated by the numbered lines in Figure 3.5.1. The synthesis loop contains a recycle line, which complicates the analysis. For simplicity, we will not consider all streams within the loop. As usual, the objective of the analysis is to specify or calculate pressure, temperature, composition, and flow rate in each line and the energy transferred into or out of each process unit. We begin by noting that the energy balances are decoupled from the mass balances for the streams selected. This means that we can solve the mole balances independent of the energy balances. If we include the determination of the flow rates of three side streams flowing into the converter, then energy balances are also needed.

The first step in the analysis is to determine if zero degrees of freedom exist in any process unit. In this case, the analysis will be simplified because of the reduction in the number of equations requiring simultaneous solution. After analyzing each process unit, we then combine the equations to determine if the process contains zero degrees of freedom. When analyzing each unit separately, we will repeat some variables and equations. For example, in line 3, the composition and flow rate variables, and the mole fraction summation, are the same for the mixer exit stream and the reactor feed stream. Later, when we combine the various processing units to determine the process degrees of freedom, we will take the duplication of variables and equations into account.

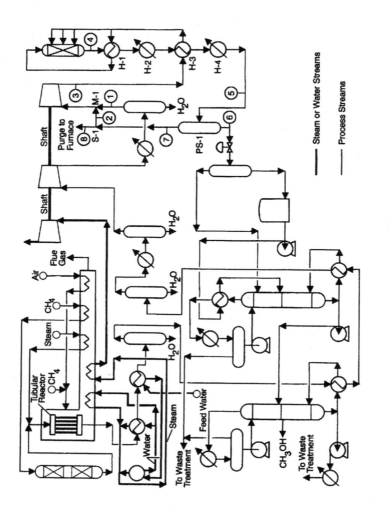

Figure 3.5.1 ICI methanol-synthesis process.

We begin the analysis by collecting all the information that is known about the process from the technical literature – journals, books, and patents. Also we can obtain information from company brochures on plant operations, pilot plant data, and laboratory data. Table 3.5.1 contains some of these data, operating conditions, and specifications.

Utilizing Table 3.5.1 we generate initial specifications for the synthesis loop, which are contained in Table 3.5.2. After completing the degrees of freedom analysis, we may have to adjust the specifications to obtain zero degrees of freedom. Market conditions determine the production rate of methanol, $y_{6,7} m_6$, as given in Table 3.5.2. The composition of the reformed gas in line 1, with a slight adjustment to include nitrogen, is taken from Fulton and Fair's [32] case-study problem. The methane and nitrogen, which are inerts, and excess hydrogen are maintained at acceptable concentrations by the purge stream. A small purge stream results in

Table 3.5.1 Process Conditions – Methanol-Synthesis Process

Reformer

Exit Temperature[a]	– 850 °C
Exit Pressure[a]	– 20 bar
Molar H_2O/CO Ratio[a]	– 3.0
Equilibrium at Reformer Exit[a]	

Converter

Exit Temperature[b]	– 270 °C
Inlet Pressure[b]	– 100 bar
Optimum Exit CH_3OH Concentration[b]	– 5 %
Pressure Drop[c]	– 5 to 6 bar

Separator

Crude Methanol Components	
Methanol	79 wt. %
Component	Concentration[d], ppm
Dimethyl Either	20-150
Carbonyl Compounds	10-35
Higher Alcohols	100-2000
Methane	None

a) Source: Reference 25
b) Source: Reference 28
c) Source: Reference 35
d) Source: Reference 32

Table 3.5.2 Specified Variables – Methanol-Synthesis Process

2nd subscript: CH_4 = 1, H_2O = 2, H_2 = 3, CO = 4, CO_2 = 5, N_2 = 6, CH_3OH = 7

Basis $y_{6,7}$ m_6 = 1000 kmol/h

Crude Methanol Concentration = 0.79 mass fraction

Variables	Mole Fraction
$y_{1,1}$	0.0085
$y_{1,2}$	≈ 0
$y_{1,3}$	0.7800
$y_{1,4}$	0.0600
$y_{1,5}$	0.1500
$y_{1,6}$	0.0015
$y_{1,7}$	≈ 0
$y_{2,2}$	≈ 0
$y_{3,1}$	0.0250
$y_{3,6}$	0.0100
$y_{4,7}$	0.0500
$y_{6,7}$	0.5816

---

high concentrations of these gases in the system and a large purge stream in low concentrations. By specifying the methane concentration, $y_{3,1}$, we fix the purge-stream flow rate. The methanol concentration at the outlet of the converter, $y_{4,7}$, is typical of the low pressure process. Finally, Fulton and Fair [32] give the methanol concentration in the crude methanol stream, $y_{6,7}$.

For a first approximation to the solution, we will assume that essentially all the methanol condenses, with only trace amounts appearing in the recycle line. We will also assume that most of the water condenses and that very small amounts of carbon monoxide, carbon dioxide, hydrogen, methane, and nitrogen dissolve in the condensate. To account for methanol and water vapor in the recycle gases and the solubility of the gases in the crude methanol, we would have to include phase equilibrium relationships in the analysis. As stated earlier, several condensable byproducts, high and low-boiling compounds in the crude methanol, are present in small amounts, as shown in Table 3.5.1. We will not consider these compounds in the synthesis-loop analysis.

At this point in the analysis we do not know if the variables are over-specified or under-specified. Table 3.5.3 gives the degrees of freedom for each process unit. As usual prime the specified variables. Except for the splitter, the analysis is straight forward. Since there is no composition change across the splitter, as stated by Equations 3.5.30 to 3.5.39, only the total mole balance is an independent equation. Also, only the sum of the mole fractions for one of the three streams is an independent equation. Table 3.5.3 shows that no process unit contains zero degrees of freedom.

Before attempting to solve the equations in Table 3.5.3, calculate the degrees of freedom for the process. First, determine the number of unique variables because some of the variables are repeated from process unit to process unit, as shown in Table 3.5.3. The process variables are equal to the sum of all the unit variables minus the repeated variables. To determine the repeated variables, examine the lines connecting the process units. Table 3.5.4 shows that the repeated variables are mole fractions and molar flow rates. From Table 3.5.5, the total number of variables for all units is 57, and the total number of repeated variables is 23. Therefore, the number of unique process variables are 34, as shown in Table 3.5.5.

Next, determine the number of independent equations by again examining each connecting line. The repeated equations are the mole fraction summations, as shown in Table 3.5.4. To determine the number of independent equations for the process, subtract the repeated equations from the sum of the equations for all the process units. The total number of equations for all process units is 39, as shown in Table 3.5.5. Although each process unit contains positive degrees of freedom, we see that the process degrees of freedom equals minus two, which means that the problem has been overspecified. Before unspecifying variables check if the number of equations are correct. By inspection – not an easy task – we find that $\sum y_{7,i} = 1$, Equation 3.5.29 in Table 3.5.3, is not independent. It can be derived by substituting Equations 3.5.30 to 3.5.34 into Equation 3.5.6, $\sum y_{2,i}$. Therefor, the number of independent equations must be reduced by 1 – from 36 to 35 – and the degrees of freedom becomes minus one.

Once you are certain that all equations are independent and no equations are missing, then unspecify one of the variables. For example, unspecify the nitrogen concentration at the converter inlet, $y_{3,6}$. Because $y_{3,6}$ is now unspecified, correct the degree of freedom analysis for both the mixer and converter. At the mixer and converter the number of variables increases by one as shown in Table 3.5.6. Thus, for the mixer $F = 12 - 7 = 5$ and for the converter $F = 14 - 9 = 5$. Because Equations 3.5.27 and 3.5.29 are not independent, the number of equations at the condenser-separator combination and the splitter are reduced by one, as shown in Table 3.5.6. Finally, because $\sum y_{7,i}$ is no longer valid, it is not a repeated equation. Thus, the repeated equations in line 7 are now zero. The revised calculation for the degrees of freedom in Table 3.5.6 shows that the process degrees of freedom is now zero.

Table 3.5.3 Summary of Mole Balances – Methanol-Synthesis Process

2nd subscripts: $CH_4 = 1$, $H_2O = 2$, $H_2 = 3$, $CO = 4$, $CO_2 = 5$, $N_2 = 6$, $CH_3OH = 7$

MIXER (M-1)

Mole Balances

$$y_{1,1}' m_1 + y_{2,1} m_2 = y_{3,1}' m_3 \tag{3.5.1}$$

$$y_{1,3}' m_1 + y_{2,3} m_2 = y_{3,3} m_3 \tag{3.5.2}$$

$$y_{1,4}' m_1 + y_{2,4} m_2 = y_{3,4} m_3 \tag{3.5.3}$$

$$y_{1,5}' m_1 + y_{2,5} m_2 = y_{3,5} m_3 \tag{3.5.4}$$

$$y_{1,6}' m_1 + y_{2,6} m_2 = y_{3,6}' m_3 \tag{3.5.5}$$

$$y_{2,1} + y_{2,3} + y_{2,4} + y_{2,5} + y_{2,6} = 1 \tag{3.5.6}$$

$$y_{3,1}' + y_{3,3} + y_{3,4} + y_{3,5} + y_{3,6}' = 1 \tag{3.5.7}$$

Variables

$y_{2,1} - y_{2,3} - y_{2,4} - y_{2,5} - y_{2,6} - y_{3,3} - y_{3,4} - y_{3,5} - m_1 - m_2 - m_3$

Degrees of Freedom

$F = 11 - 7 = 4$

CONVERTER (R-3)

(A) $2H_2 + CO \rightarrow CH_3OH$

(B) $3H_2 + CO_2 \rightarrow CH_3OH + H_2O$

Mole Balances

$$y_{3,1}' m_3 = y_{4,1} m_4 \tag{3.5.8}$$

$$x_B y_{3,5} m_3 = y_{4,2} m_4 \tag{3.5.9}$$

$$y_{3,3} m_3 = y_{4,3} m_4 + 2 x_A y_{3,4} m_3 + 3 x_B y_{3,5} m_3 \tag{3.5.10}$$

$$y_{3,4} m_3 = y_{4,4} m_4 + x_A y_{3,4} m_3 \tag{3.5.11}$$

$$y_{3,5} m_3 = y_{4,5} m_4 + x_B y_{3,5} m_3 \tag{3.5.12}$$

Table 3.5.3 continued

$$y_{3,6}' \, m_3 = y_{4,6} \, m_4 \tag{3.5.13}$$

$$x_A \, y_{3,4} \, m_3 + x_B \, y_{3,5} \, m_3 = y_{4,7}' \, m_4 \tag{3.5.14}$$

$$y_{3,1}' + y_{3,3} + y_{3,4} + y_{3,5} + y_{3,6}' = 1 \tag{3.5.15}$$

$$y_{4,1} + y_{4,2} + y_{4,3} + y_{4,4} + y_{4,5} + y_{4,6} + y_{4,7}' = 1 \tag{3.5.16}$$

Variables

x_A - x_B - $y_{3,3}$ - $y_{3,4}$ - $y_{3,5}$, - $y_{4,1}$ - $y_{4,2}$ - $y_{4,3}$ - $y_{4,4}$ - $y_{4,5}$ - $y_{4,6}$ - m_3 - m_4

Degrees of Freedom

$F = 13 - 9 = 4$

CONDENSER-SEPARATOR
(the system includes H-1, 2, 3, 4 and PS-1)

Mole Balances

$$y_{4,1} \, m_4 = y_{7,1} \, m_7 \tag{3.5.17}$$

$$y_{4,2} \, m_4 = y_{6,2} \, m_6 \tag{3.5.18}$$

$$y_{4,3} \, m_4 = y_{7,3} \, m_7 \tag{3.5.19}$$

$$y_{4,4} \, m_4 = y_{7,4} \, m_7 \tag{3.5.20}$$

$$y_{4,5} \, m_4 = y_{7,5} \, m_7 \tag{3.5.21}$$

$$y_{4,6} \, m_4 = y_{7,6} \, m_7 \tag{3.5.22}$$

$$y_{4,7}' \, m_4 = y_{6,7}' \, m_6 \tag{3.5.23}$$

$$y_{6,7}' \, m_6 = 1000 \tag{3.5.24}$$

$$y_{4,1} + y_{4,2} + y_{4,3} + y_{4,4} + y_{4,5} + y_{4,6} + y_{4,7}' = 1 \tag{3.5.25}$$

$$y_{6,2} + y_{6,7}' = 1 \tag{3.5.26}$$

$$y_{7,1} + y_{7,3} + y_{7,4} + y_{7,5} + y_{7,6} = 1 \tag{3.5.27}$$

Table 3.5.3 Continued

Variables

$y_{4,1} - y_{4,2} - y_{4,3} - y_{4,4} - y_{4,5} - y_{4,6} - y_{6,2} - y_{7,1} - y_{7,3} - y_{7,4} - y_{7,5} - y_{7,6} - m_4 - m_6 - m_7$

Degrees of Freedom

$F = 15 - 11 = 4$

SPLITTER (S-1)

Mole Balances

$$m_7 = m_2 + m_8 \tag{3.5.28}$$

$$y_{7,1} + y_{7,3} + y_{7,4} + y_{7,5} + y_{7,6} = 1 \tag{3.5.29}$$

$$y_{7,1} = y_{2,1} \tag{3.5.30}$$

$$y_{7,3} = y_{2,3} \tag{3.5.31}$$

$$y_{7,4} = y_{2,4} \tag{3.5.32}$$

$$y_{7,5} = y_{2,5} \tag{3.5.33}$$

$$y_{7,6} = y_{2,6} \tag{3.5.34}$$

$$y_{7,1} = y_{8,1} \tag{3.5.35}$$

$$y_{7,3} = y_{8,3} \tag{3.5.36}$$

$$y_{7,4} = y_{8,4} \tag{3.5.37}$$

$$y_{7,5} = y_{8,5} \tag{3.5.38}$$

$$y_{7,6} = y_{8,6} \tag{3.5.39}$$

Variables

$y_{7,1} - y_{7,3} - y_{7,4} - y_{7,5} - y_{7,6} - y_{8,1} - y_{8,3} - y_{8,4} - y_{8,5} - y_{8,6} - y_{2,1} - y_{2,3} - y_{2,4} - y_{2,5} - y_{2,6} - m_2 - m_7 - m_8$

Degrees of Freedom

$F = 18 - 12 = 6$

Table 3.5.4 Repeated Variables and Equations – Methanol-Synthesis Process

Line Number	Repeated Variables V_R	Repeated Equations R_R
2	$y_{2,1} - y_{2,3} - y_{2,4} - y_{2,5} - y_{2,6} - m_2$	0
3	$y_{3,3} - y_{3,4} - y_{3,5} - m_3$	$\sum y_{3,i} = 1$
4	$y_{4,1} - y_{4,2} - y_{4,3} - y_{4,4} - y_{4,5} - y_{4,6} - m_4$	$\sum y_{4,i} = 1$
7	$y_{7,1} - y_{7,3} - y_{7,4} - y_{7,5} - y_{7,6} - m_7$	$\sum y_{7,i} = 1$

Table 3.5.5 Degrees of Freedom Calculation – Methanol-Synthesis Process

Process Unit	Unit Variables V_U	Unit Equations R_U
Mixer	11	7
Converter	13	9
Condenser-Separator	15	11
Splitter	18	12
Total	57	39

Line Number	Repeated Variables V_R	Repeated Equations R_R
2	6	0
3	4	1
4	7	1
7	6	1
Total	23	3

Process Degrees of Freedom

$F_P = (V_U - V_R) - (R_U - R_R) = (57 - 23) - (39 - 3) = 34 - 36 = -2$

Table 3.5.6 Revised Degrees of Freedom Calculation—Methanol-
Synthesis Process

Process Unit	Unit Variables V_U	Unit Equations R_U
Mixer	12	7
Converter	14	9
Condenser-Separator	15	10
Splitter	18	11
Total	59	37

Line Number	Repeated Variables V_R	Repeated Equations R_R
2	6	0
3	5	1
4	7	1
7	6	0
Total	24	2

Process Degrees of Freedom

$$F_P = (V_U - V_R) - (R_U - R_R) = (59 - 24) - (37 - 2) = 35 - 35 = 0$$

Now that the problem is formulated we turn our attention to solving the
equations. One solution method that we could use is the sequential modular
method. For this method, select one of the process units as the starting point for
the calculation. Then, assume values for some of variables to reduce the degrees of
freedom to zero for that unit. Next, precede unit-by-unit through the flow sheet

until you can calculate the assumed variables to compared with the original guesses. Westerberg et al. [16] have reviewed the sequential modular method, as well as other methods, in detail. This particular method has the advantage that the calculation procedure can be visualized physically. Also, at any particular time the number of equations that require simultaneous solution is considerably reduced.

For this problem we can solve the reduced set of equations simultaneously, using POLYMATH (Version 4.0) [19] or by some other suitable mathematical software. Since POLYMATH cannot solve more than 32 simultaneous, nonlinear equations and explicit algebraic expressions, we must reduce the number of equations listed in Table 3.5.3.

First, drop all the repeated equations listed in Table 3.5.3. By substituting Equations 3.5.30 to 3.5.34 into Equations 3.5.19 to 3.5.22, we eliminate the mole fraction variables in line seven. We do not need Equations 3.5.35 to 3.5.39 for the solution, so they can be dropped. Table 3.5.2 lists the specified variables, except for the nitrogen mole fraction, $y_{3,6}$, which is now unspecified. Table 3.5.7 lists the reduced set of equations.

Before solving the Equations in Table 3.5.7, we must select initial guess values for all the variables. Selecting guess values for variables to start a calculation is always a problem. For some initial values of the variables, the solution may not converge. One strategy for obtaining correct initial guesses is to examine each variable for limits. For example, values of mole fraction must be limited to the range from zero to one. Temperatures in heat exchangers are limited by the freezing point of the fluids and the stability of the fluids at high temperatures. Obtaining stable initial guess values is an iterative procedure. Table 3.5.8 lists the composition and flow rates from the POLYMATH solution.

To complete the process circuit analysis, we now assign pressures and temperatures in lines 1 to 8. The pressures in the various streams given in Table 3.5.8, are determined after specifying 100 bar at the reactor inlet, an optimum synthesis pressure [30]. Then, we assign pressure drops, based on experience, of 0.34 bar across each heat exchanger [8] and 5.0 bar across the converter. The pressure drop across the gas-liquid phase separator, PS-1, and piping is small compared to the other system pressure drops. Starting at 100 bar at the converter inlet we can now specify pressures in lines 1 to 8, except line 6. The pressure at line 6 should be high enough to overcome the pressure drop across the upper plates of the first column, 0.1 bar, plus the pressure across the two condensers. Therefore, the total pressure drop is 0.1 + 2 (0.34) or 0.78 bar which is the pressure at line 6. The copper-oxide catalyst sinters significantly at high temperatures, i.e., there is growth of the copper-oxide crystals. Consequently, there will be a corresponding reduction in surface area and catalytic activity. Thus, limit the gas temperature to 270 °C [8]. Because the compressor work increases with increasing volumetric flow rate, we must keep the temperature at the compressor inlet low. If we assume a temperature of 40 °C in lines 1 and 2, then the temperatures in lines 5, 7 and 8 will also be 40 °C. The temperature in line 3 can be determined by an energy balance across the compressor, which will be considered in Chapter 5. Finally, we can find the temperature in line 6 by making a flash calculation. Table 3.5.8 gives the pressures and temperatures in each line. This completes the analysis of the methanol-synthesis flow loop.

Table 3.5.7 Revised Summary of Mole Balances – Methanol-Synthesis Process

MIXER (M-1)

Mole Balances

$$y_{1,1}' \, m_1 + y_{2,1} \, m_2 = y_{3,1}' \, m_3 \tag{3.5.1}$$

$$y_{1,3}' \, m_1 + y_{2,3} \, m_2 = y_{3,3} \, m_3 \tag{3.5.2}$$

$$y_{1,4}' \, m_1 + y_{2,4} \, m_2 = y_{3,4} \, m_3 \tag{3.5.3}$$

$$y_{1,5}' \, m_1 + y_{2,5} \, m_2 = y_{3,5} \, m_3 \tag{3.5.4}$$

$$y_{1,6}' \, m_1 + y_{2,6} \, m_2 = y_{3,6} \, m_3 \tag{3.5.5}$$

$$y_{2,1} + y_{2,3} + y_{2,4} + y_{2,5} + y_{2,6} = 1 \tag{3.5.6}$$

CONVERTER (R-3)

Mole Balances

$$y_{3,1} \, m_3 = y_{4,1} \, m_4 \tag{3.5.9}$$

$$x_B \, y_{3,5} \, m_3 = y_{4,2} \, m_4 \tag{3.5.10}$$

$$y_{3,3} \, m_3 = y_{4,3} \, m_4 + 2 \, x_A \, y_{3,4} \, m_3 + 3 \, x_B \, y_{3,5} \, m_3 \tag{3.5.11}$$

$$y_{3,4} \, m_3 = y_{4,4} \, m_4 + x_A \, y_{3,4} \, m_3 \tag{3.5.12}$$

$$y_{3,5} \, m_3 = y_{4,5} \, m_4 + x_B \, y_{3,5} \, m_3 \tag{3.5.13}$$

$$y_{3,6} \, m_3 = y_{4,6} \, m_4 \tag{3.5.14}$$

$$x_A \, y_{3,4} \, m_3 + x_B \, y_{3,5} \, m_3 = y_{4,7}' \, m_4 \tag{3.5.15}$$

Table 3.5.7 Continued

$$y_{3,1}' + y_{3,3} + y_{3,4} + y_{3,5} + y_{3,6} = 1 \quad (3.5.16)$$

CONDENSER-SEPARATOR
(the system includes H-1, 2, 3, 4 and PS-1)

Mole Balances

$$y_{4,1}\, m_4 = y_{2,1}\, m_7 \quad (3.5.19)$$

$$y_{4,2}\, m_4 = y_{6,2}\, m_6 \quad (3.5.21)$$

$$y_{4,3}\, m_4 = y_{2,3}\, m_7 \quad (3.5.22)$$

$$y_{4,4}\, m_4 = y_{2,4}\, m_7 \quad (3.5.23)$$

$$y_{4,5}\, m_4 = y_{2,5}\, m_7 \quad (3.5.23)$$

$$y_{4,6}\, m_4 = y_{2,6}\, m_7 \quad (3.5.24)$$

$$y_{4,7}'\, m_4 = y_{6,7}\, m_6 \quad (3.5.25)$$

$$y_{6,7}'\, m_6 = 1000 \quad (3.5.26)$$

$$y_{4,1} + y_{4,2} + y_{4,3} + y_{4,4} + y_{4,5} + y_{4,6} + y_{4,7}' = 1 \quad (3.5.27)$$

$$y_{6,2} + y_{6,7}' = 1 \quad (3.5.28)$$

SPLITTER

$$m_7 = m_2 + m_8 \quad (3.5.30)$$

Variables

x_A - x_B - m_1 - m_2 - m_3 - m_4 - m_6 - m_7 - m_8 - $y_{2,1}$ - $y_{2,3}$ - $y_{2,4}$ - $y_{2,5}$ - $y_{2,6}$ - $y_{3,1}$, $y_{3,3}$ - $y_{3,4}$ -$y_{3,5}$ - $y_{4,1}$ - $y_{4,2}$ - $y_{4,3}$ - $y_{4,4}$ - $y_{4,5}$ - $y_{4,6}$ - $y_{6,2}$

Degrees of Freedom

$$F = 25 - 25 = 0$$

Table 3.5.8 Summary of Stream Properties—Methanol Synthesis Process

Line No.	Temperature °C	Pressure[a] bar	Flow Rate[b] kgmol/h	Composition, Mole Fraction						
				CH_4	H_2O	H_2	CO	CO_2	N_2	CH_3OH
1	40	94.32	5184	0.0085	–	0.7800	0.0600	0.1500	0.00150	–
2	40	94.32	16816	0.03009	–	0.9041	0.0208	0.03972	0.005309	–
3	–	100.68	22000	0.02500	–	0.8748	0.03004	0.06571	0.004412	–
4	270	95.68	20000	0.0275	0.03597	0.8264	0.01901	0.03631	0.004853	0.0500
5	40	94.32	20000	0.0275	0.02500	0.8264	0.01901	0.03631	0.004853	0.0500
6	40	0.78	1719	–	0.4184	–	–	–	–	0.5816
7	–	94.32	18281	0.03009	–	0.9041	0.0208	0.03972	0.005309	–
8	40	94.32	1465	0.03009	–	0.9041	0.0208	0.03972	0.005309	–

a) Multiply by 0.9869 to convert to atmospheres.
b) Multiply by 2.205 to convert to lbmol/h.

Nomenclature

English

A   area

a   constant in the heat-capacity equation or in Redlic-Kwong's equation

b   constant in the heat-capacity equation or in Redlic-Kwong's equation

c   constant in the heat-capacity equation

c_P heat capacity at constant pressure

C   cost

C_D direct cost

C_G general cost

C_I indirect cost

C_T total cost

d   constant in the heat-capacity equation or pipe diameter

F   degrees of freedom

f   friction factor or function of

g   acceleration of gravity

h   enthalpy or heat-transfer coefficient

Δh_R enthalpy of reaction

Δh_{VR} enthalpy of vaporization at a reference temperature

Δh_{VW} enthalpy of vaporization at the wet bulb temperature

k   mass-transfer coefficient

K   degrees Kelvin or phase equilibrium constant or constant of integration

K_P chemical equilibrium constant

m   molar flow rate or mass flow rate

N   number of moles

p   partial pressure or vapor pressure

P   property or total pressure

Q   heat transferred or volumetric flow rate

R   gas constant or number of relationships (tabular, graphical or algebraic)

Re   Reynolds group

t   temperature or time

t_P temperature of the process fluid

t_R reference temperature

t_W water or wet-bulb temperature

T   absolute temperature

U   overall heat-transfer coefficient

v   specific volume or velocity

V   variable or vessel volume

W   work done

x   conversion

y   mole or mass fraction

z   elevation

Greek

ε   roughness

ρ   molar density

Subscripts

B   at the bubble-point temperature

D   at the dew-point temperature

i   ith component

L   liquid phase

LM   logarithmic mean

p   process or constant pressure

P   property

R   reference

S   saturated vapor or liquid

V   vapor phase

w   wet bulb or water

REFERENCES

1. Polya, G., How to Solve It, 2nd ed., Doubleday, Garden City, NY, 1957.
2. Reid, R.C., Prausnitz, J. M., Sherwood, T.K., Properties of Gases and Liquids, 3rd ed., McGraw-Hill Book Co., New York, NY, 1977.
3. OLvjic, Z , Compute Friction Factors Fast for Flow in Pipes, Chem. Eng., 88, 25, 91, 1981.
4. Nathan, M. F., Choosing a Process for Choride Removal, Chem. Eng., 85, 3, 93, Jan. 1978.
5. LeDuff, C., Welder Is Killed in an Explosion at the Brooklyn Navy Yard, NY Times, New York, NY, Oct. 1998.
6. Constance, J. D., Solve Gas Purging Problems Graphically, Chem. Eng., 87, 26, 65, 1980
7. Reid, W. T., et al., Heat Generation and Transport, Chemical Engineers Handbook, 5th ed. Perry, R. H., Chilton, C. H., (Eds.), p. 9.19, McGraw-Hill, New York, NY, 1973.
8. Anonymous, Methanation Unit Design, AICHE Student Contest Problem, American Institute of Chemical Engineers, New York, NY, 1982
9. Aagaard, V., Ledegaard, B. H., Liquefied Natural Gas Storage, Barcelona, CN Post NR 87, Christiani & Wielsen, Kobenhaven, Denmark, Nov., 1969.
10. Gunther, R.,Verbrennung und Feuerrungen, Springer-Verlag, Berlin, 1974.
11. Woodson, R., Cooling Towers, Sci. Amer., 224, 5, 70, 1971.
12. Walas, S. M., Chemical Process Equipment, Selection and Design, Butterworths, Boston, MA, 1988.
13. Chaar, L., Gallagher, J. S., Kelly, G. S., NBSINRC, Hemisphere Publishing, New York, NY, 1984.
14. Bagnoli, E., Fuller, F. H., Johnson, V. J., Morris, R. W., Evaporative Cooling, Chemical Engineers Handbook, Perry, R. H., Chilton, C. H., 5th ed., p. 12-12, McGraw-Hill, New York, NY, 1973.
15. Bird, R. B., Stewart, W. E., Lightfoot, E. N., Transport Phenomena, John Wiley & Sons, New York, NY, 1960.
16. Westerberg, A.W., Hutchinson, H. P., Motard, R. L., Winter, P., Process Flow-Sheeting, Cambridge University Press, Cambridge, England, 1979.
17. Rudd, D. F., Watson, C. C., Strategy of Process Engineering, John Wiley & Sons, New York, NY, 1968.

18. Meyers, A.L., Seider, W.D., Introduction to Chemical Engineering and Computer Calculations, Prentice Hall, Englewood Cliffs, NJ, 1976.
19. Shacham, M., Cutlip, M.B., POLYMATH (Version 4.0), CACHE Corp., Austin, TX, 1996.
20. Shacham, M., Shacham O., Finding Boundaries of the Domain of Definition for Functions Along a One- Dimensional Ray, Acm. Trans. Math. Softw., 16, 3, 258, 1990.
21. Swearingen, J. S., Turboexpanders and Processes that Use Them, Chem. Eng. Progr., 68, 7, 95, 1972.
22. King, C. J., Separation Processes, 2nd ed., McGraw-Hill, New York, NY, 1980.
23. Greek, B. F., Slump Hits Gas-Based Petrochemicals, Chem. & Eng. News, p. 18, March 29, 1982.
24. Diem, H., Formaldehyde Routes Bring Cost, Production Benefits, Chem. Eng., 85, 5, 83, 1978.
25. Wade, L. E., Gengelbach, R. B., Trumbley, J. L., Hallbauer, W. L., Methanol, Kirk-Othmer Encyclopedia of Chemical Technology, 3rd ed., p. 16, John Wiley & Sons, New York, NY, 1981.
26. Anderson, E. V., Large Volume Fuel Market Still Eludes Methanol, Chem. & Eng. News, p. 9, July 16, 1984.
27. Anonymous, Steady Growth Ahead for Methanol in Europe, Chem. & Eng. News, p. 30, Nov. 17, 1980.
28. Horsley, J. B., Rogerson, P. L., Scott, R. H., The Design and Performance of the ICI 100 Atmosphere Methanol Plant, 74th Annual Meeting, American Institute of Chemical Engineers, New York, NY, March 1973.
29. Rogerson, P. L., The ICI Low Pressure Methanol Process, Het In genieursblad, 40e joargang, nr. 21, 659, 1971.
30. Pinto, A., Rogerson, P. L., Optimizing the ICI Low Pressure Methanol, Chem. Eng., 84, 4, 102, 1977.
31. Anonymous, The ICI Low Pressure Methanol Process, Agricultural Division, Imperial Chemical Industries, Bilingham, England, No date.
32. Fulton, W., Fair, J.R., Manufacture of Methanol and Substitute Natural Gas, Case Study 17, Chem. Eng. Dept., Washington University, St. Louis, MO, Sept. 1, 1974.

4

Process Heat Transfer

Heat transfer is a frequently occurring process operation. Within a process two general types of heat exchange occur. One type is the exchange of heat between two process streams. The heat exchanger where this occurs is frequently called an interchanger. In the second type, heat exchange occurs between the process and the surroundings, which requires a heat-transfer fluid. Water is the most common fluid. If the temperature is sufficiently high, then it may be economical to recover work from a process stream by generating high pressure steam and then expanding the steam through a turbine. This occurs in processes for synthesizing methanol where superheated steam is generated when cooling the reformer exit stream.

After the process analysis is completed, the heat-exchange requirements of the process will be specified. The next step is to calculate the heat-exchanger surface area which will allow you to calculate its installed cost. The cost calculation proceeds according to the following steps:

1. select a heat-transfer fluid
2. evaluate and select a heat-exchanger type
3. locate the shell-side and tube-side fluids
4. specify the terminal temperatures of the fluid streams
5. determine the overall heat-transfer coefficient
6. calculate the heat-exchanger surface area
7. estimate the total installed cost

To calculate the heat-transfer surface area requires a calculation procedure. The approach used here will be to use a simple procedure. A detailed procedure requires specifying a tube length, diameter, and layout. Although this detail will eventually be needed, at the preliminary stage of a process design we are only interested in an approximate estimate of the cost. Kern [1] gives detailed heat-exchanger design procedures, which, according to Frank [29], are too conserva-tive. Some up-to-date procedures can be found in Reference 15 and in the engi-neering literature.

Heat-Transfer Fluids

Before selecting a heat-transfer fluid, examine the process for any possibility of interchanging heat between process streams to conserve energy. Frequently, one process stream needs to be heated and another process stream cooled. After this possibility has been exhausted, select a heat-transfer fluid to cool or heat the proc-ess stream. A variety of heat-transfer fluids are available, ranging from the cryo-genic to the high-temperature region as shown in Table 4.1.

Because air and water are common heat-transfer fluids, we must frequently select one or the other. For an air-cooled heat exchanger, Frank [7] recommends that if the process-fluid temperature is

1. > 65 °C (149 °F) use an air-cooled heat exchanger
2. < 50 °C (122 °F) use water

Between 50°C and 65°C an economic analysis is required, but for a preliminary analysis this will not be necessary.

The factors that must be considered in evaluating and selecting a heat transfer fluid are:

1. operating temperature range
2. environmental effects
3. toxicity
4. flammability
5. thermal stability
6. corrosivity
7. viscosity

The primary consideration is to match the process temperature requirements with the recommended operating temperature range of the heat-transfer fluid. Ta-ble 4.1 lists the range for several heat-transfer fluids. Steam is normally consid-ered first for high temperatures, but above 180°C (356°F) the steam pressure in-creases rapidly with increasing temperature. Consequently, piping and vessel costs will also rise rapidly. Thus, other high-temperature heat-transfer fluids must be considered. A low vapor pressure at a high temperature is the major reason for choosing an organic fluid over steam. Pressurized water could be used from 300 to 400°C (572 to 752 °F), but high pressures are required to maintain the water in

Table 4.1 Selected Heat-Transfer Fluids

Heat-Transfer Fluid	Operating Temperature Range, °C	Reference
Refrigerants		
Methane		2
Ethane and Ethylene	–60 to –115	2
Propane and Propylene	5 to –46	2
Butanes	–12 to 16	2
Ammonia	–32 to 27	2
Fluorocarbon (R-12)[a]	–29 to 27	2
Water + Ethylene Glycol (50%/50%)	–50 to 90	
Water		
Water (wells, rivers, lakes)	32 to 49	3
Chilled Water	1.7 to 16	4
Cooling Tower Water	30	
High Temperature Water	300 to 400	6
Air	65 to 260	7, 8
Steam		
Low Pressure (2.7 bar)[b]	126	
Low Pressure (4.6 bar)[b]	148	
Organic Oils[c]	–50 to 430	9
Silicone Oils	–23 to 399	11
Molten Salts		
25% $AlCl_3$, 75% $AlBr_3$	75 to 500	9
40% $NaNO_2$, 7% $NaNO_3$, 53% KNO_3	204 to 454	10
Liquid Metals		
56% Na, 44% K or 22% Na, 78% K	204 to 454	6
Mercury	316 to 538	6
Combustion Gases	> 500	9

a) Dichlorodifluoromethane
b) Typical steam pressures
c) For example: diphenyl-diphenyl oxide, hydrogenated terphenyl, aliphatic oil, aromatic oil

the liquid state. Singh [9] recommends using the nitrate salt mixture listed in Table 4.1 in the temperature range of 204 to 454 °C (367 to 850 °F). Above 500 °C (932 °F) combustion gases and liquid metals are possibilities. Although mercury was considered in the past for power plants, the risk is too great. The other liquid metals are used for cooling nuclear reactors. Temperatures from 50 to 1000 °C (90 to 1830 °F) can also be achieved by electrical heating.

Since accidental chemical spills occur occasionally, the effect of the heat-transfer fluid on the environment and health must be considered. Since the use of chemicals may be governed by laws, the process engineer must comply. In 1979, the EPA banned the use of polychlorinated biphenyls (PCBs) because of the concern over environmental contamination [12].

The factors numbered three to six can be reduced to economic considerations. Ultimately, the heat-transfer fluid selected will depend on the total cost, both capital and operating costs. For example, if a heat-transfer fluid meets the first two requirements, but it is more toxic than other possibilities, then the heat-transfer system will have to contain extra safety features, increasing its cost. The heat-transfer fluid will then need to have other compensating features to reduce the cost of transferring heat.

Organic heat-transfer fluids require stringent leakage control because they are all flammable from 180 to 540 °C (356 to 1000 °F) [10], and most of the fluids irritate eyes and skin [9]. Although a nitrate salt mixture is nonflammable, it is a strong oxidizing agent and thus should not contact flammable materials.

Organic heat-transfer fluids can degrade somewhat, either by oxidation or thermal cracking. The primary cause is thermal degradation. In thermal degradation, chemical bonds are broken forming new smaller compounds that lower the flash point of the fluid. At the flash point, flammable fluids will momentarily ignite on application of a flame or spark. Organic fluids will also degrade to form active compounds. The compounds will then polymerize to form large molecules thereby increasing the fluid viscosity, which reduces heat transfer. Heat-transfer fluids are usually heated in a furnace and then distributed to several heat exchangers in a process. At high temperatures thermal degradation accelerates, forming coke at the heater surface in furnaces, which eventually leads to heater failure. Even the most stable fluids will eventually degrade so that some means must be provided for removal of the degradation products in the design of the system. Alternatively, the fluid could be replaced periodically and the spent fluid sent back to the producer for recovery.

Generally, a heat-transfer fluid should be noncorrosive to carbon steel because of its low cost. Carbon steel may be used with all the organic fluids, and with molten salts up to 450°C (842 °F) [6]. With the sodium-potassium alloys, carbon, and low-alloy steels can be used up to 540°C (1000 °F), but above 540°C stainless steels should be used [6]. Stainless steels contain 12 to 30% Cr and 0 to 22% Ni, whereas a steel containing small amounts of nickel and chromium, typically 1.85% Ni and 0.80% Cr, is referred to as a low alloy steel [6]. Cryogenic fluids require special steels. For example, liquid methane requires steels containing 9% nickel. To aid in the selection of a heat-transfer fluid, Woods [28] has constructed a temperature-pressure chart for several fluids.

Heat-Exchanger Evaluation and Selection

The process engineer must be familiar with the types of equipment that are available for the various process units. Because the evaluation and selection of equipment occur frequently, we will first establish general criteria that applies to most equipment. These criteria are to determine:

1. operating principles
2. equipment type
3. sealing
4. thermal expansion
5. maintenance
6. materials of construction – shell, tubes, and seals
7. temperature-pressure rating
8. economics

There may also be other special considerations that do not fit in the above criteria.

The most commonly used heat exchangers are the coil and double pipe for small heat-exchange areas and the shell-and-tube design for large areas. Devore et al. [13] recommend that if:

1. $A < 2m^2$ (21.5 ft^2) select a coiled heat exchanger
2. $2 \ m^2 < A < 50 \ m^2$ (538 ft^2) select a double-pipe heat exchanger
3. $A > 50 \ m^2$ select a shell-and-tube heat exchanger

The coiled heat exchanger is very compact, and it is frequently used when space is limited. The decision between the heat-exchanger types is not as distinct as indicated. At the boundary of each category, a detailed analysis is required to arrive at the most economical choice. Walas [5] discusses other heat-exchanger designs.

The most frequently used heat exchanger is the shell-and-tube heat exchanger, which is available in several designs. Figure 4.1 shows some of the more common ones. Each heat exchanger consists of entrance and exit piping, called nozzles, and hundreds of lengths of tubing contained in a shell. Usually, the outside diameter of the tubes are 0.75, 1.0, 1.5, and 2.0 in (1.9, 2.5, 3.8, 5.1 cm) [14]. The tubes are arranged in parallel and joined to metal plates, called tube sheets, as shown in Figure 4.1. The tubes are joined to the tube sheet by either welding or expanding the ends of the tube – called rolling. These methods of joining make very reliable seals. Tube diameters less than 0.75 in (1.9 cm) are difficult to clean and therefore should be used with clean fluids. The tubes are arranged in standard patterns, as shown in Figure 4.2. Although the triangular pitch is a more compact arrangement, resulting in a larger surface area per unit volume of heat exchanger, the other tube layouts are more accessible for cleaning. Also, the square pitch has a lower shell pressure drop than the triangular pitch, if the flow is in the direction indicated in Figure 4.2. Normally, tube lengths are 8, 12, 16, and 20 ft (2.44, 3.66, 4.88, 6.10 m) [5].

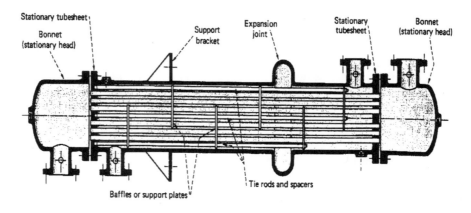

Fixed Tube Sheet

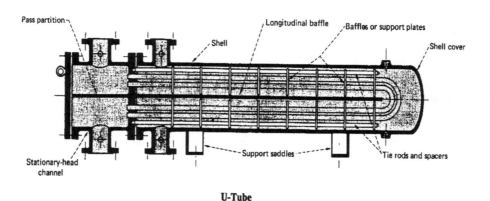

U-Tube

Figure 4.1 Shell-and-tube heat-exchanger designs. From Ref. 14 with permission.

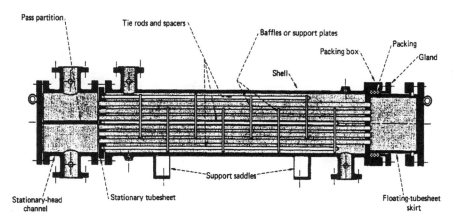

Floating Head, Outside-Packed Stuffing Box

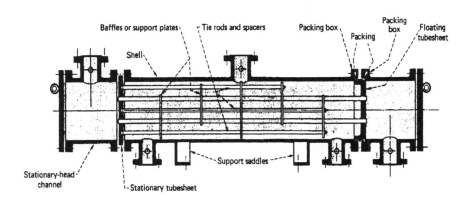

Floating Head, Outside-Packed Lantern Ring

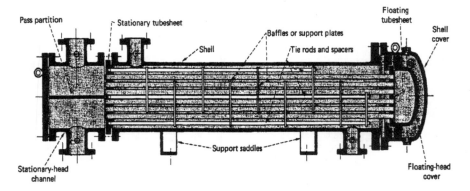

Inside-Split Backing Ring

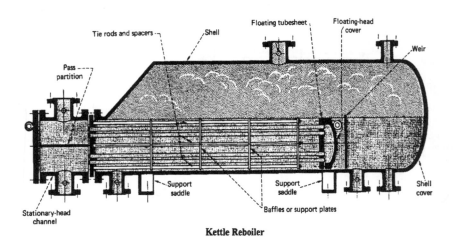

Kettle Reboiler

Figure 4.1 Continued.

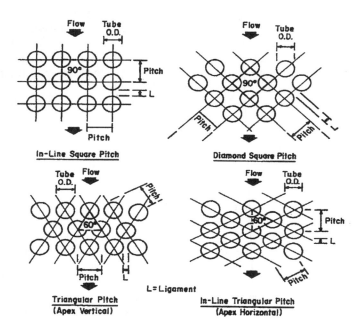

Figure 4.2 Shell-and-tube heat-exchanger tube layouts. (Source Ref. 15.)

For each heat exchanger shown in Figure 4.1, except the reboiler, the fluid enters the shell side in one nozzle, is forced to flow across the tubes by the baffles, and finally leaves in another nozzle. The baffles create turbulence, increasing the shell-side heat-transfer coefficient, and support the tubes to prevent sagging and flow induced vibrations. If the tube-side fluid flows through all of the tubes in one pass, it may be difficult to obtain a high fluid velocity and therefore an acceptable heat-transfer coefficient. Thus, the fluid is forced to flow through a fraction of the tubes in one pass, and then the fluid reverses direction to make at least one more pass. This is illustrated in Figure 4.1 for the inside-split-backing-ring heat exchanger, where a pass partition divides the tubes into two sections. A heat exchanger with one shell pass and two or more tube passes is referred to as a 1-2 heat exchanger. It is thus seen that the flow is not purely countercurrent. In the shell side there is crossflow, and in the tube side the flow is countercurrent to the shell fluid in one direction and then cocurrent to the shell fluid in the other direction.

The seal type selected depends on the pressure and temperature in the shell and tubes. Three types of seals employed in shell-and-tube heat exchangers, shown in Figure 4.3, are the flat gasket, the outside-packed stuffing box, and the outside-packed lantern ring. The latter two seals are sliding seals, which allow for movement between the sealing surfaces, thus relieving thermal stresses.

The maintenance required is cleaning, because of fouling of the heat-transfer surface, replacing seals, and replacing or plugging leaky tubes. Because most heat exchangers are overdesigned, some tubes could be plugged rather than replaced. Cleaning can either be done chemically or mechanically.

Scale formation is referred to as fouling and may be caused by the following mechanisms [25]:

1. precipitation of a salt from solution – frequently calcium carbonate in water
2. chemical reaction – such as polymerization of a monomer or corrosion, which are accelerated by a warm surface
3. growth of a microorganisms
4. depositing of suspended matter

The fixed-tube-sheet heat exchanger, shown in Figure 4.1, is the most popular design. This heat exchanger has straight tubes sealed in tube sheets, which are welded to the shell. Because the shell side is inaccessible for cleaning, we must use clean fluids – such as steam, refrigerants, gases, and organic heat-transfer fluids [16]. Differential thermal expansion must be considered when selecting a heat exchanger. Because the shell and tubes may be made of different materials to reduce the cost, differential expansion could be considerable. Without an expansion joint in the shell, the temperature difference between the shell and tube fluids is limited to 80°C (144 °F) [17]. With an expansion joint, as shown in Figure 4.1, a higher temperature difference is possible, but then the shell pressure is limited to only 8.0 bar (7.90 atm) [17].

In the U-tube heat exchanger, shown in Figure 4.1, the tubes are free to expand within the shell to prevent thermal stresses. Because the tubes are bent, only one tube sheet is needed, minimizing the number of connections. This feature plus the gasket-type seal make this heat-exchanger suitable for high pressure applications. Maintenance, however, is more difficult than for other shell-and-tube heat exchangers because any leaky inner tubes cannot be replaced, and must be plugged. Mechanical cleaning in the tubes is also difficult because of the U-bends, but chemical cleaning is possible. Also, hydraulic tube cleaners can clean both the straight and curve part of the tubes [6].

Another way of relieving thermal stresses is to use an outside-packed stuffing box or an outside-packed lantern ring, shown in Figure 4.1, and also in detail in Figure 4.3. For both designs, one tube sheet is free to slide along the packing. For the outside-packed stuffing box, the shell-side pressure is limited to 42.4 bar (41.8 atm) and the temperature to 320°C (608 °F) [16]. If the packing leaks, the shell and tube-side fluids will not mix. To clean the shell side of both heat exchangers, requires removing both ends and then sliding the tube bundle out of the shell. The tubes and shell can be cleaned mechanically and the seals easily replaced.

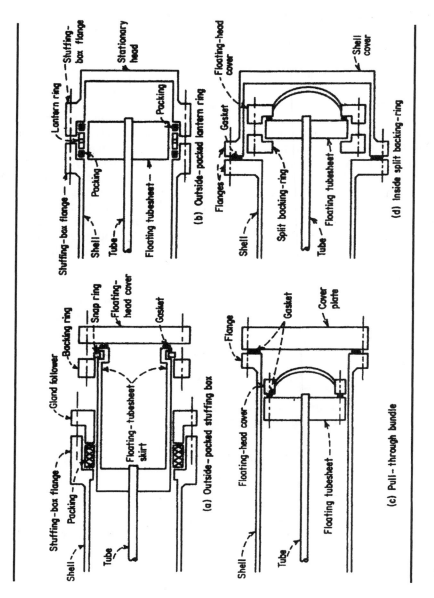

Figure 4.3 Floating-head heat-exchanger seal designs. From Ref.16 with permission.

For the outside-packed lantern ring, the shell and tube-side fluids will not mix within the shell. If the packing leaks, then the liquid will flow through the weep holes in the lantern ring and drop to the floor. This design will not be satisfactory for dangerous liquids unless a means for collecting the liquid safely is devised. This particular design is limited to 11.4 bar (11.3 atm) and 160°C (320 °F).

When higher shell-side temperatures and pressures than are attainable with a packing-type seal are required, then the inside-split backing-ring design is used. This design uses only gaskets as shown in Figure 4.1. To remove the tube bundle for maintenance requires removing the front end, and the split ring, and the float-ing-head cover at the back end. Because no seal can be guaranteed to be leak proof, there is the possibility that shell-side and tube-side fluids could mix so that this design is limited to fluids that can mix without creating a hazard.

The final heat-exchanger design considered is the kettle-type reboiler, shown in Figure 4.1. The boiling fluid, which could be a refrigerant or other proc-ess fluids, is placed on the shell side. In this design, the shell is enlarged to allow some separation of entrained liquid droplets in the vapor. Also, the tube bundle can be removed for maintenance. As was the case for the split-ring design, the kettle reboiler should not be used if mixing of the shell-side and tube-side fluids creates a hazard. The tubes in the kettle reboiler are free to expand in the shell.

If mixing of the shell-side and tube-side fluids cannot be tolerated, then use the double tube-sheet design shown in Figure 4.4 for extra protection. Because

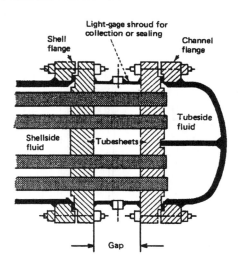

Figure 4.4 Double-tube sheet heat-exchanger design. From Ref. 18 with permission.

leaks could occur at the tube sheets, either the shell or tube-side fluid will collect in the space between both tube sheets. It is unlikely that both tube sheets will leak simultaneously.

In Table 4.2, the shell-and-tube heat exchangers just discussed are compared. Table 4.2 illustrates a general approach for evaluating and selecting equipment. To compare various designs, first, list the important design features of heat exchangers in the left column. Then, list the available heat exchanger designs in the column headings.

Table 4.2 Comparison of Shell-and-Tube, Heat-Exchanger Designs (Source Ref. 16 with permission..)

Design Features	Fixed Tubesheet	Return Bend (U-Tube)	Outside-Packed Stuffing Box	Outside-Packed Lantern Ring	Pull-Through Bundle	Inside Split Backing Ring
Is tube bundle removable?	No	Yes	Yes	Yes	Yes	Yes
Can spare bundles be used?	No	Yes	Yes	Yes	Yes	Yes
How is differential thermal expansion relieved?	Expansion joint in shell	Individual tubes free to expand	Floating head	Floating head	Floating head	Floating head
Can individual tubes be replaced?	Yes	Only those in outside rows without special designs	Yes	Yes	Yes	Yes
Can tubes be chemically cleaned, both inside and outside?	Yes	Yes	Yes	Yes	Yes	Yes
Can tubes be physically cleaned on inside?	Yes	With special tools	Yes	Yes	Yes	Yes
Can tubes be physically cleaned on outside?	No	With square or wide triangular pitch	With square or wide triangular pitch	With square or wide triangular pitch	With square or wide triangular pitch	With square or wide triangular pitch
Are internal gaskets and bolting required?	No	No	No	No	Yes	Yes
Are double tubesheets practical?	Yes	Yes	Yes	No	No	No
What number of tubeside passes are available?	Number limited by number of tubes	Number limited by number of U-tubes	Number limited by number of tubes	One or two	Number limited by number of tubes. Odd number of passes requires packed joint or expansion joint	Number limited by number of tubes. Odd number of passes requires packed joint or expansion joint.
Relative cost in ascending order, least expensive = 1	2	1	4	3	5	6

For safety and for ease of manufacture, organizations are established to develop standards and to facilitate the exchange of design information. The mechanical design of heat exchangers is governed by the Tubular Exchanger Manufacturers Association (TEMA) [19], the American Petroleum Institute (API) [21], and the American Society of Mechanical Engineers (ASME) [20]. These organizations publish standards and update them regularly.

Fluid Location

Locate the fluid on the tube side if the fluid is:

1. more corrosive
2. less viscous
3. more fouling
4. at a higher pressure
5. hotter
6. at a higher flow rate

and also if the fluid requires a low pressure drop. Generally, the more "obnoxious" fluid is placed on the tube side because:

1. the tube side is relatively easy to clean
2. tubes are easier to replace or plugged if damaged
3. high heat-transfer coefficients can be obtained at a low pressure drop
4. a high-pressure fluid is more economically contained in tubes because of their smaller diameter compared to the shell

Cooling water, for example, will be placed on the tube side because of its tendency to form a scale. Water usually contains dissolved salts, like calcium carbonate, which may deposit on the tube wall. A condensing fluid will be placed in the shell side to prevent the liquid film from growing too large, reducing the heat-transfer coefficient, or in the tube side if subcooling of the liquid is desirable. In the shell side, turbulence occurs at a lower Reynolds number than in the tube side because of the baffles. Thus, the shell side is the best location for very viscous fluids.

Heat-Exchanger Sizing

The well-known formula for sizing heat exchangers is

$$Q = U_o A_o (\Delta t)_{LM} \tag{4.1}$$

where the subscript, o, signifies that the overall heat-transfer coefficient is based on the outside tube area. Sizing a heat exchanger entails calculating the area required to transfer a specified amount of heat. This formula, which may be used for both countercurrent and cocurrent flow, is derived in a number of texts (for example, see Reference 4.22). Although countercurrent flow is the most efficient, cocurrent flow is used when it is necessary to limit the final temperature of a heat sensitive material. Cocurrent flow is also used when a rapid change in temperature is needed (quenching) [8].

The logarithmic-mean temperature difference, $(\Delta t)_{LM,}$ is defined by

$$(\Delta t)_{LM} = \frac{(t_4 - t_1) - (t_3 - t_2)}{\ln \dfrac{(t_4 - t_1)}{(t_3 - t_2)}} \qquad (4.2)$$

where the subscripts correspond to the streams in Figure 4.5.

To derive Equation 4.2 the assumptions made are:

1. constant overall heat-transfer coefficient
2. constant heat capacity
3. isothermal phase change
4. adiabatic operation

The first assumption is that the overall heat-transfer coefficient, U_o, is constant. It may vary along the length of the heat exchanger because the changing temperature affects fluid properties. Assumptions two and three mean that the cooling or heating curves are linear for both fluids. The curves are plots of temperature versus the amount of heat transfer up to any particular point in the heat exchanger. Nonisothermal phase changes occur when processing multicomponent mixtures, and will frequently result in nonlinear curves as illustrated in Figure 4.6. If, however, the nonlinear curves are divided up into short enough segments so that they are essen-

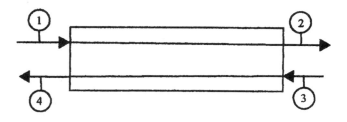

Figure 4.5 Countercurrent-flow heat exchanger.

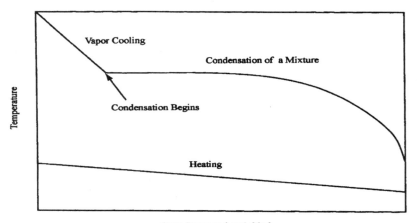

Figure 4.6 Heating and cooling curves for a heat exchanger.

tially linear, then the logarithmic-mean temperature difference can be used for each segment as shown below.

The total surface area,

$$A_o = A_1 + A_2 + \ldots + A_n \tag{4.3}$$

Substituting Equation 4.1 into Equation 4.3 for each segment we find that

$$A_o = \frac{Q_1}{U_o\,(\Delta t)_{LM1}} + \frac{Q_2}{U_o\,(\Delta t)_{LM2}} + \ldots + \frac{Q_n}{U_o(\Delta t)_{LM\,n}} \tag{4.4}$$

If the segments are chosen so that

$$Q_1 = Q_2 = \ldots = Q/n \tag{4.5}$$

then

$$A_o = \frac{Q/n}{U_o} \left[\frac{1}{(\Delta t)_{LM1}} + \frac{1}{(\Delta t)_{LM2}} \ldots + \frac{1}{(\Delta t)_{LMn}} \right] \tag{4.6}$$

where n is the number of segments. After solving Equation 4.6 for Q, we find that

$$Q = U_o A_o \frac{n}{1/(\Delta t)_{LM1} + 1/(\Delta t)_{LM2} + \ldots + 1/(\Delta t)_{LMn}} \tag{4.7}$$

The expression to the right of A_o is an effective logarithmic-mean temperature difference. Thus,

$$Q = U_o A_o (\Delta t)_{LM,eff}$$

Correction Factor for Non-countercurrent Flow

It was seen from the discussion of heat exchangers that the fluid streams are not strictly countercurrent. Baffles on the shell side induce crossflow, and in a two-tube-pass heat exchanger both countercurrent and cocurrent flow occur. To account for deviations from countercurrent flow, the logarithmic-mean temperature difference is multiplied by a correction factor, F. Thus,

$$Q = U_o A_o F (\Delta t)_{LM} \tag{4.9}$$

An equation for the correction factor can be derived with the following assumptions:

1. adiabatic operation
2. well mixed shell-side fluid
3. the heat-transfer surface area is the same for each tube pass
4. constant overall heat-transfer coefficient
5. constant heat capacity
6. no phase change for either fluid

The correction factor for a one-shell-pass and a two-tube-pass heat exchanger (a 1-2 heat exchanger), which is derived by Kern [1], is

$$F = \frac{(R^2 + 1)^{1/2}}{R - 1} \frac{\ln \dfrac{1 - S}{1 - R S}}{\ln \dfrac{2 - S [(R + 1) - (R^2 + 1)^{1/2}]}{2 - S [(R + 1) + (R^2 + 1)^{1/2}]}} \tag{4.10}$$

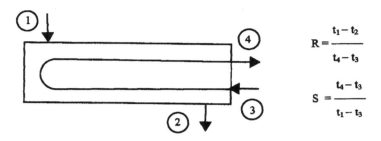

$$R = \frac{t_1 - t_2}{t_4 - t_3}$$

$$S = \frac{t_4 - t_3}{t_1 - t_3}$$

Figure 4.7 Definition of parameters for the logarithmic-mean-temperature correction factor.

where R and S are defined in Figure 4.7. According to Kern [1], the values of F for the worst case are less than two percent apart when comparing 1-2 and 1-8 heat exchangers. Thus, for any heat exchanger having one shell pass and two or more even-numbered tube passes in countercurrent-cocurrent flow, Equation 4.10 is satisfactory. For other flow arrangements, correction factors can be found in the Chemical Engineering Handbook [6]. If the flow is perfectly countercurrent, then F = 1. According to Coulson et al. [17], an economic heat exchanger design cannot be attained for a value of F less than 0.85, whereas Kern [1] and Goyal [23] recommend a minimum value of 0.75. Taborex [24], however, shows that the minimum value of F varies with R and S. We will use 0.85.

Overall Heat-Transfer Coefficients

The overall heat-transfer coefficient, defined by Equation 4.11, is derived in a number of texts (see for example Reference 4.22). If the heat transferred is based on the outside area of the tube, then the overall heat-transfer coefficient is

$$U_o = \frac{1}{\dfrac{x_{fi}\, D_o}{k_{fi}\, D_i} + \dfrac{D_o}{h_i\, D_i} + \dfrac{x_w\, D_o}{k_w\, D_{LM}} + \dfrac{1}{h_o} + \dfrac{x_{fo}}{k_{fo}}} \qquad (4.11)$$

where the logarithmic-mean diameter, D_{LM}, is defined by

$$D_{LM} = \frac{D_o - D_i}{\ln (D_o / D_i)} \tag{4.12}$$

Each term in the denominator of Equation 4.11 is the reciprocal of a heat-transfer coefficient, and thus represents a resistance to heat transfer. The first term in the denominator represents the resistance to heat conduction across a scale formed on the inside surface of the tube, where the thickness and the thermal conductivity of the scale is rarely known. The thermal conductivity and the thickness of scale are not reported in the literature, but its reciprocal is designated by $R_{f\,i}$, the resistance to heat transfer caused by the tube-side scale, where

$$R_{fi} = x_{fi}\, D_o / k_{fi}\, D_i \tag{4.13}$$

Also, $R_{f\,o}$, the resistance to heat transfer (fouling resistance or fouling factor) caused by the shell-side scale is equal to the last term in the denominator of Equation 4.11.

$$R_{fo} = x_{f\,o} / k_{f\,o} \tag{4.14}$$

The scale thickness will vary with time. When a heat exchanger is first installed, it is clean. With use the scale thickness increases. If a fouling resistance is specified, the time required to form the scale is indirectly specified, usually 1 to 1 ½ years [1]. When this period of time is reached, the heat exchanger must be taken out of service and cleaned. The longer the time before cleaning (service time), the greater the required heat-transfer area and cost of the heat exchanger and hence capital cost, but the cost of cleaning and operating cost will be less. On the other hand, if the service time is reduced, the heat-exchanger cost will decrease, but the cleaning cost will increase. Therefore, there is an optimum service time which minimizes the total cost. This optimization problem has been studied by Crittenden and Khater [26].

The second term in the denominator of Equation 4.11 represents the convective resistance to heat transfer caused by the inside fluid film on the scale surface. The third term is the conductive resistance caused by the tube wall, which is usually small, because the thermal conductivity of many metals is large. We will neglect the conductive resistance to heat transfer, unless the thermal conductivity is very small and tube wall thickness large. The fourth term is the convective resistance to heat transfer of the outside fluid film on the scale surface. After substituting Equations 4.13 and 4.14 into Equation 4.11,

$$U_o = \frac{1}{R_{fi} + \dfrac{D_o}{D_i}\dfrac{1}{h_i} + \dfrac{1}{h_o} + R_{fo}} \tag{4.15}$$

Individual heat-transfer coefficients and the fouling resistance or fouling factor, are listed
in Table 4.3. The heat transfer coefficients in Table 4.3 are divided according to whether the fluid is inorganic or organic, a gas or a liquid, and whether it is heated or cooled, with or without a phase change. The inorganic fluids are water and ammonia, whereas the organic fluids are divided into three categories, light, medium, or heavy, depending on their viscosity. If the fluid is a gas, pressure will also affect the transfer properties. The footnotes in Table 4.3 define a light, medium, and heavy organic fluid.

For boiling liquids, the heat flux cannot be too large or a vapor blanket will form on the heat transfer surface, effectively insulating the surface and thus reducing the heat-transfer coefficient. Gases or vapors have a much lower heat-transfer coefficient than a boiling liquid. If heat is supplied by a condensing vapor or hot liquid, a considerable reduction in heat transfer will occur if a vapor blanket forms. If the fluid is being heated electrically, however, heat transfer will remain essentially the same, and the heater surface temperature will rise until the heater "burns out". To avoid this problem, Walas [3] recommends designing a heat exchanger for a boiling heat flux of less than 1.3×10^5 W/m^2 (4.12×10^4 Btu/h-ft^2).

Terminal Temperatures of the Fluid Streams

Before calculating the logarithmic-mean temperature difference, determine the terminal temperature of each fluid stream. Three of the four terminal temperatures are usually specified, and the fourth can be found by optimizing the fixed and operating costs for the heat exchanger. If we consider cooling a process stream, then the stream temperature at the inlet and outlet of the heat exchanger will usually be known. The stream leaves one process unit and enters the heat exchanger. Then, the stream is cooled to a specified temperature, depending on the requirements of the next process unit. Also, if the coolant is water, which is generally the case, its temperature varies throughout the year. Take the worst case, which is approximately 30 $^\circ$C (86 $^\circ$F) in the New York area. The next step is to calculate the exit water temperature, which is discussed in Example 4.1.

Table 4.3 Approximate Heat Transfer Coefficients for Shell-and-Tube Heat Exchangers (Source Ref. 27 with permission.)

To convert $w/m^2 = K$ to $Btu/h\text{-}ft^{2\text{-}\circ}F$ multiply by 0.1761

Fluid Conditions		h, W/m^2 K [a,b]	Fouling resistance, m^2 K/W[a]
Sensible heat transfer			
Water[c]			
	Liquid	5 000 - 7 500	1 x 10^{-4} – 2.5 x 10^{-4}
Ammonia			
	Liquid	6 000 - 8 000	0-1 x 10^{-4}
Light organics[d]			
	Liquid	1 500 – 2 000	1 x 10^{-4} – 2 x 10^{-4}
Medium organics[e]			
	Liquid	750 –1 500	1.5 x 10^{-4} –4 x 10^{-4}
Heavy organics[f]			
	Liquid, Heating	250 -750	2 x 10^{-4} –3 x 10^{-3}
	Cooling	150 - 400	2 x 10^{-4} –3 x 10^{-3}
Very Heavy Organics[g]			
	Liquid, Heating	100 - 300	4 x 10^{-4} –3 x 10^{-3}
	Cooling	60 -150	4 x 10^{-4} –3 x 10^{-3}
Gas[h]			
	Pressure 100-200kN/m^2 abs	80 -125	0-1 x 10^{-4}
Gas[h]			
	Pressure 1 MN/m^2 abs	250 - 400	0-1 x 10^{-4}
Gas[h]	Pressure 10 MN/m^2 abs		
		500 - 800	0-1 x 10^{-4}
Condensing heat transfer			
Steam, ammonia	Pressure 10kN/m^2 Abs, no noncondensables[i]		
		8 000 – 12 000	0-1 x 10^{-4}
Steam, ammonia	Pressure 10kN/m^2 abs 1% noncondensables[i]		
		4 000 – 6 000	0-1 x 10^{-4}
Steam, ammonia	Pressure 10kN/m^2 abs 4% noncondensables		
	Pressure 100 kN/m^2 abs, no condensables[i,k,l]	2 000 – 3 000	0-1 x 10^{-4}
Steam, ammonia	Pressure 1 MN/m^2 abs no condensables[i,j,k,l]	10 000 - 15 000	0-1 x 10^{-4}
Steam, ammonia	Pure component, pressure 10 kN/m^2 abs, no non-		

Table 4.3 continued

	condensables[j]	15 000 – 25 000	0-1 x 10⁻⁴
Light organics[d]	Pressure 10 kN/m² abs, 4% noncondesables[k]	1 500 - 2 000	0-1 x 10⁻⁴
Light organics[d]	Pure component, pressure 100 kN/m² abs, no non-condensables[j]		
Light organics[d]	Pure component, pressure 1 MN/m² abs	750 – 1 000	0-1 x 10⁻⁴
Light organics[d]	Pure component or narrow condensing range, pressure 100kN/m² abs[m,n]	2 000 - 4 000	0-1 x 10⁻⁴
Medium organics[e]	Narrow condensing range, pressure 100kN/m² abs[m,n]	3 000 – 7 000	0-1 x 10⁻⁴
Heavy organics	Medium condensing range, pressure 100kN/m² abs[l,m,o]	1 500 – 4 000	1 x 10⁻⁴ –3 x 10⁻⁴
Light multicomponent mixtures, all condensable[d]	Medium condensing range, pressure 100kN/m² abs[l,m,o]	600 – 2 000	2 x 10⁻⁴ –5 x 10⁻⁴
Medium multicomponent mixtures, all condensable	Medium condensing range, pressure 100kN/m² abs[l,m,o]	1 000 – 2 500	0-2 x 10⁻⁴
Heavy multicomponent mixtures, all condensable[f]	Pressure < 0.5 MN/m² abs, ΔT_{SH}, max = 25K		
Vaporizing heat transfer[h,i]	Pressure > 0.5 MN/m² abs, pressure < 10 MN/m² abs, ΔT_{SH}, max = 20K	600 – 1 500	1 x 10⁻⁴ –4 x 10⁻⁴
Water[r]	Pressure < 3 MN/m² abs, ΔT_{SH}, max = 20K	300 - 600	2 x 10⁻⁴ –8 x 10⁻⁴
Water[r]	Pure component, pressure < 2MN/m² abs, ΔT_{SH}, max = 20 K		
Ammonia	Narrow boiling range[r], pressure < 2 MN/m² abs, ΔT_{SH}, max = 15 K	3 000 – 10 000	1 x 10⁻⁴ –2 x 10⁻⁴
Light organics[d]	Pure component, pressure < 2 MN/m² abs, ΔT_{SH}, max = 20	4 000 – 15 000	1 x 10⁻⁴ –2 x 10⁻⁴
	Narrow boiling range[r],		

Table 4.3 continued

Light organics[d]	pressure < 2 MN/m² abs, $^{\Delta T}SH$, max = [15 K]	3 000 – 5 000	1 N x 10⁻⁴ –2 x 10⁻⁴
Medium organics[e]	Pure component, pressure < 2 MN/m² abs, $^{\Delta T}SH$, max = [20 K]	1 000 – 4 000	1 x 10⁻⁴ –2 x 10⁻⁴
	Narrow boiling ranges[g], pressure < 2 MN/m² abs, $^{\Delta T}SH$, max = [15 K]	750 – 3 000	1 x 10⁻⁴ –3 x 10⁻⁴
Medium organics[e]	Narrow boiling range[g], pressure < 2 MN/m² abs, $^{\Delta T}SH$, max = [15 K]	1 000 – 3 500	1 x 10⁻⁴ –3 x 10⁻⁴
Heavy organics[f]		600 – 2 500	1 x 10⁻⁴ –3 x 10⁻⁴
Heavy organics[g]		750 – 2 500	2 x 10⁻⁴ –5 x 10⁻⁴
Very heavy organics[h]		400 – 1 500	2 x 10⁻⁴ –8 x 10⁻⁴
		300 – 1 000	2 x 10⁻⁴ –1 x 10⁻³

[a] Heat transfer coefficients and fouling resistances are based on area in contact with fluid. Ranges shown are typical, not all-encompassing. Temperatures are assumed to be in normal processing range; allowances should be made for very high or low temperatures.

[b] Allowable pressure drops on each side are assumed to be about 50 –100 kN/m² except for (1) low-pressure gas and two-phase flows, where the pressure drop is assumed to be about 5% of the absolute pressure; and (2) very viscous organics, where the allowable pressure drop is assumed to be about 150-250 kN/m².

[c] Aqueous solutions give approximately the same coefficients as water.

[d] "Light organics" include fluids with liquid viscosities less than about 0.5 x 10⁻³ N/m², such as hydrocarbons through C, , gasoline, light alcohols and ketones, etc.

[e] "Medium organics" include fluids with liquid viscosities between about 0.5 x 10⁻³ and 2.5 x 10⁻³ Ns/m², such as kerosene, straw oil, hot gas oil, absorber oil, and light crudes.

[f] "Heavy organics" include fluids with liquid viscosities greater than 2.5 x 10⁻³ Ns/m², but not more than 50 x 10⁻³ Ns/m², such as cold gas oil, lube oils, fuel oils, and heavy and reduced crudes.

[g] "Very heavy organics" include tars, asphalts, polymer melts, greases, etc. having liquid viscosities greater than about 50 x 10⁻³ Ns/m². Estimation of coefficients for these materials is very uncertain and depends strongly on the temperature difference, because natural convection is often a significant contribution to heat transfer in heating, whereas congelation on the surface and particularly between fins can occur in cooling. Since many of these materials are thermally unstable, high surface temperatures can lead to extremely severe fouling.

[h] Values given for gases apply to such substances as air, nitrogen, carbon dioxide, light hydrocarbon mixtures (no condensation), etc. Because of the very high thermal conductivities and specific heats of hydrogen and helium, gas

Table 4.3 continued

mixtures containing appreciable fractions of these components will generally have substantially higher heat transfer coefficients.

^{i} Superheat of a pure vapor is removed at the same coefficient as for condensation of the saturated vapor if the exit coolant temperature is less than the saturation temperature (at the pressure existing in the vapor phase) and if the (constant) saturation temperature is used in calculating the mean temperature difference. But see note k for vapor mixtures with or without noncondensable gas.

^{j} Steam is not to be condensed on conventional low-finned tubes; its high surface tension causes bridging and retention of the condensate and a severe reduction of the coefficient below that of the plain tube.

^{k} The coefficients cited for condensation in the presence of noncondensable gases or for multicomponent mixtures are only for very rough estimation purposes because of the presence of mass transfer resistances in the vapor (and to some extent, in the liquid) phase. Also, for these cases, the vapor-phases temperature is not constant, and the coefficient given is to be used with the mean temperature differences estimated using vapor-phase inlet and exit temperatures, together with the coolant temperatures.

^{l} As a rough approximation, the same relative reduction in low-pressure condensing coefficients due to noncondensable gases can also be applied to higher pressures.

^{m} Absolute pressure and noncondensables affect condensing coefficients for medium and heavy organics in approximately the same proportion as for light organics. Because of thermal degradation, fouling may become quite severe for the heavier condensates. For large fractions of noncondensable gas, interpolate between pure component condensation and gas cooling coefficients.

^{n} "Narrow condensing range" implies that the temperature difference between dew point and bubble point is less than the smallest temperature difference between vapor and coolant at any place in the condenser.

^{o} "Medium condensing range" implies that the temperature difference between dew point and bubble point is greater than the smallest temperature difference between vapor and coolant, but less than the temperature difference between inlet vapor and outlet coolant.

^{p} Boiling and vaporizing heat transfer coefficients depend very strongly on the nature of the surface and the structure of the two-phase

flow past the surface in addition to all of the other variables that are significant for convective heat transfer in other modes. The flow velocity and structure are very much governed by the geometry of the equipment and its connecting piping. Also, there is a maximum heat flux from the surface that can be achieved with reasonable temperature differences between surface and saturation temperature of the boiling fluid; any attempt to exceed this maximum heat flux by increasing the surface temperature leads to partial or total coverage of the surface by a film of vapor and a sharp *decrease* in the heat flux.

Therefore, the heat transfer coefficients given in this table are only for very rough estimating purposes and assume the use of plain or low-finned tubes without special nucleation enhancement. $\Delta T_{SH, max}$ is the maximum allowable temperature difference between surface and saturation temperature of the boiling surface. No attempt is made in this table to distinguish among the various types of vapor-generation equipment, since the major heat transfer distinction to be made is the propensity of the process steam to foul. Severely fouling streams will usually call for a vertical thermosiphon or a forced-convection (tube-side) reboiler for ease of cleaning.

^{q} Subcooling heat load is transferred at the same coefficient as latent heat load in kettle reboilers, using the saturation temperature in the mean temperature difference. For horizontal and vertical thermosiphons, a separate calculation is required for the sensible heat transfer area, using appropriate sensible heat transfer coefficients and the liquid temperature profile for the mean temperature difference.

^{r} Aqueous solutions vaporize with nearly the same coefficient as pure water if attention is given to boiling-point elevation and if the solution does not become saturated and care is taken to avoid dry wall conditions.

^{s} For boiling of mixtures, the saturation temperature (bubble point) of the final liquid phase (after the desired vaporization has taken place) is to be used to calculate the mean temperature difference. A narrow-boiling-range mixture is defined as one for which the difference between the bubble point of the incoming liquid and the bubble point of the exit liquid is less than the temperature difference between the exit hot stream and the bubble point of the exit boiling liquid. Wide-boiling-range mixtures require a case-by-case analysis and cannot be reliably estimated by these simple procedures.

Example 4.1 Optimum Cooling-Water Exit Temperature

The exit water temperature could be calculated by minimizing the total cost of operating a heat exchanger. This optimization problem is approached by listing all the relationships and variables to determine if there are any degrees of freedom. Table 4.4.1 lists the equations for the optimization. The mass flow rate of cooling water into the heat exchanger equals the mass flow rate of water out, as given by Equation 4.4.1, where the subscript, w, refers to water. Also, we must calculate the amount of heat transferred from the process stream to the water stream, so that an energy balance is written for the tube side instead of over the entire heat exchanger, which would eliminate Q. Because the kinetic energy and potential energy changes are usually insignificant, and the work term is zero, the energy equation reduces to Equations 4.4.2, which states that the heat transferred to the water is equal to the change in enthalpy of the water.

Because the cost of a heat exchanger depends on its size, and because its size will depend on the heat-transfer rate, a rate equation must be introduced. The rate equation is given by Equation 4.4.3. The logarithmic-mean temperature difference in Equation 4.4.3 is given by Equation 4.4.4. Because perfect countercurrent flow can never be achieved in an actual heat exchanger, the logarithmic-mean temperature difference correction factor, F, is needed. For simplicity, Equation 4.10, discussed earlier, is expressed as Equation 4.4.5, which states that F depends only on the terminal temperatures, once a particular heat exchanger is selected.

Several cooler sizes will cool a process fluid to a specified temperature, but there is only one that is the most economical. If the cooling-water exit temperature increases, less water is needed and its cost will be less. To achieve a high exit-water temperature, however, requires more heat-exchanger surface area and consequently a more costly heat exchanger. Equation 4.4.6, the total annual cost, expresses the trade-off between the cost of cooling water and the cost of a heat exchanger. The total cost consists of the sum of the first term, which is the cooling-water cost, and the second term, which is the installed cost of the heat exchanger and the maintenance cost. In Equation 4.4.6, C_W equals the cost of water per pound, C_C capital cost of the heat exchanger per square foot, and C_M the maintenance cost per square foot. Because we want to obtain the optimum cooling-water exit temperature and therefore, an optimally-sized heat exchanger, the total cost should be the minimum. Therefore, the derivative of the total cost with respect to the exit water temperature $dC_T / dt_{2,w}$ is set equal to zero.

Finally, to complete the formulation of the problem, we need system property data. For this particular problem, enthalpy, a thermodynamic property, is required for the energy balance and the overall heat-transfer coefficient, a transfer property, is needed for the rate equation. These system property relationships are given by Equations 4.4.8 to 4.4.10. The economic balance also requires cost data.

Table 4.4.1 Summary of Equations for Calculating the Optimum Cooling-Water Exit Temperature

First subscript: Process stream = 1,2,3 or 4
Second subscript: Component = p or w

Mass Balances

$$m_{1,w} = m_{2,w} \tag{4.4.1}$$

Energy Balances

$$Q = h_{2,w}\, m_{2,w} - h_{1,w}\, m_{1,w} \tag{4.4.2}$$

Rate Equations

$$Q' = U_o \, A_o \, F \, (\Delta t)_{LM} \tag{4.4.3}$$
$$(\Delta t)_{LM} = f(t_{1,w}', \, t_{2,w}, \, t_{3,P}', \, t_{4,P}') \; — \; \text{Equation 4.2} \tag{4.4.4}$$
$$F = f(t_{1,w}', \, t_{2,w}, \, t_{3,P}', \, t_{4,P}') \; — \; \text{Equation 4.10} \tag{4.4.5}$$

Economic Relations

$$C_T = m_{1,w} \, \theta' \, C_W' + A \, (C_C' + C_M') \tag{4.4.6}$$
$$dC_T/dt_{2,w} = 0 \tag{4.4.7}$$

Thermodynamic Properties

$$h_{1,w} = f(t_{1,w}') = C_{PW} \, (t_{1,w}' - t_B') \tag{4.4.8}$$
$$h_{2,w} = f(t_{2,w}) = C_{PW} \, (t_{2,w} - t_B') \tag{4.4.9}$$

Transfer Properties

$$U_o = f(\text{heat-exchanger type}', \text{ shell fluid}', \text{ tube fluid}') \tag{4.4.10}$$

Variables

$$m_{1,w} - m_{2,w} - h_{1,w} - h_{2,w} - t_{2,w} - A_o - F - U_o - (\Delta t)_{LM} - C_T$$

Degrees of Freedom

$$F = V - R = 10 - 10 = 0$$

Because the degrees of freedom are zero, the problem is completely formulated and we can now solve the equations listed in Table 4.4.1. Next, express the total cost equation in terms of a single variable, which is the exit water temperature, $t_{2,w}$, so that it can be differentiated. It is reasonable to assume that in the temperature range of interest the heat capacity of the cooling water will not vary appreciably. Thus, from Equations 4.4.1 and 4.4.2,

$$Q = (h_{2,w} - h_{1,w}) \, m_{1,w} \tag{4.4.11}$$

and after substituting Equations 4.4.8 and 4.4.9, where t_B is a base temperature, into Equation 4.4.11

$$(h_{2,w} - h_{1,w}) = \; = m_{1,w} \, C_{PW} \, (t_{2,w} - t_{1,w}) \tag{4.4.12}$$

Therefore,

$$Q = m_{1,w} \, C_{PW} \, (t_{2,w} - t_{1,w}) \tag{4.4.13}$$

Use Equations 4.4.13 and 4.4.3 to eliminate $m_{1,W}$ and A_o from the total cost equation, Equation 4.4.6. Thus,

$$C_T = \frac{Q\,\theta\,C_W}{C_{PW}\,(t_{2,W} - t_{1,W})} + \frac{Q\,(C_C + C_M)}{U_o\,F\,(\Delta t)_{LM}} \qquad (4.4.14)$$

With the exception of F, all the parameters in both terms in Equation 4.4.14 are constants. In order to obtain a first approximation for the exit-water temperature and to simplify the derivation, assume that F is constant. To obtain an economically-viable heat exchanger, let F = 0.85.

After substituting the logarithmic-mean temperature difference, Equation 4.4.4, into Equation 4.4.14 we find that

$$C_T = \frac{F\,Q\,\theta\,C_W}{C_{PW}\,(t_{2,W} - t_{1,W})} + \frac{Q\,(C_C + C_M)}{U_o\,F\,[\,(t_{4,P} - t_{1,W}) - (t_{3,P} - t_{2,W})\,]} \ln \frac{(t_{4,P} - t_{1,W})}{(t_{3,P} - t_{2,W})} \qquad (4.4.15)$$

Equation 4.4.15 can now be differentiated with respect to $t_{2,W}$. After setting the derivative equal to zero, and rearranging the equation, it is found that

$$\frac{U_o\,\theta\,C_W\,(\Delta t_1 - \Delta t_2)^2}{C_{PW}\,(C_F + C_M)\,(t_{2,W} - t_{1,W})^2} = \frac{(\Delta t_1 - \Delta t_2)}{\Delta t_2} - \ln \frac{\Delta t_1}{\Delta t_2} \qquad (4.4.16)$$

where $\Delta t_1 = t_{4,P} - t_{1,W}$ and $\Delta t_2 = t_{3,P} - t_{2,W}$. Equation 4.4.16 is dimensionless. The optimum cooling-water temperature, $t_{2,W}$, is obtained from Equation 4.4.16 by iteration.

APPROACH TEMPERATURE DIFFERENCES

Frequently, an approximate value of the optimum exit-water temperature is all that is required, and a rule-of-thumb will be satisfactory. Table 4.4 lists the approach temperature difference, which is the difference between the two terminal temperatures of two passing streams, for several heat exchangers. Several approach temperature differences were taken from Ulrich [8]. For refrigerants, Ulrich's range of 10 to 50°C is on the high side. Frank [7] recommends a range of 3 to 5°C whereas Walas [3] recommends a value of 5.6°C or less.

Table 4.4 Summary of Heat-Exchanger Approach Temperature Differences and Pressure Drops

Heat-Exchanger	Heat-Transfer Fluid	Approach Temperature Difference[a], °C	Pressure Drop[a], bar	
			Shell	Tube
Chiller (H)[c]	Brine	3.0 – 5.0[b]		
Cooler (H)	Water Air	5.0 – 50.0 5.0 – 50.0	 0.0012	
Condenser (H or V) (I)	Water Air	10.0 –50.0 5.0 – 50.0	0.1 0.0012	
Heater (H)	d	10.0 – 50.0	0.1	
Superheater (H)	d	50.0 – 100.0	0.05 – 0.6	0.05 – 0.6
Reboiler or Vaporizer Kettle (H) Thermosyphon (V)	 d d	 10.0 – 50.0 20.0 – 60.0	 negligible[e] 0.1	 0.1 0.2 – 0.6
Interchanger (H)	Process Fluid	10.0 – 50.0		

Liquids	
Average Viscosity, cp	Pressure Drop Shell or Tube, bar (psi)
< 1.0 1.0 – 10.0 > 10.0	0.34 (5.0) 0.48 (7.0) 0.70 (10)
Gases	
Pressure, bar	Pressure Drop Shell or Tube, bar
High Vacuum < 1.7 > 1.7	0.004 – 0.008 0.035 5.0 to 10.0% of the inlet pressure

a. Source: Reference 8 except where indicated. Multiply by 0.9869 to obtain atm.
   Multiply by 0.9869 to obtain atmospheres.
b. Source: Reference 7.
c. The letters in the parenthesis is the normal installation
   position, H for horizontal, V for vertical, and I for inclined.
d. Steam, organic, hot gases
e. Source: Reference 1.

Also, listed in Table 4.4 are pressure-drop ranges for heat exchangers for making preliminary estimates. The pressure drop depends on whether the fluid is a gas or a liquid, or if the fluid is condensing or vaporizing. For gases, the pressure drop depends on the total pressure. Below atmospheric pressure, the pressure drop is critical and should be small because of the cost of vacuum pumps.

SIZING HEAT EXCHANGERS

There are two general classes of problems encountered by a process engineer. One class is the design problem, which requires calculating the size of a process unit. The other class is the rating problem, which requires determining if an existing process unit will satisfy process conditions. For a heat exchanger, the sizing problem is calculating surface area for transferring a specified amount of heat. Then, a heat exchanger can be designed in detail to give the calculated area. For the rating problem, the heat-transfer area is fixed. The heat exchanger may be available in a plant, at a used equipment dealer, or supplied by a manufacturer, who usually produces standard heat exchangers in discrete sizes. Rating a process unit is a frequently occurring problem. We will consider the design problem first.

We have now developed sufficient background material to outline a sizing procedure for a preliminary estimate of the heat-exchanger surface area. Equations for sizing heat exchangers are summarized in Table 4.5. Table 4.6 outlines the calculation procedure. Because heat transfer coefficients and fouling factors are contained in Table 4.3, we represent this mathematically by using functional notation as shown by Equations 4.6.10 to 4.6.13 in Table 4.5.

Table 4.5 Summary of Equations for Sizing Shell-and-Tube Heat Exchangers

First subscript: Process stream = 1,2,3 or 4
Second subscript: Component =1 or 2

Mass Balance

$$m_3' = m_4 \qquad\qquad (4.5.1)$$

Energy Balance

$$Q = h_{3,2}\, m_4 - h_{4,2}\, m_3'$$ (4.5.2)

Rate Equations

$$Q = U_o\, A_o\, F\, (\Delta t)_{LM}$$ (4.5.3)

$$F = f\,(t_1', t_2, t_3', t_4') \;\text{—— Equation 4.10}$$ (4.5.4)

$$(\Delta t)_{LM} = f\,(t_1', t_2, t_3', t_4') \;\text{—— Equation 4.2}$$ (4.5.5)

Thermodynamic Properties

$$h_{3,2} = f(t_3')$$ (4.5.6)

$$h_{4,2} = f(t_4')$$ (4.5.7)

Economic Relation

$$t_3' - t_2 = \text{approach } \Delta t \;\text{—— from Table 4.4}$$ (4.5.8)

Transfer Properties

$$U_o = 1/[R_{fi} + 1/h_i + 1/h_o + R_{fo}]$$ (4.5.9)

$$r_{fi} = f(\text{fluid}', \text{type of phase change}') \;\text{—— Table 4.3}$$ (4.5.10)

$$h_i = f(\text{fluid}', \text{type of phase change}') \;\text{—— Table 4.3}$$ (4.5.11)

$$h_o = f(\text{fluid}', \text{type of phase change}') \;\text{—— Table 4.3}$$ (4.5.12)

$$R_{fo} = f(\text{fluid}', \text{type of phase change}') \;\text{—— Table 4.3}$$ (4.5.13)

Variables

$$m_4 - t_2 - (\Delta t)_{LM} - h_{3,2} - h_{4,2} - Q - U_o - A - F - R_{f1} - h_1 - h_o - R_{fo}$$

Degrees of Freedom

$$F = 13 - 13 = 0$$

Example 4.2 Sizing a Distilled-Water Interchanger

Distilled water at 34 °C is cooled to 30 °C by a raw-water feed at 23 °C flowing to an evaporator. Estimate the heat-transfer area required to cool 79,500 kg/h (8.16×10^5 lb/h) of distilled water using a 1-2 heat exchanger.

The Equations listed in Table 4.5 can be solved one at a time. Table 4.6 outlines the calculation procedure. From Equations 4.5.1, 4.5.2, 4.5.6, and 4.5.7 in Table 4.5, and noting that the enthalpy difference is equal to C_P ($t_{3,2} - t_{4,2}$), we find that the heat transferred,

$$Q = m_3\,(h_{3,2} - h_{4,2}) = m_3\,C_p\,(t_{3,2} - t_{4,2})$$

where the first subscript 3 refers to the entering distilled water stream, and 4 refers to the exit distilled water stream. The second subscript 2 refers to distilled water, and the subscript 1 refers to raw water.

$$Q = (79500 \text{ kg/h})\,(1h/3600s)\,(4.187 \times 10^3 \text{ J/kg-°C})\,(34 - 30)\text{ °C}$$

$$= 3.699 \times 10^5 \text{ J/s } (1.26 \times 10^6 \text{ Btu/h})$$

Table 4.6 Calculation Procedure for Sizing Shell-and-Tube Heat Exchangers

1. Calculate the heat transferred from Equations 4.5.1, 4.5.2, 4.5.6, and 4.5.7.

2. Select approximate values of the individual heat-transfer coefficients and fouling resistance from Equations 4.5.10 to 4.5.13.

3. Calculate the overall heat-transfer coefficient from Equation 4.5.9.

4. Calculate the exit temperature, t_2, from Equation 4.5.8 for the approach temperature difference.

5. Calculate the logarithmic-mean temperature differences from Equation 4.5.5.

6. Calculate the logarithmic-mean correction factor, F, from Equation 4.5.4.

7. Calculate the required surface area, A_o, of the tubes from Equation 4.5.3.

Use Table 4.3 to obtain approximate values of the individual heat-transfer coefficients and fouling resistances. Then, calculate the overall heat-transfer coefficient from Equation 4.5.9 after selecting a conservative heat-transfer coefficient of 5000 W/m²-K for water on both the shell and tube sides. Also, select a high value of 2.5×10^{-4} for the fouling resistance for raw water and a low value of 1×10^{-4} for the fouling resistance for distilled water, which is clean. Thus, the overall heat-transfer coefficient,

$$U_o = 1/[2.5 \times 10^{-4} + (1/5.0 \times 10^3) + (1/5.0 \times 10^3) + 1.0 \times 10^{-4}]$$

$$= 1.333 \times 10^3 \text{ W/m}^2\text{-K (235 Btu/h-ft}^2\text{-}^\circ\text{F)}$$

This value for the overall heat-transfer coefficient appears to be on the high side. Ludwig (4.15) reports a range of coefficients of 170 to 225 Btu/h-ft^2-$^\circ$F (965 to 1280 W/m^2-K) for raw water in the tubes and treated water in the shell. We will use the value of the coefficient calculated above, and then, correct the area calculation by using a large safety factor.

For a cooler, select from Table 4.4 an approach temperature difference of 5.0 $^\circ$C, which is an economic rule-of-thumb. This approach is selected rather than the upper limit of 50.0 $^\circ$C to conserve heat, but the surface area will be larger for the 5.0 $^\circ$C approach. From Equation 4.5.8, the exit raw-water temperature, t_2, equals 29 $^\circ$C. Because the raw water has a tendency to scale, it is located on the tube side. At a water temperature of about 50 $^\circ$C and above, scale formation increases so that the exit water temperature should never exceed 50 $^\circ$C (122 $^\circ$F).

From Equation 4.5.5, the logarithmic-mean temperature difference is

$$(\Delta t)_{LM} = \frac{(30 - 23) - (34 - 29)}{\ln \dfrac{(30 - 23)}{(34 - 29)}} = 5.944 \ ^\circ\text{C (10.7 }^\circ\text{F)}$$

Next, calculate the logarithmic-mean temperature difference correction factor, F, from Equation 4.5.4. Calculate F either from Equation 4.10 or use plots of Equation 4.10 given in the chemical engineering handbook [1]. In either case, first calculate the parameters R and S. R and S are defined in Figure 4.7.

$$R = \frac{34 - 30}{29 - 23} = 0.6667$$

$$S = \frac{29 - 23}{34 - 23} = 0.5455$$

By solving Equation 4.54 using Polymath, F = 0.9471. Because F = 0.9471 is greater than the minimum recommended value of 0.85, the design is acceptable. Finally, using Equation 4.5.3, we find that the required surface area,

$$A = (3.669\text{x}10^5) / 1.33\text{x}10^3 \, (0.9471) \, (5.944) = 49.0 \text{ m}^2 \, (527.2 \text{ ft}^2)$$

Because of the uncertainty in the overall heat-transfer coefficient, allow for a safety factor of 20%, which results in an area of 58.80 m^2. Round off the area to 60 m^2 (64.6 ft^2).

RATING HEAT EXCHANGERS

The objective of a rating problem is to determine if an existing process unit will satisfy process conditions. To arrive at an approximate calculation procedure for rating a heat exchanger, first define a clean overall heat-transfer coefficient, i.e., in the absence of any fouling. Therefore, R_{fi} and R_{fo} = 0 in Equation 4.15.

$$U_{oC} = \cfrac{1}{\cfrac{D_o}{D_i h_i} + \cfrac{1}{h_o}} \tag{4.16}$$

For many situations, $D_o / D_i \approx 1$.

Substitute Equation 4.16 into Equation 4.15 and let R_{fi} and R_{fo} equal $(R_{fi})_A$ and $(R_{fo})_A$, the available fouling resistances. Then, the overall heat-transfer coefficient,

$$U_o = 1/[(R_{fi})_A + 1/U_{oC} + (R_{fo})_A] \tag{4.17}$$

If the individual fouling resistances are added to obtain the total fouling resistance, R_{oA}, then

$$U_o = 1/(R_{oA} - 1/U_{oC}) \tag{4.18}$$

Rearranging Equation 4.18, the total available fouling resistance,

$$R_{oA} = \cfrac{U_{oC} - U_o}{U_{oC} U_o} \tag{4.19}$$

where the overall heat-transfer coefficient for the existing heat exchanger is calculated from Equation 4.20.

$$Q = U_o A_o F (\Delta t)_{LM} \tag{4.20}$$

Thus, R_{oA} can be calculated from Equation 4.19 after calculating U_{oC} from Equation 4.16 and U_o from Equation 4.20.

Next, add the fouling resistances caused by the inside and outside scale,

$$R_{oR} = (R_{fi})_R + (R_{fo})_R \tag{4.21}$$

where R_{oR} is the required combined fouling resistance. R_{oR} is calculated using individual fouling resistances obtained from Table 4.3, assuming that one to one-and-a-half years of service before cleaning is optimum.

For an existing heat exchanger to be adequate for new process conditions,

$$R_{oA} \geq R_{oR} \tag{4.22}$$

A value of R_{oA} larger than R_{oR} means that the heat exchanger will last longer than the optimal time before cleaning. For any value of R_{oA} smaller than R_{oR}, the heat exchanger will operate at less than the optimum time.

Tabele 4.7 lists the equations for rating heat exchangers and Table 4.8 outlines thecalculating procedure.

Example 4.3 Rating an Ammonia Condenser

It is required to condense 650 kg/h (1430 lb/h) of ammonia vapor at 14.8 bar (14.6 atm) using water. The available heat exchanger is a 1-2 heat exchanger with 46 m^2 (495 ft^2) of surface area. The enthalpy of vaporization is 261.4 kcal/kg. Is this heat exchanger adequate for this service? Show why or why not.

Follow the solution procedure.

From Table 4.3, the following conservative values of the heat-transfer coefficients and fouling resistances are selected. Because water is dirtier than ammonia, locate the water on the tube side. Also, a condensing vapor is usually located on the shell side.

$h_i = 5000$ W/m^2-K
$h_o = 8000$ W/m^2-K
$R_{fi} = 2.5 \times 10^{-4}$ m^2-k/W
$R_{fo} = 1 \times 10^{-4}$ m^2-k/W

Assuming that there is no subcooling of the condensed ammonia, from Equations 4.7.1 and 4.7.2,

$Q = 261.4$ kcal/kg (650 kg/hr) $= 1.7 \times 10^5$ kcal/h (6.75×10^5 Btu/h)

To calculate the logarithmic-mean temperature difference, the terminal temperatures of the condenser must be fixed. Because the condensation is essentially isobaric, the inlet and outlet temperatures of the ammonia stream are 41.4°C (106.5 °F). From Table 4.1, the inlet cooling-water temperature is 30°C (86.0 °F) if cooling-tower water is used. Also, for thermodynamic considerations the exit water temperature must be less than 41.4°C, and it is calculated from Equation 4.7.6. If the lower value of the approach temperature difference of 5 °C (9.0 °F) is selected from Table 4.4, a low cooling-water flow rate will be needed. Thus, exit water temperature is 36.4°C. Therefore, from Equation 4.7.5, the logarithmic-mean temperature difference,

$$(\Delta t)_{LM} = \frac{41.4 - 36.4 - (41.4 - 30.0)}{\ln \dfrac{5.0}{11.4}} = 7.765 \ °C \ (14.0 \ °F)$$

For isothermal condensation, the logarithmic-mean temperature difference correction factor, F, equals one. Therefore, from Equation 4.7.3 for the existing heat exchanger, the available overall heat-transfer coefficient,

$$U_o = \frac{Q}{A \ F \ (\Delta t)_{LM}}$$

$$U_o = \frac{1.7 \times 10^5 \ \text{kcal}}{h} \quad \frac{1}{46 \ m^2} \quad \frac{1}{7.765 \ °C} \quad \frac{4.183 \times 10^3 \quad J}{1} \quad \frac{1 \quad h}{\text{kcal} \quad 3600 \ s}$$

or $U_o = 97.5$ Btu/h-ft^2-°F.

From Equation 4.7.11, the clean overall heat-transfer coefficient,

$$U_{oC} = \frac{h_i \ h_o}{h_i + h_o} = \frac{5000 \ (8000)}{5000 + 8000} = 3.077 \times 10^3 \ \text{W/m}^2\text{-K} \ (542 \ \text{Btu/h-ft}^2\text{-°F})$$

Then, from Equation 4.7.10, the available fouling resistance,

$$R_{oA} = \frac{U_{oC} - U_o}{} = \frac{3077 - 553.5}{} = 1.482 \times 10^{-3} \ \text{m}^2\text{-°C/W} \ (2.60 \times 10^{-4} \ \text{h-ft}^2\text{-°F/Btu})$$

U_{oC} U_o 3077 (553.5)
From Equation 4.7.9, the required fouling resistance for the condenser,

$R_{oR} = R_{fi} + R_{fo} = 2.5 \times 10^{-4} + 1.0 \times 10^{-4} = 3.5 \times 10^{-4}$ m^2-K/W (6.16x10^{-5} h-ft^2-°F/Btu)

Therefore,

$R_{oA} > R_{oR}$

Thus, the condenser will be adequate for the service. Because the available fouling resistance is greater than the required fouling resistance, the condenser will last longer than the specified time before cleaning.

Table 4.7 Summary of Equations for Rating a Heat Exchanger

First subscript: Process stream = 1,2,3 or 4
Second subscript: Component =1 or 2

Mass Balance

$m_3' = m_4$ (4.7.1)

Energy Equation

$Q = h_{3,2} m_3' - h_{4,2} m$ (4.7.2)

Rate Equation

$Q = U_o A_o F (\Delta t)_{LM}$ (4.7.3)

$F = f(t_1', t_2, t_3', t_4')$ — Equation 4.10   (4.7.4)

$(\Delta t)_{LM} = f(t_1', t_2, t_3', t_4')$ — Equation 4.2   (4.7.5)

Economic Relations

$t_3' - t_2 =$ approach Δt — Table 4.5   (4.7.6)

Thermodynamic Properties

$h_{3,2} = f(t_3')$ (4.7.7)

Table 4.7 Continued

Transfer Properties

$$R_{oR} = (R_{fi})_R + (R_{fo})_R \tag{4.7.9}$$

$$R_{oA} = \frac{U_{oC} - U_o}{U_o \, U_{oC}} \tag{4.7.10}$$

$$U_{oC} = \frac{h_i \, h_o}{h_i + h_o} \tag{4.7.11}$$

$$R_{fi} = f(\text{fluid}', \text{type of phase change}') \; — \; \text{Table 4.3} \tag{4.7.12}$$

$$h_i = f(\text{fluid}', \text{type of phase change}') \; — \; \text{Table 4.3} \tag{4.7.13}$$

$$h_o = f(\text{fluid}', \text{type of phase change}') \; — \; \text{Table 4.3} \tag{4.7.14}$$

$$R_{fo} = f(\text{fluid}', \text{type of phase change}') \; - \; \text{Table 4.3} \tag{4.7.15}$$

Variables

$$m_4 \text{ - } t_2 \text{ - } (\Delta t)_{LM} \text{ - } h_{3,2} \text{ - } h_{4,2} - Q \text{ - } U_o \text{ - } U_{oC} - F \text{ - } R_{oA} \text{ - } R_{oR} \text{ - } (R_{fi})_R \text{ - } (R_{fo})_R - h_i \text{ - } h_o$$

Degrees of Freedom

$$F = 15 - 15 = 0$$

Table 4.8 Calculation Procedure for Rating Heat Exchangers

1. Calculate the heat transferred from Equations 4.7.1, 4.7.2, 4.7.7 and 4.7.8.

2. Calculate the approximate values of the heat-transfer coefficients from Equations 4.7.12 to 4.7.15.

Table 4.8 Continued

3. Calculate the coolant exit temperature, t_3, from Equation 4.7.6.

4. Calculate the logarithmic-mean temperature difference, $(\Delta t)_{LM,}$ from Equation 4.7.5.

5. Calculate the logarithmic-mean temperature-difference correction factor, F, from Equation 4.7.4.

6. Calculate the overall heat-transfer coefficient, U_o, from Equation 4.7.3.

7. Calculate the clean overall heat-transfer coefficient, U_{oC}, from Equation 4.7.11.

8. Calculate the available fouling resistance, R_{oA}, from Equation 4.7.10 and the require fouling resistance, R_{oR}, from Equation 4.7.9.

9. If $R_{oA} \geq R_{oR}$, the condenser will be adequate for the process.

NOMENCLATURE

English

A   surface area

A_o outside surface area of a tube

C_C capital cost per unit area

C_M maintenance cost per unit area

C_P heat capacity at constant pressure

C_T total cost

C_W cost of water per unit mass

D   tube diameter

F   logarithmic mean temperature correction factor or degrees of freedom

h   heat transfer coefficient or enthalpy

k   thermal conductivity

m   mass flow rate

Q   heat transfer rate

R	number of independent relations
R_f	fouling resistance
t	temperature
t_B	base temperature
U	overall heat-transfer coefficient
U_o	oveall heat-transfer coefficient base on the outside area of a tube
V	number of variables
x_f	scale thickness
x_W	wall thickness

Greek

$(\Delta t)_{LM}$	logarithmic-mean temperature difference
$(\Delta t)_{LM,eff}$	effective logarithmic-mean temperature difference
θ	annual number of hours of operation

Subscripts

A	available
C	clean
f	refers to a solid deposit
i	inside of the tube or tube side
LM	logarithmic mean
o	outside of the tube or shell side
p	process fluid
R	required
W	water

REFERENCES

1. Kern, P.Q., Process Heat Transfer, McGraw-Hill, New York, NY, 1959.
2. Mehra, Y.R., Refrigeration Systems for Low Temperature Processes

Chem. Eng., <u>89</u>, 14, 94, 1982.

3. Walas, S. M., Rules of Thumb, Chem. Eng., <u>94</u>, 4, 75, 1987.
4. Brochure, Vacuum Cooling, Croll-Reynolds, Westfield, NJ, No date.
5. Walas, S.M., Chemical Process Equipment, Butterworths, Boston, MA, 1988.
6. Shilling,R.L., Bell, K.J., Bernhagen, P.M., Flynn, T.M., Goldschmidt, V. M., Hrnjak, P.S., Standford, F.C., Timmerhaus, K.D., Heat-Transfer Equipment, eds. Perry, R.H., Green, D.W., Perry's Chemical Engineers Handbook, 7th ed., McGraw-Hill, New York, NY, 1997.
7. Frank,O., Simplified Design Procedures for Tubular Heat Exchangers, Practical Aspects of Heat Transfer, Chem. Eng. Prog., Tech. Manual, Am. Inst. of Chem. Eng., New York., NY, 1978.
8. Ulrich, G.P., A Guide to Chemical Engineering Process Design and Economics, John Wiley & Sons, New York, NY, 1984.
9. Singh, J., Selectinq Heat-Transfer Fluids for High-Temperature Service, Chem. Eng., <u>88</u>, 11, 53, 1981.
10. Fried, J.R., Heat-Transfer Agents for High-Temperature Systems, Chem. Eng., <u>80</u>, 12, 89, 1973.
11. Worthy, W., Silicone Heat-Transfer Fluids Use Grows, Chem. & Eng. News, p.32, Dec. 8, 1986.
12. Kirschen, N.A., PCBs in Transformer Fluids, Amer. Lab., <u>13</u>, 12, 65, 1981.
13. Devore, A., Vago, G.J., Picozzi, G.J., Heat Exchangers, Specifying and Selecting, Chem. Eng., <u>87</u>, 20, 133, 1980.
14. Mehra, D.K., Shell and Tube Heat Exchangers, Chem. Eng., <u>90</u>, 15, 47, 1983.
15. Ludwig, E.E., Applied Process Design for Chemical and Petrochemical Plants, Vol. III, 3rd ed., Butterworth-Heinemann, Woburn, MA, 2001.
16. Lord, R.C., Minton, P.E., Slusser, R.P., Design of Heat Exchangers, Chem. Eng., <u>77</u>, 2, 45, 1970.
17. Coulson, J.M., Richardson, J.F., Sinnott, E.K. Chemical Engineering, Vol. 6, An Introduction to Chemical Engineering Design, Pergamon Press, New York, NY, 1983.
18. Yokell, S., Double-Tubesheet Heat-Exchanger Design Stops Shell-Tube Leakage, Chem. Eng., <u>80</u>, 11,133, 1973.
19. Anonymous, Standards of Tubular Exchanger Manufacturers Association, 7th ed., Tubular Exchangers Manufacturers Association, Tarrytown, NY, 1988.
20. Anonymous, ASME Boiler and Pressure Vessel Code, Section VIII, Div. 1, American Society of Mechanical Engineers, New York, NY, 1980.
21. Anonymous, Heat Exchangers for General Refinery Service, API Standard 660, American Petroleum Institute, Washington, D.C.
22. McCabe, W.L., Smith, J.C., Harriott, P., Unit Operations of Chemical Engineering, 6th ed., McGraw-Hill, New York, NY, 2001.

23. Goyal, O.P., Guidelines on Exchangers, Hydrocarbon Process., <u>64</u>, 8, 55, 1985.
24. Taborex, J., Evolution of Heat Exchanger Design Technologies, Heat Transfer Eng., <u>1</u>, 11 15, 1979.
25. Kundsen, J.G., Fouling of Heat Exchangers: Are We Solving the Problems?, Chem. Eng. Progr., <u>80</u>, 2, 63, 1984.
26. Crittenden, B.D., Khater, E.H., Economic Fouling Resistance Selection, in Knudsen, J.G., ed. Fouling of Heat Transfer Equipment, Hemisphere Publishing Corp., New York, NY, 1981.
27. Thermal and Hydraulic Design of Heat Exchangers, K.J. Bell, Approximate Sizing of Shell-and-Tube Heat Exchangers, p 3.1.4-1, Heat Exchanger Design Handbook, Vol. 3, Begell House, New York, NY, 1998.
28. Woods, D.R, Process Design and Engineering Practice, PTR Prentice Hall, Englewood Cliffs, NJ, 1995.
29. Frank. O., Personal Communication, Consulting Engineer, Convent Station, NJ, Jan. 2002.

5

Compressors, Pumps, and Turbines

Compressors are required to transfer gases from one process unit to another and to compress them to carry out chemical reactions, separations, and to liquefy gases. Compressors cover the range from vacuum to high pressure and are called vacuum pumps, fans, blowers and compressors, according to their operating pressure range. Roughly vacuum pumps compress gases from about 0.00133 to 1.01 bar (0.0193 to 14.7 psia) [3], fans from 1.01 to 1.15 bar (14.7 to 16.7 psia) [1], blowers from 1.15 to 1.70 bar (16.7 to 24.7 psia) [1], and compressors above 1.70 bar (24.7 psia) [1]. These regions of application are not distinct, and may overlap. The word pump is usually reserved for transferring liquids, but in the vacuum region the compressor is called a vacuum pump.

Compressors are divided into two main classes, positive displacement and dynamic. Positive-displacement compressors compress essentially the same volume of gas in a chamber regardless of the discharge pressure. In a dynamic compressor, a gas is first accelerated to a high velocity to increase its kinetic energy. Then, the compressor converts kinetic energy into pressure by reducing the gas velocity, according to the macroscopic energy balance.

VACUUM PUMPS

In the vacuum region, pressures down to 0.00133 bar (0.0193 psia) are of interest to process engineers for process operations such as distillation, drying and evaporation. Some applications below 0.00132 bar (0.193 psia) are molten metal degassing, molecular distillation, and freeze drying.

The most commonly used vacuum pumps are steam-jet ejectors and several positive-displacement pumps, which are shown in Figures 5.1 and 5.2. Some of the characteristics of vacuum pumps are given in Table 5.1. A prime consideration when selecting a vacuum pump is the compatibility of a gas with a seal fluid. To avoid these problems, there is a trend toward using dry pumps where a seal fluid or lubricant is not used [60].

In an ejector, steam enters the nozzle at the pressure, P_1, shown in Figure 5.1. The nozzle increases the velocity of the steam, reducing the pressure to P_2 at the suction to evacuated a vessel. Then, the steam and suction fluid are compressed in the diffuser section where the kinetic energy of the mixed fluid is converted to the pressure P_3. Both condensable and noncondensable gases, usually air, are entrained by the steam. When staging ejectors, the load on the downstream ejectors is considerably reduced if intercondensers are used to remove condensable gases. Table 5.1 shows some of the characteristics of staged ejectors. An advantage of ejectors is that there are no moving parts. A disadvantage is that the ejector is designed to meet specific conditions [2], and it is inflexible under widely varying conditions.

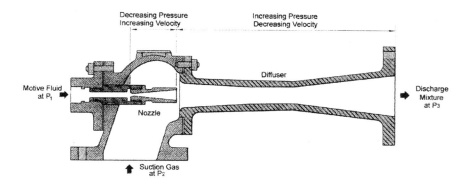

Figure 5.1 A steam-jet ejector. From Ref. 3 with permission.

Table 5.1 Vacuum-Pump Characteristics

Vacuum Pump Type[a]	Inlet Flow Rate[b] m³/h	Single-Stage Compression Ratio	Minimum Suction Pressure[c] bar
Steam-Jet Ejector			
One Stage	$17.0 - 1.7 \times 10^6$	6.0	0.1
Two Stages			0.016
Three Stages			1.3×10^{-3}
Four Stages			2.7×10^{-4}
Five Stages			2.7×10^{-5}
Six Stages			4.0×10^{-6}
Positive Displacement			
Rotary Piston			
One Stage[d]	$5.1 - 1.4 \times 10^3$	1.0×10^5	2.7×10^{-5}
Two Stages			1.3×10^{-6}
Rotary Vane			
Dry Operation	$30.0 - 1.0 \times 10^4$	1.0×10^5	6.7×10^{-2}
Oil Sealed	$85.0 - 1.4 \times 10^3$		1.3×10^{-3}
Oil Sealed[e] (1 stage)	$5.1 - 85.0$		2.7×10^{-5}
Oil Sealed[e] (2 stages)			1.3×10^{-6}
Rotary Blower			
One Stage	$51.0 - 5.1 \times 10^4$	2.3	0.4
Two Stages			8.0×10^{-2}
Liquid Ring			
Water Sealed (16 °C)			
One Stage	$51.0 - 5.1 \times 10^4$	10.0	0.1
Two Stages			5.3×10^{-2}
Oil Sealed			1.3×10^{-2}

a) Source: Reference 3
b) To convert to ft³/min multiply by 0.589.
c) Discharge pressure limited to 1.22 bar (1.20)
   to convert to atm multiply by 0.987.
d) Maximum temperature = 370 K (666 °R)
e) Spring-loaded vanes

A rotary-piston pump is an oil-sealed, positive-displacement vacuum pump. The oil both lubricates the pump and seals the discharge from the suction side of the pump. As the piston rotates, gas enters a chamber, as shown in Figure 5.2. Then, the inlet port closes, and the gas is compressed in the chamber until the discharge valve opens, exhausting the gas to the atmosphere. Possible contamination of the oil with condensable vapors, usually water, is a problem. One way condensation can be avoided is by reducing the partial pressure of the condensable gases by allowing air to leak into the cylinder, which is called a gas ballast.

A rotary-vane vacuum pump is also a positive-displacement vacuum pump. The vanes slide in slots and are forced against the wall of a stationary cylinder by springs for laboratory pumps or by centrifugal force for process pumps. Sealing is accomplished either by oil or a dry seal using nonmetallic vanes which continuously wear thereby forming a tight seal. As can be seen in Figure 5.2, a gas enters the pump, is trapped between two vanes, is compressed as the volume of the chamber is reduced, and finally exhausted at the discharge port. The rotary-vane vacuum pump is sensitive to contamination, which can reduce its performance rapidly.

In the rotary blower, shown in Figure 5.2, gases are trapped in between two interlocking rotors which rotate in opposite directions. The blower requires no seal fluid. Because of the required clearances between the rotors of 0.025 to 0.25 mm (9.84×10^{-4} to 9.84×10^{-3} in), backflow reduces the blower capacity [3]. Also, overheating limits the pressure increase.

In the liquid-ring pump, shown in Figure 5.2, a seal liquid, usually water, is thrown against the casing by a rotating impeller forming a liquid ring. Gas drawn from an inlet port is compressed in the chamber between the rotor blades as the impeller rotates on an axis that is offset from the casing. Some of the seal liquid is entrained with the exhausted gas. If the gas contains a condensable component, the pump behaves like a direct contact condenser. Provisions can be made for separating the condensable component from the seal liquid which is then recirculated. Because of its ability to handle condensable vapors, the liquid-ring pump is ideally suited for filtering operations. Another advantage is that the seal liquid is a heat sink, limiting the temperature rise of the compressed gases [3]. A disadvantage of this pump is that it uses twice as much energy as an oil-sealed- rotary-vane or a rotary-piston pump of the same capacity [4].

The performance of a vacuum pump is depicted by a plot of flow rate against suction pressure, which is called the characteristic curve. Physically, a vacuum pump must operate at some point on the curve, depending on the design of the system. In Figure 5.3, the characteristic curves for a rotary piston, a liquid-ring pump and steam-jet ejector are plotted. For a perfect positive-displacement pump, the curve should be flat over the whole pressure range. Instead, for the rotary-piston pump, the curve increases slightly with increasing suction pressure because of reduced leakage. The curve for the ejector increases

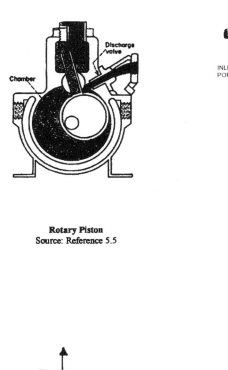

Rotary Piston
Source: Reference 5.5

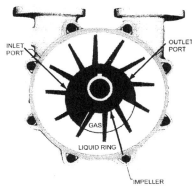

Liquid Ring
Source: Reference 5.59

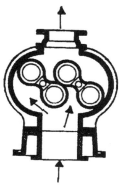

Rotary Lobe
Source: Reference 5.4

Rotary Vane
Source: Reference 5.6

Figure 5.2 Positive-displacement vacuum pumps. References with permission.

at low pressures and then decreases at high pressures. For the liquid-ring vacuum pump, the curve increases at low pressures, resembling the ejector, and
then flattens out at high pressures, resembling a positive-displacement pump.
Isentropic efficiencies for some vacuum pumps are plotted against the suction
pressure, as shown in Figure 5.4.

To size a vacuum pump requires calculating the volumetric flow rate and
frictional pressure loss in the vacuum system. The volumetric flow rate consists
of condensable and noncondensable gases. The noncondensable gases originate
from the material being processed and from air leaking into the system. Assuming that reasonable care is taken when sealing a vacuum system, Ryans and
Croll [3] have devised a procedure for estimating acceptable leakage rates
through various pump seals, valves, and sight glasses. To estimate the flow rate
of condensable gases, it is assumed that the noncondensable gases are saturated
with the condensable vapors. Once the total flow rate of gases and the required
pressure are known, the vacuum pump power can be calculated according to a
method used for compressors described later.

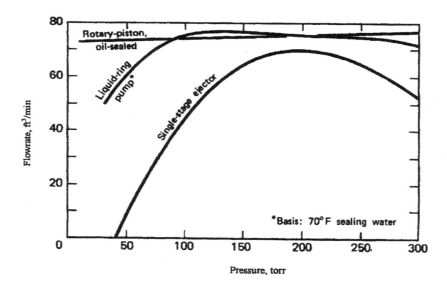

Figure 5.3 Characteristic curves of vacuum pumps. From Ref. 3 with
permission.

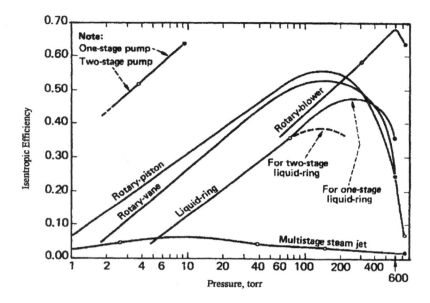

Figure 5.4 Isentropic Efficiencies of vaccum pumps. (Source Ref. 3 with permission)

FANS

Fans are designed to move a large volume of gas near atmospheric pressure. Because the clearance between the impeller and casing are large, the pressure developed is low, between 1.01 to 1.15 bar (14.7 to 16.7 psia). Fans, which are all of the dynamic type, are classified according to the direction of air flow. In a centrifugal fan, gas flows along the fan shaft, turns ninety degrees by the impeller, which imparts kinetic energy to the gas as it flows radially outward. Then, the gas is converted to pressure as it leaves the fan parallel to the shaft. In an axial flow fan, gas enters and leaves the fan parallel to the shaft. These fan types are shown in Figure 5.5.

Centrifugal fans are classified according to their blade geometry – radial, forward curved, backward curved, and air foil. The radial fan's major characteristic is its ability to compress gases to a higher pressure but delivers lower flow rates than the other fan types. Its characteristic curve is shown in Figure 5.6. The blades are self cleaning, tending to fling off particles and thus can be used to pneumatically convey solids. Other applications are listed in Table 5.2

The backward-incline fan design consists of the single-thickness blade and the air-foil blade. The single thickness blade can be used for pneumatic con-

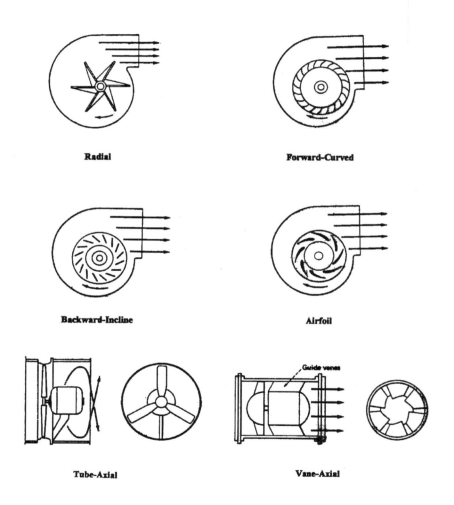

Figure 5.5 Fan types. From Ref. 7 with permission.

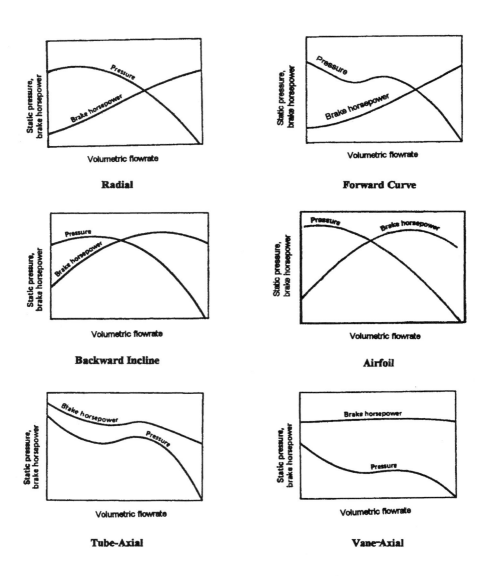

Figure 5.6 Fan characteristic curves. From Ref. 7 with permission.

Table 5.2 Applications for Several Fan Designs   (Source Ref. 5.1 with permission)

Application	Tube-Axial	Vane-Axial	Radial	Forward-Curved	Backward-Inclined	Airfoil
Conveying systems			X		X	
Supplying air for oil and gas burners or combustion furnaces	X	X	X	X	X	X
Boosting gas pressures			X		X	X
Ventilating process plants	X	X			X	X
Boilers, forced-draft		X			X	X
Boilers, induced-draft			X	X		
Kiln exhaust			X	X		
Kiln supply		X			X	X
Cooling towers	X					
Dust collectors and electrostatic precipitators			X	X		
Process drying	X	X	X		X	X
Reactor off-gases or stack emissions			X	X		

veying of solids as shown in Table 5.2. The air-foil type has aerodynamically-shaped blades to reduce flow resistance, resulting in a high efficiency. Entrained particles will damage the blades, and thus this fan is not suitable for pneumatic conveying. An attractive feature of both fans is that they have nonoverloading power curves as shown in Figure 5.6. This means that as the flow rate increases, the required power increases, reaches a maximum, and then decreases instead of continuously increasing. When operating under conditions where the flow rate varies, this characteristic is an asset.

The forward-curved-blade fan is designed for low to medium flow rates at low pressures. Because of the cupped shaped blades, solids tend to be held in the fan, and thus this fan is also not suitable for pneumatic conveying of solids. In the characteristic curve for the fan, shown in Figure 5.6, there is a region of instability to the left of the pressure peak. Thus, the fan must be operated to the right of that region. The horsepower increases continuously with increasing flow rate.

Axial fans consist of the tube-axial fan and the vane-axial fan, which are designed for a wide range of flow rates at low pressures. These fans consist of a propeller enclosed in a duct. They are limited to applications where the gas does not contain entrained solids. In a tube-axial fan, the discharged flow follows a helical path creating turbulence. To reduce turbulence and increase the fan efficiency, the vane-axial fan contains flow straightening vanes (Figure 5.5). The tube-axial fan has an unusual power-flow curve, as can be seen in Figure 5.6. The required power initially decreases with increasing flow rate and then increases, reaching a maximum before decreasing again. Also, the pressure curve has an unstable region so that the fan must be operated to the right of the maximum.

Table 5.3 Fan Efficiencies

Fan Type	Fan Efficiency[a] η_F, %
Radial[b]	65-70
Backward inclined	
Single Thickness[c]	84
Air Foil[b]	90
Forward Curved	70-75
Axial	
Tube[b]	75-80
Vane[c]	85

a) Includes fluid and mechanical
   frictional losses
b) Source: Reference 7
c) Source: Reference 8

A final consideration is the fan operating temperature and environmental conditions. Most axial fans contain the motor, bearings, and drive components within the duct. Thus, the gases must be noncorrosive, at a low temperature (−34 to 82 °C) (−29.2 to 180 °F) [1], nonflammable and without any particulate matter. At low or high temperatures, most steels lose their strength [8]. Thompson and Trickler [8] discuss some elements of duct-system design. Table 5.3 summarizes efficiencies for approximate sizing of the fans that have been described.

Fan Power

The most frequently used relationship in the design of flow systems is the macroscopic mechanical-energy balance. This equation is obtained by integrating the microscopic mechanical-energy balance over the volume of the system as shown by Bird et al. [9]. The balance is given by

$$\frac{\Delta(v^2/\alpha)}{2 g_c} + \frac{g}{g_c} \Delta z + \int_1^2 \frac{dp}{\rho} + W + E = 0 \tag{5.1}$$

Each term has the dimensions of energy per unit of mass – in this case, ft-lb_F/lb_M. The factor, α, in the kinetic energy term, $\Delta v^2/2\alpha g_C$, corrects for the velocity profile across a duct. For laminar flow in a circular duct, the velocity profile is parabolic, and $\alpha = 1/2$. If the velocity profile is flat, $\alpha = 1$. For very rough pipes and turbulent flow, α may reach a value of 0.77 [10]. In many engineering applications, it suffices to let $\alpha = 1$ for turbulent flow.

The second term in the mechanical-energy balance, Equation 5.1, is the change in potential energy and requires no comment. The third term is "pressure work" and its evaluation depends on whether the fluid is compressible or incompressible. Because the increase in pressure across the fan is small, we treat the flow as essentially incompressible. Thus, the fluid density may be removed from the integral sign and the mechanical energy balance becomes

$$\frac{\Delta(v^2/\alpha)}{2\,g_C} + \frac{\Delta p}{g_C}\Delta z + \frac{\Delta P}{\rho} + W + E = 0 \tag{5.2}$$

The last two terms are the work done by the system, W, and the friction loss, E. The system is defined by the fan inlet and discharge. Because the density of a gas at atmospheric pressure is small, Δz can be neglected. Since W is defined as the work done by the system, the work done on the gas by the fan is $-W_H$. Thus, Equation 5.2 becomes

$$\frac{\Delta(v^2)}{2\,g_C} + \frac{g}{\rho} - W_H + E = 0 \tag{5.3}$$

The frictional loss term, E, can be included in an hydraulic efficiency which accounts for the gas frictional losses in the fan according to

$$\eta_H = \frac{W_H - E}{W_H} \tag{5.4}$$

$$W_H = \frac{1}{\eta_H}\left[\frac{\Delta(v^2)}{2\,g_C} + \frac{\Delta P}{\rho}\right] \tag{5.5}$$

where η_H is a hydraulic efficiency that accounts for pressure losses caused by fluid friction in the fan.

In addition to the work lost by fluid friction, some work is lost because of mechanical friction in the seals and bearings. This lost work is accounted for by a mechanical efficiency. Thus, the fan work,

$$W_F = \frac{1}{\eta_F} \left[\frac{\Delta(v^2)}{2 \, g_C} + \frac{\Delta P}{\rho} \right] \qquad (5.6)$$

where $\eta_F = \eta_M \, \eta_H$, and W_F is the work delivered to the shaft of the fan.

Because power is the rate of doing work, the fan shaft power – frequently called brake power – is calculated from Equation 5.7.

$$P_F = m \, W_F \qquad (5.7)$$

Example 5.1 Calculation of Fan Power

A fan will pneumatically convey 1360 kg/h (2300 lb/h) of a powdered resin from a storage bin to a mixer [11]. The duct diameter is 15.2 cm (6 in). Assume an electrical-motor efficiency of 95%. If the air flow rate needed to convey the resin is 1670 m³/h (983 ft³/min) at 300 K (540 °R) and 1.013 bar (14.7 psia). The pressure drop in the duct system is 0.0893 bar (1.29 psi), what is the required fan power?

The fan shaft work is calculated from Equation 5.6. From Table 5.2, it is seen that a radial fan is acceptable for the conveying system. From Table 5.3 a conservative value for the radial fan efficiency of 65% is selected. The air velocity is

$$v = \frac{V}{A} = 1670 \, \frac{m^3}{h} \; \frac{4}{\pi \, (0.152)^2 \, m^2} \; \frac{1 \; h}{3600 \; s} = 25.56 \; m/s \; (83.9 \; ft/s)$$

From the ideal gas law, the air density,

$$\rho = \frac{P \, M}{R \, T} = \frac{1.013 \; bar}{1} \; \frac{1 \times 10^5 \; N}{1 \; m^2\text{-}bar} \; \frac{1 \; kgmol\text{-}K}{8314 \; N\text{-}m} \; \frac{1}{300 \; K} \; \frac{29 \; kg}{1 \; kgmol}$$

$$= 1.179 \; kg/m^3 \; (0.0736 \; lb/ft^3)$$

Because SI units are used, g_C is not needed in Equation 5.6. From Equation 5.6, the fan work,

$$W_F = \frac{1}{0.65} \left[\frac{(25.56)^2 \text{ kg-m}^2}{2 \text{ kg-s}^2} + \frac{0.0893 \text{ bar}}{1} \quad \frac{1 \times 10^5 \text{ N}}{1 \text{ m}^2\text{-bar}} \quad \frac{1 \text{ m}^3}{1.179 \text{ kg}} \right] =$$

$= 1.216 \times 10^4$ N-m/kg or 1.216×10^4 J/kg (5.23 Btu/lb)

The shaft power,

$$P_F = m\,W_F = \frac{1670 \text{ m}^3}{1 \text{ h}} \quad \frac{1 \text{ h}}{3600 \text{ s}} \quad \frac{1.179 \text{ kg}}{1 \text{ m}^3} \quad \frac{1.216 \times 10^4 \text{ J}}{1 \text{ kg}} - \frac{1 \text{ kW-s}}{1000 \text{ J}}$$

$= 6.651$ kW

$P_F = 6.651$ kW / 0.7457 kW/hp = 8.919 hp

The electric motor efficiency is 0.95. The motor horsepower,

$P_E = 8.919/0.95 = 9.388$ hp

Therefore, select a standard 10 hp (7.46 kW) motor, which gives a safety factor of 6.52%.

COMPRESSORS

Figure 5.7 shows that positive-displacement compressors, like vacuum pumps, are divided into two main classes: reciprocating and rotary. Table 5.4 lists characteristics of these compressors. Ludwig [14] discusses compression equipment and calculation methods in detail.

Positive-Displacement Compressors

Reciprocating compressors consist of direct-acting and diaphragm types. The direct-acting compressor consists of one or more cylinders, each with a piston or plunger that moves back and forth. A gas enters or leaves a cylinder through valves that are activated by the difference in pressure in the cylinder and intake or discharge. When the pressure in the cylinder drops below the inlet pressure, a valve opens allowing gas to flow into the cylinder. After compressing the gas to a pressure above the discharge pressure, the discharge valve opens allowing gas to flow out. This is illustrated in Figure 5.8 for a double-acting reciprocating compressor, i.e., the gas is compressed during both the forward and backward stroke of the piston. The valves in Figure 5.8 are not shown in any detail. If the piston is just a straight rod, called a plunger, the compressor cannot be double acting. An advantage of a

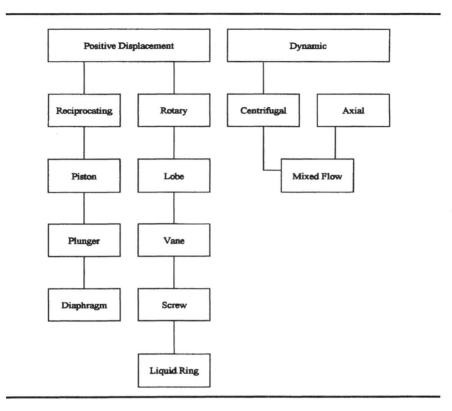

Figure 5.7 Compressor classification chart.

double-acting compressor is that the discharge flow will be smoother than a single acting compressor. The reciprocating compressor is a fixed capacity machine as long as the driver speed is constant. By altering the speed of the driver, the compressor capacity can be changed. A packing-type seal contained in the stuffing box, shown in Figure 5.8, seals the piston rod from the atmosphere.

In a diaphragm compressor, a piston acts indirectly by applying pressure to a hydraulic oil, which flexes a thin metal diaphragm to compress the gas. It is used for small flow rates, below the range for reciprocating compressors, and is limited by the construction of the diaphragm. An advantage of the diaphragm compressor is that leakage of either the gas or oil into the gas is prevented. Thus, the diaphragm compressor is ideal for compressing flammable, corrosive, or toxic gases at high pressures. A disadvantage is the high maintenance cost, mainly because the diaphragm has to be replaced after about 2000 h of operation [13].

Table 5.4 Compressor Characteristics

Compressor Type	Inlet Flow Rate[f] 1000 m³/h	Compression Ratio	Maximum Temperature[g] K	Overall Efficiency[b] η
Positive Displacment[a]				
Reciprocating		3.0 – 4.0	450 – 510	0.75 – 0.85
Diaphragm[d]	0.0051 – 0.051	20.0		
Rotary				
Helical Scew	34.0	2.0 – 4.0	450 – 510	0.75
Spiral Axial	22.0	3.0	450 – 510	0.70
Straight Lobe[c]	52.0	1.7	450 – 510	0.68
Sliding Vane	10.0	2.0 – 4.0	450 - 510	0.72
Liquid Ring	22.0	5.0	450 - 510	0.50
Dynamic				
Centrifugal	85.0 – 340	6.0 – 8.0	111 - 505[e]	
Axial	1.3 - 1000	12.0 - 24	590	

a) Source Reference 2 except where indicated
b) η = isentropic efficency
c) Contains two lobes
d) Source: Reference 13
e) Source: Reference 22
f) To convert to ft³/min multiply by 0.5885
g) To convert to °R multiply by 1.8.

The rotary-compressor types have been discussed when the vacuum pumps were described, except for the screw pump. A rotary-screw compressor contains a male and female rotor, which are shown in Figure 5.9. The rotation of the rotors causes an axial progression of successive sealed cavities, which compresses the gas [14]. One of the major advantages of a screw compressor is that it can handle polymer-forming gases and gases containing significant amounts of entrained liquids. Also, the compression chamber is dry so that lubricating oils will not contaminate the compressed gases, which is necessary in food and drug-production processes.

Dynamic Compressors

Dynamic compressors, like fans, are divided into two classes, centrifugal and axial, according to the direction of gas flow through the machine. A compression stage for a centrifugal compressor, shown in Figure 5.10, consists of a row of

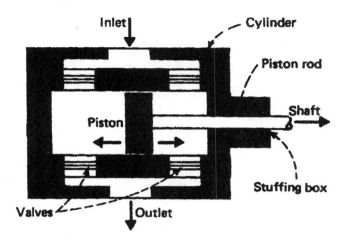

Figure 5.8 A double-acting reciprocating compressor. (Source Ref. 6 with permission).

Figure 5.9 Screw-compressor rotors. (Source Ref. 15).

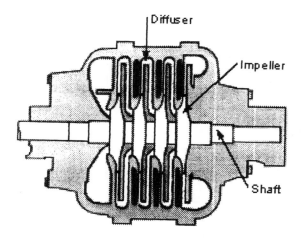

Figure 5.10 A centrifugal compressor containing four impellers. (Source Ref. 16).

blades attached to an impeller, a diffuser, and a diaphragm. The impeller increases the kinetic energy of the gas as it flows radially outward. Then, the diffuser, which is an expanding passage, converts the kinetic energy into pressure. The diffuser and the diaphragm direct the flow to the center of the next impeller. Curved guide vanes, located before each impeller, guides the gas into the impeller at the proper angle. If the pressure rise across the compressor is too large, increasing the gas temperature, intercooling may be necessary.

In an axial-flow compressor, shown in Figure 5.11, the gas flows through an annular passage parallel to the compressor axis. The cross sectional area of the annular passage decreases towards the outlet as the gas density decreases. One compression stage consists of one row of rotating and one row of stationary blades. As the gas flows through the compressor, the rotating blades increase both the pressure and kinetic energy of the gas. In a row of stationary blades, kinetic energy is converted into pressure. The stationary blades also guide the gas flow into the next row of rotating blades. Generally, half the pressure rise is accomplished in the rotating blades and the other half in the stationary blades [18]. Axial-flow compressors are more efficient and are used for higher flow rates than centrifugal compressors. Since axial compressors are more sensitive to deposits, corrosion, and erosion, they are used for very clean, noncorrosive gases. Axial compressors are designed without any intercooling.

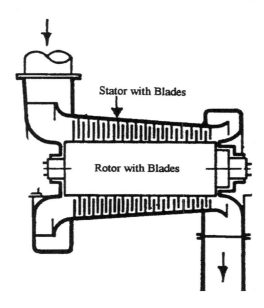

Figure 5.11 An axial-flow compressor.

In Figure 5.12, the characteristic curve illustrates the performance of a compressor rotating at a definite speed. In addition to showing the pressure-capacity characteristic, the curve also shows important operating limits. The most important one is the "surge limit" or minimum-flow point below which the compressor operation becomes unstable. If the flow rate is reduced, the pressure developed by the compressor decreases. Then, the pressure in the discharge line becomes greater, and the gas flows back into the compressor. As soon as the pressure in the discharge line drops to below that developed by the compressor, the gas again flows into the discharge line. Then, the cycle repeats. The oscillating pressure and flow rate will cause audible vibrations and shocks, and could damage the compressor blades, seals, and other components. Therefore, the compressor requires an antisurge control system to limit the flow rate at a minimum point, safely away from the surge limit. The surge limit usually is clearly marked, but, if not, it should be understood that the left end of the curve terminates at the surge limit. The lower right end of the curve usually terminates before reaching a limiting condition referred to as the "choke limit", where the gas flows at the speed of sound. If the curve were extrapolated as shown by the dashed line, it

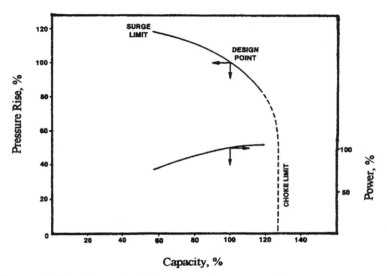

Figure 5.12 Characteristic curve for a centrifugal compressor. From Ref. 19 with permission.

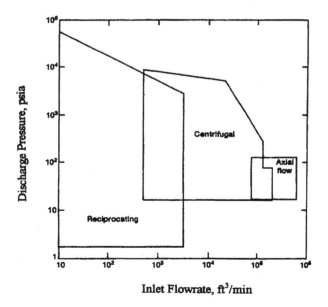

Figure 5.13 Compressor selection chart. From Ref. 12 with permission.

would become vertical at the choke limit, indicating that the flow rate has reached a maximum. Controls to prevent operation too near the choke limit usually are not required. The design point is selected to allow for an increase or decrease in the flow rate if the process conditions vary.

In Figure 5.13, the operating range of the various compressors are shown for comparison, except for the rotary compressors which are expected to occupy a region between the reciprocating and centrifugal compressors. Figure 5.13 can help to guide the process engineer in selecting a compressor design.

COMPRESSOR POWER

To size a compressor requires calculating the power needed for compression. This can be done by assuming an isentropic compression and then correcting the result by dividing by an isentropic efficiency. The power can also be calculated by assuming a polytropic compression, and then correcting the result by dividing by a polytropic efficiency. Both methods will be considered. The isentropic method is also used for blowers and vacuum pumps, but the polytropic method could also be used if data were available. First, we need to derive relationships to calculate the compressor power.

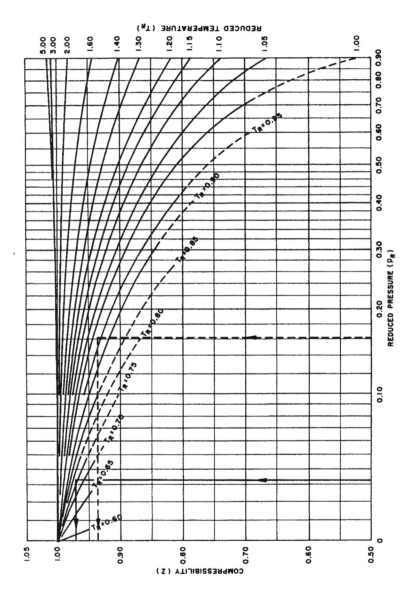

Figure 5.14 Compressibility-factor chart. From Ref. 56).

To obtain an equation for calculating the work of compression, first apply Bernoulli's equation, Equation 5.1, across the compressor. The first term, the kinetic energy term, is small compared to the other terms in the balance. The second term is the change in potential energy, and it is also small. The last two terms are the work done by the system and the friction loss. First, we consider frictionless flow. Thus, the compressor work,

$$W = -\int_1^2 \frac{dp}{\rho} \tag{5.8}$$

Isentropic Compression

For isentropic compression of an ideal gas, the dependence of pressure on temperature is given by

$$\frac{T_2}{T_1} = \left(\frac{P_2}{P_1} \right)^{(k-1)/k} \tag{5.9}$$

where k is the ratio of the heat capacity at constant pressure to the heat capacity at constant volume. Equation 5.9 is derived in several thermodynamic texts and by Bird et al. [9]. Table 5.5 contains values of k for several gases.

By integrating Equation 5.8 over an isentropic path using Equation (5.9), it can be shown that the work of compression for an ideal gas,

$$W_S = \frac{R T_1}{(k-1)/k} \left[\left(\frac{P_2}{P_1} \right)^{(k-1)/k} - 1 \right] \tag{5.10}$$

where k is assumed constant.

For a real gas, we define the compressibility factor, z, by

$$P V = z n R T \tag{5.11}$$

If the gas is ideal, $z = 1$. In Figure 5.14, the compressibility factor is plotted as a function of reduce pressure and temperature. The compressibility factor in Equation 5.11 will vary as the temperature and pressure changes from the compressor inlet to the compressor outlet.

Table 5.5: Properties of Gases

(Most values taken from Natural Gas Processors Suppliers Association Engineering Data Book—1972, Ninth Edition)

Gas or Vapor	Hydrocarbon Reference Symbols	Chemical Formula	Molecular Mass	Specific Heat Ratio $k = c_p/c_v$ at 15.5°C	Critical Conditions Absolute Pressure P_c (bar)	Critical Conditions Absolute Temperature T_c (K)	*$C_{p,m}$ at 0°C	*$C_{p,m}$ at 100°C
Acetylene	C₂=	C₂H₂	26.04	1.24	62.4	309.4	42.16	48.16
Air		N₂+O₂	28.97	1.40	37.7	132.8	29.05	29.32
Ammonia		NH₃	17.03	1.31	112.8	406.1	34.65	37.93
Argon		A	39.94	1.66	48.6	151.1	20.79	20.79
Benzene		C₆H₆	78.11	1.12	49.2	562.8	74.18	103.52
Iso-Butane	iC₄	C₄H₁₀	58.12	1.10	36.5	408.3	89.75	116.89
n-Butane	nC₄	C₄H₁₀	58.12	1.09	38.0	425.6	93.03	117.92
Iso-Butylene	iC₄=	C₄H₈	56.10	1.10	40.0	418.3	83.36	104.96
Butylene	nC₄=	C₄H₈	56.10	1.11	40.2	420.0	83.40	105.06
Carbon Dioxide		CO₂	44.01	1.30	74.0	304.4	36.04	40.08
Carbon Monoxide		CO	28.01	1.40	35.2	134.4	29.10	29.31
Carbureted Water Gas (1)	-		19.48	1.35	31.3	130.6	31.58	33.78
Chlorine		Cl₂	70.91	1.36	77.2	417.2	35.29	35.53
Coke Oven Gas (1)		-	10.71	1.35	28.1	109.4	31.95	34.21
n-Decane	nC₁₀	C₁₀H₂₂	142.28	1.03	22.1	619.4	218.35	280.41
Ethane	C₂	C₂H₆	30.07	1.19	48.8	305.6	49.49	62.14
Ethyl Alcohol		C₂H₅OH	46.07	1.13	63.9	516.7	69.92	81.97
Ethyl Chloride		C₂H₄Cl	64.52	1.19	52.7	460.6	59.61	70.16
Ethylene	C₂—	C₂H₄	28.05	1.24	51.2	283.3	40.90	51.11
Flue Gas (1)			30.00	1.38	38.8	146.7	30.17	30.98
Helium		He	4.00	1.66	2.3	5.0	20.79	20.79
n-Heptane	nC₇	C₇H₁₆	100.20	1.05	27.4	540.6	161.20	202.74
n-Hexane	nC₆	C₆H₁₄	86.17	1.06	30.3	508.3	138.09	174.27
Hydrogen		H₂	2.02	1.41	13.0	33.3	28.67	29.03
Hydrogen Sulphide		H₂S	34.08	1.32	90.0	373.9	33.71	35.07
Methane	C₁	CH₄	16.04	1.31	46.4	191.1	34.50	40.13
Methyl Alcohol		CH₃OH	32.04	1.20	79.8	513.3	42.67	55.32
Methyl Chloride		CH₃Cl	50.49	1.20	66.7	416.7	45.60	49.82
Natural Gas (1)		-	18.82	1.27	46.5	210.6	34.66	39.54
Nitrogen		N₂	28.02	1.40	33.9	126.7	29.10	29.31
n-Nonane	nC₉	C₉H₂₀	128.25	1.04	23.8	596.1	197.07	253.10
Iso-Pentane	iC₅	C₅H₁₂	72.15	1.08	33.3	461.1	112.09	145.56
n-Pentane	nC₅	C₅H₁₂	72.15	1.07	33.7	470.6	115.21	145.94
Pentylene	C₅—	C₅H₁₀	70.13	1.08	40.4	474.4	102.11	130.37
n-Octane	nC₈	C₈H₁₈	114.22	1.05	25.0	569.4	176.17	226.17
Oxygen		O₂	32.00	1.40	50.3	154.4	29.17	29.92
Propane	C₃	C₃H₈	44.09	1.13	42.5	370.0	68.34	88.88
Propylene	C₃—	C₃H₆	42.08	1.15	46.1	365.6	60.16	75.70
Blast Furnace Gas (1)		-	29.6	1.39	—	—	29.97	30.64
Cat Cracker Gas (1)		-	28.83	1.20	46.5	286.1	46.16	57.31
Sulphur Dioxide		SO₂	64.06	1.24	78.7	430.6	38.05	40.00
Water Vapor		H₂O	18.02	1.33	221.2	647.8	33.31	34.07

(1) Approximate values based on average composition.

*Use straight line interpolation or extrapolation to approximate $C_{p,m}$ [in kJ/(kmol•K)] at actual inlet T. (For greater accuracy, average T should be used.)

Source: Ref. 56 with permission.

Then, for a real gas the isentropic work of compression is approximated by

$$W_S = \frac{z\,R\,T_1}{(k-1)/k} \left[\left(\frac{P_2}{P_1} \right)^{(k-1)/k} - 1 \right] \tag{5.12}$$

where z is taken as an average of the inlet and discharge compressibility factors.

To obtain the actual work of compressing a gas, W_A, divide the isentropic work by an isentropic efficiency.

$$W_A = \frac{W_S}{\eta_A} \tag{5.13}$$

Equation 5.9 cannot be use to obtain the gas-discharge temperature for a real compression because it was derived for an isentropic compression. Instead, use the macroscopic energy balance which applies for any process. Thus,

$$\Delta h = c_P\,(T_2 - T_1) = W_A \tag{5.14}$$

where it is assumed that c_P is a constant.

Polytropic Compression

When adiabatic conditions are not attained, the process is called a polytropic change of state. Such a change of state is described by the Equation 5.15.

$$P\,V^n = \text{a constant} \tag{5.15}$$

Thus,

$$P\,V^n = P_1\,V_1^n \tag{5.16}$$

The exponent n depends on the amount of cooling, mechanical friction, and fluid friction during compression. It is determined experimentally for any particular compressor. By integrating Equation 5.8 over a polytropic path, using Equation 5.16, it can be shown that the polytropic work of compression for an ideal gas,

$$W_P = \frac{R\,T_1}{(n-1)/n} \left[\left(\frac{P_2}{P_1} \right)^{(n-1)/n} - 1 \right] \tag{5.17}$$

Again, for a real gas use the compressibility factor.

$$W_P = \frac{z\,R\,T_1}{(n-1)/n}\left[\left(\frac{P_2}{P_1}\right)^{(n-1)/n} - 1\right]$$ (5.18)

To obtain the actual work of compression, divide the polytropic work of compression by the polytropic efficiency.

$$W_A = \frac{W_P}{\eta_P}$$ (5.19)

The polytropic efficiency is defined by Equation 5.20.

$$\eta_P = \frac{(k-1)/k}{(n-1)/n}$$ (5.20)

Figure 5.15 contains plots of the polytropic efficiency for centrifugal and axial compressors as a function of the volumetric flow rate at the compressor inlet.

In addition to the isentropic and polytropic efficiencies, there are other efficiencies that affect the actual power that is delivered to the gas. The isentropic and polytropic efficiencies are hydraulic efficencies because some of the work done on the gas is consumed by fluid friction. Other frictional energy losses are caused by the compressor seal, the bearings supporting the shaft, and any gears needed to reduce or increase the rotational speed. Table 5.6 lists these efficiencies. The engineering literature either reports these efficiencies separately or combined. The seal, bearing, and gear efficiencies may be combined into a mechanical efficiency. Then, the hydraulic and mechanical efficiencies are combined into an overall efficiency for the machine, which is designated as the compressor efficiency, η_C.

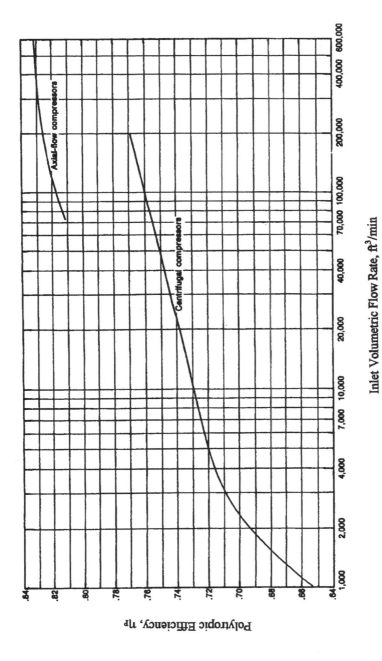

Figure 5.15 Polytropic efficiency for centrifugal and axial-flow compressors. (Source Ref. 12 with permission).

After calculating the isentropic work, then calculate the shaft work, brake work, or compressor work, i.e, the work that is delivered to the shaft of the compressor. The compressor work,

$$W_C = \frac{W_S}{\eta_A \, \eta_S \, \eta_B \, \eta_G} = \frac{W_S}{\eta_C} \qquad (5.21)$$

If the polytropic work is calculated, then the compressor work,

$$W_C = \frac{W_P}{\eta_P \, \eta_S \, \eta_B \, \eta_G} \qquad (5.22)$$

where W_C is called the compressor, shaft, or brake work.

Using either method, the compressor work should be approximately the same. Finally, the compressor horsepower,

$$P_C = \frac{m \, W_C}{550} \qquad (5.23)$$

If the compressor driver is an electric motor, then divide the compressor power by an electric-motor efficiency to size the electric motor.

Table 5.6 Compressor Gear, Bearing, and Seal Efficiencies

Component	Efficiency
Gears	0.95 to 0.98
Bearings	0.95 to 0.99
Mechanical Seals[a]	0.98 to 0.995

a) For seal power losses from 5 kW to 25kW

Optimum Compression Ratio

When compressing a gas to a high pressure, the compressor may be divided into two or more stages. The pressure ratio per stage for various compressors are given in Table 5.4. The pressure ratio for a compression stage is determined by mechanical considerations and is the concern of mechanical engineers. If the gas temperature rises too high, then the gas must be cooled after one or more compression stages. This is illustrated in Figure 5.16 for a centrifugal compressor, where the gas is removed, cooled in an external heat exchanger, called an intercooler, and returned to the compressor. The objective is to calculate the number of intercoolers because they affect the work of compression.

The minimum work of compression is obtained if the compression is isothermal. Figure 5.17 illustrates this, where the isothermal work is compared with the adiabatic work by comparing the area under to the left of the curves. Thus, the compressor should be operated isothermally, but practically it is difficult to remove heat fast enough to obtain isothermal operation because the surface area needed for heat transfer cannot be contained inside the compressor. Isothermal operation can be approached by removing the gases from the compressor periodically and cooling the gases in an intercooler, as illustrated in Figure 5.17.

After specifying the compressor inlet and discharge pressures, then the problem is to find the pressure ratio for each stage of compression. A stage may contain one or more impellers. We define a stage of compression as one or more impellers in series with no intercooler between the impellers. Thus, the compressor in Figure 5.16 contains two stages and each stage contains three impellers. If part of the gases condenses after cooling, then there will also be a

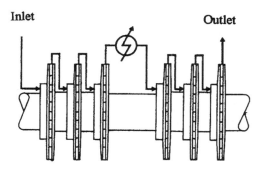

Figure 5.16 A centrifugal compressor with intercooling. (Source Ref. 21).

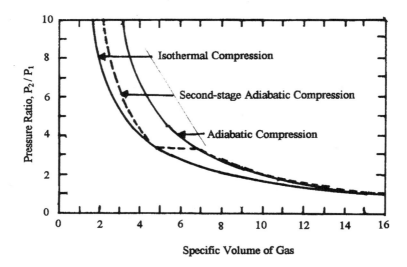

Figure 5.17 The effect of operating mode on compressor work.

phase separator after the condenser to separate the gas from the liquid. The total work of compression is equal to the sum of the work for each stage of compression. Assuming that the compressed gases are cooled to the inlet temperature of the compressor after each stage, we can use Equation 5.18 for each stage. If it is also assumed that the pressure drop across intercoolers, phase separators, and piping is negligible, then the total compressor work,

$$W_P = \frac{z\,R\,T_1}{(n-1)/n} \left[\left(\frac{P_2}{P_1} \right)^{(n-1)/n} + \left(\frac{P_3}{P_2} \right)^{(n-1)/n} + \ldots \right] \qquad (5.24)$$

where z is the average of the inlet and discharge compressibility factors.

Because the work of compression should be a minimum, differentiate Equation 5.24 with respect to P_2, and then set the derivative equal to zero.

$$\frac{\delta W_P}{\delta P_2} = \frac{z\,R\,T_1}{(n-1)/n} \left[\left(\frac{P_2}{P_1} \right)^{(n-1)/n} + \left(\frac{P_3}{P_2} \right)^{(n-1)/n} + \frac{P_3}{P_2} \right] = 0 \qquad (5.25)$$

Next, solve for the pressure ratio for the first stage, P_2/P_1.

$$\frac{P_2}{P_1} = \frac{P_3}{P_2} \tag{5.26}$$

Similarly, after differentiating Equation 5.24 with respect to P_3, and setting the derivative equal to zero, the pressure ratio for the second stage,

$$\frac{P_3}{P_2} = \frac{P_4}{P_3} \tag{5.27}$$

Therefore,

$$\frac{P_2}{P_1} = \frac{P_3}{P_2} = \frac{P_4}{P_3} = \cdots \frac{P_N}{P_{N-1}} \tag{5.28}$$

Thus, the pressure ratio for each stage should be equal to obtain the minimum work of compression. Then, the pressure ratio for each stage,

$$\frac{P_{S+1}}{P_S} = \left(\frac{P_D}{P_I}\right)^{1/N} \tag{5.29}$$

Next, consider the case where the pressure drop across the intercooler and connecting piping is significant. If the gas requires cooling between stages, the pressure drop across the cooler, separator, and piping can be approximated by $\Delta P = 0.1\, P_D^{0.7}$ for centrifugal compressors except when compressing air [58], where P_D is the discharge pressure for a stage. For air compressors $\Delta P = 0.05\, P_D^{0.7}$ [58], and for reciprocating compressors $\Delta P = 0.3\, P_D^{0.7}$ [58]. Now, we can develop a procedure for calculating the compressor work if intercooling is necessary.

Assuming a two stage compressor, the pressure at the inlet of the second stage is $P_3 = P_2 - 0.1\, P_2^{0.7}$.

$$W_P = \frac{z R T_1}{(n-1)/n}\left[\left(\frac{P_2}{P_1}\right)^{(n-1)/n} + \left(\frac{P_4}{P_2 - 0.1P_2^{0.7}}\right)^{(n-1)/n} + \cdots \right] \tag{5.30}$$

Let $\theta = (n-1)/n$ and then differentiate Equation 5.30 with respect to P_2. Then, set the derivative equal to zero to obtain the minimum work of compression. Thus.

$$\left(\frac{P_2}{P_1}\right)^{\theta-1}\frac{1}{P_1} = \left(\frac{P_4}{P_2 - 0.1 P_2^{0.7}}\right)^{\theta-1}\left[\frac{P_4}{(P_2 - 0.1 P_2^{0.7})^2}\right](1 - 0.07 P_2^{-0.3}) \tag{5.31}$$

Multiply both sides of Equation 5.31 by P_2 and rearrange the result to obtain

$$\left(\frac{P_2}{P_1}\right)^{\theta} = \left(\frac{P_4}{P_2 - 0.1\,P_2{}^{0.7}}\right)^{\theta} \frac{1 - 0.07\,P_2{}^{-0.3}}{1 - 0.1\,P_2{}^{-0.3}} \qquad (5.32)$$

Because

$$\frac{1 - 0.07\,P_2{}^{-0.3}}{1 - 0.1\,P_2{}^{-0.3}} \approx 1 \qquad (5.33)$$

we find for the first stage that

$$\frac{P_2}{P_1} = \frac{P_4}{P_2 - 0.1\,P_2{}^{0.7}} \qquad (5.34)$$

Also, the pressure at the inlet to the second stage is given by $P_3 = P_2 - 0.1\,P_2{}^{0.7}$. Thus, $P_2/P_1 = P_4/P_3$.

Similarly, for the third stage,

$$\frac{P_4}{P_3} = \frac{P_6}{P_4 - 0.1\,P_4{}^{0.7}} \qquad (5.35)$$

and the pressure at the inlet to the third stage is given by $P_5 = P_4 - 0.1\,P_4{}^{0.7}$. Thus, $P_4/P_3 = P_6/P_5$. Because $P_2 \neq P_3$ and $P_4 \neq P_5$ etc., the pressure ratio across any stage and across the entire compressor are not simply related to the number of stages as given by Equation 5.29.

The maximum allowed temperature determines the number of intercoolers. This limit is determined by the stability of seals, lubricants, and other materials that contact the gas. The gas temperature may have to be even lower than this limit if the gases are corrosive; undergo chemical reactions at high temperatures, possibly exploding; or react with the lubricating oil. High compressor operating temperatures lead to high power consumption and may promote polymerization of gases such as ethylene, acetylene and butadiene. In this case, the gas temperature should be limited to 107 °C (225 °F) [2] If the stability of the materials are the only constraint, then use the temperature limits listed in Table 5.4. If the discharge temperature exceeds this limit, then the pressure ratio across a stage must be reduced. Ulrich [23] recommends that the temperature be no greater than 200 °C

(392 °F). For low temperatures, Moens [22] reported a temperature as low as −162 °C (−260 °F).

Compressor Sizing

Table 5.7 lists the equations for the sizing a centrifugal compressor and Table 5.8 outlines the calculation procedure. The equations listed in Table 5.7 assumes three stages of compression. Equation 5.7.1 in Table 5.8 sums up the work for three stages of compression, where z is the average of the inlet and outlet compressibility factors. Equations 5.7.1 to 5.7.3 can be adjusted to include more or less stages of compression. The other equations remain the same. To determine the stages of compression and the number of intercoolers, first assume one stage of compression, and then check if the discharge-temperature limit is exceeded. If it is, then assume two stages of compression with intercooling after the first stage, and again check if the temperature limit is exceeded. Repeat the process until the gas temperature is below the maximum acceptable value after each stage of compression. The discharge temperature can be calculated from Equation 5.14, which was discussed earlier. If $R = c_p - c_V$, is substituted into Equation 5.14, the result is Equation 5.7.6 in Table 5.7,

Table 5.7 Summary of Equations for Sizing a Compressor

$$
W_{PN} = \frac{z\,R'\,T_1'}{(n-1)/n} \left[\left(\frac{P_2}{P_1'}\right)^{(n-1)/n} - 1 + \left(\frac{P_4}{P_2 - 0.1\,P_2^{0.7}}\right)^{(n-1)/n} - 1 \right.
$$

$$
\left. + \left(\frac{P_6'}{P_4 - 0.1\,P_4^{0.7}}\right)^{(n-1)/n} - 1 \right] \tag{5.7.1}
$$

for two stages of compression:

$$
\frac{P_2}{P_1'} = \frac{P_4}{P_2 - 0.1\,P_2^{0.7}} \tag{5.7.2}
$$

for three stages of compression:

$$
\frac{P_4}{P_3} = \frac{P_6'}{P_4 - 0.1\,P_4^{0.7}} \tag{5.7.3}
$$

$$
P_2/P_1 = P_4/P_3 \tag{5.7.4}
$$

Table 5.7 Continued

$$W_{CN} = \frac{W_{PN}}{\eta_P \, \eta_S' \, \eta_B' \, \eta_G'} \tag{5.7.5}$$

$$W_{CN} = \frac{R'}{(k-1)/k} \, (T_D - T_1') \tag{5.7.6}$$

$$\frac{n-1}{n} = \frac{k-1}{\eta_P k} \tag{5.7.7}$$

$$\eta_P = f(V_1) \; — \; \text{Figure 5.15} \tag{5.7.8}$$

$$V_1 = v_1 \, m_1' \tag{5.7.9}$$

$$P_{CP} = W_{CN} \, m_1' \tag{5.7.10}$$

Thermodynamic Properties

$$z = (z_1 + z_D)/2 \tag{5.7.11}$$

$$z_1 = f(T_{R1}, P_{R1}) \; — \; (\text{Figure 5.14}) \tag{5.7.12}$$

$$z_D = f(T_{RD}, P_{RD}) \; — \; (\text{Figure 5.14}) \tag{5.7.13}$$

$$T_{R1} = T_1'/T_C \tag{5.7.14}$$

$$T_{RD} = T_D'/T_C \tag{5.7.15}$$

$$P_{R1} = P_1'/P_C \tag{5.7.16}$$

$$P_{RD} = P_D'/P_C \tag{5.7.17}$$

$$T_C = \sum_i y_i' \, T_{Ci} \tag{5.7.18}$$

$$P_C = \sum_i y_i' \, P_{Ci} \tag{5.7.19}$$

$$k = \sum_i y_i' \, k_i \tag{5.7.20}$$

$$T_{Ci} = f(\text{chemical compound}') \; — \; \text{from Table 5.5} \tag{5.7.21}$$

$$P_{Ci} = f(\text{chemical compound}') \; — \; \text{from Table 5.5} \tag{5.7.22}$$

Table 5.7 Continued

$$k_i = f(\text{chemical compound'}) \quad\text{— from Table 5.5} \qquad (5.7.23)$$

$$v_1 = \frac{z_1 R' T_1'}{P_1'} \qquad (5.7.24)$$

Variables

W_{PN} - W_{CN} - z - z_1 - z_D - P_2 - P_3 - P_4 - P_C - P_{Ci} - P_{CP} - T_C - T_{Ci} - P_{R1} - P_{RD} - T_D - T_{R1} - T_{RD} - k - k_i - v_1 - V_1 - η_P - n

Degrees of Freedom

$F = 24 - 24 = 0$

Table 5.8 Compressor Sizing Procedure

1. Calculate T_C, P_C and k from Equations 5.7.18 to 5.7.23.

2. Calculate z_1 from Equations 5.7.12, 5.7.14, and 5.7.16.

3. Calculate V_1, v_1, η_P and $(n-1)/n$ from Equations 5.7.7 to 5.7.9, and 5.7.24.

4. Assume one stage of compression ($N = 1$, $P_D = P_2$).

5. Calculate W_{P1} from Equation 5.7.1.

6. Calculate the discharge temperature, T_2 ($T_D = T_2$), from Equations 5.7.5 and 5.7.6.

7. If $T_2 > T_{max}$, assume two stages of compression ($N = 2$).

8. Calculate P_2 ($P_D = P_4$) from Equation 5.7.2.

9. Calculate W_{P2} from Equation 5.7.1.

10. Calculate the discharge temperature, T_4 ($T_D = T_4$), from Equations 5.7.5 and 5.7.6.

11. Caculate the average compressibility factor z, from Eq. 5.7.11.

Table 5.8 Continued

12. If $T_4 > T_{max}$, assume three stages of compression (N = 3).

13. Calculate P_2 and P_4 ($P_D = P_6$) by solving Equations 5.7.2, 5.7.3, and 5.7.4 simultaneously.

14. Calculate W_{P3} from Equation 5.7.1.

15. Calculate the discharge temperature, T_6 ($T_D = T_6$), from Equations 5.7.5 and 5.7.6.

16. If $T_6 < T_{max}$, calculate z_D ($z_D = z_6$) from Equations 5.7.13, 5.6.15, and 5.7.17.

17. Calculate z from Equation 5.7.11.

18. Recalculate W_{P3} from Equation 5.7.1 using the new value of z.

19. Recalculate W_{CN} from Equation 5.7.5. T_6 will change and could be recalculated from Equation 5.7.6, but in most cases this will not be necessary.

20. Calculate P_{CP} from Equation 5.7.10.

Example 5.2 Calculation of Compressor Power

This problem was taken from Reference 5.56. Assume that the electric motor efficiency is 94%. Calculate the power required for an electric motor drive for a compressor to compress a process gas containing propane, butane and methane from 5 °C (41°F) and from 1.4 to 7.0 bar (20.3 to 101 psia). The composition of the gas in mole percent is: C_3H_8 = 89.0, n-C_4H_{10} = 6.0, and C_2H_6 = 5.0. The flow rate is 1090 kgmol/h (2403 lbmol/h).

Follow the procedure outlined in Table 5.8. First, calculate the mole fraction averages of the heat capacity ratio, critical temperature, and critical pressure from Equations 5.7.18 to 5.7.23. Critical pressures and temperatures are given in Table 5.5.

k = 0.89 (1.13) + 0.06 (1.09) + 0.05 (1.19) = 1.131

T_C = 0.89 (370.0) + 0.06 (425.6) + 0.05 (305.6) = 370.1 K (666 °R)

P_C = 0.89 (42.5) + 0.06 (38.0) + 0.05 (48.8) = 42.55 bar (617 psia)

To determine the compressibility factor at the compressor inlet, z_1, first calculate the reduced temperature and pressure.

$$T_{R1} = \frac{T_1}{T_C} = \frac{278.2}{370.1} = 0.7517$$

$$P_{R1} = \frac{P_1}{P_C} = \frac{1.4}{42.55} = 0.03290$$

From Figure 5.14, $z_1 = 0.97$.
At the compressor inlet, the specific volume of the gas (Equation 5.7.24),

$$v_1 = \frac{z_1 R T_1}{P_1} = \frac{0.97}{1} \frac{0.08314}{1} \frac{\text{bar-m}^3}{\text{kgmol-K}} \frac{1}{1.4 \text{ bar}} \frac{278.2 \text{ K}}{1}$$

$$= 16.03 \text{ m}^3/\text{kgmol} \ (257 \text{ ft}^3/\text{lbmol})$$

At the inlet conditions, the volumetric flow rate of the gas from Equation 5.7.9,

$$V_1 = \frac{16.03}{1} \frac{\text{m}^3}{\text{kgmol}} \frac{1090.0 \text{ kgmol}}{1 \text{ h}} = 1.747 \times 10^4 \text{ m}^3/\text{h}$$

or

$$V_1 = \frac{1.747 \times 10^4 \text{ m}^3}{1 \text{ h}} \frac{1 \text{ h}}{60 \text{ min}} \frac{35.31 \text{ ft}^3}{1 \text{ m}^3} = 1.028 \times 10^4 \text{ ft}^3/\text{min}$$

From Figure 5.15 at V_1 the polytropic or hydraulic efficiency, $\eta_P = 0.73$. Therefore, from Equation 5.7.7,

$$\frac{n-1}{n} = \frac{(k-1)/k}{\eta_P} = \frac{(1.131-1)/1.13}{0.73} = 0.1587$$

Assume one stage of compression, and calculate the polytropic work of compression given by Equation 5.7.1. Because the discharge temperature is unknown, the average compressibility factor for Equation 7.7.11 cannot be calculated until work done is calculated. To start the calculation, use the compressibility factor at the inlet for one stage of compression, N = 1, in Equation 5.7.1. Therefore,

$$W_{P1} = \frac{z_1 \, R \, T_1}{(n-1)/n} \left[\left(\frac{P_2}{P_1} \right)^{(n-1)/n} - 1 \right]$$

$$W_{P1} = \frac{0.97}{0.1587} \quad \frac{8314.0}{1} \quad \frac{J}{kgmol\text{-}K} \quad \frac{278.2 \, K}{1} \left[\left(\frac{7.0}{1.4} \right)^{0.1587} - 1 \right]$$

$W_{P1} = 4.114 \times 10^6$ J/kgmol (1.770×10^3 Btu/lbmol)

Now, calculate the discharge temperature from Equation 5.7.6.

$(k-1)/k = (1.131 - 1)/1.131 = 0.1158$

$$T_2 = \frac{0.1158}{1} \quad \frac{4.114 \times 10^6}{1} \quad \frac{J}{kgmol} \quad \frac{1}{8314.0} \quad \frac{kgmol\text{-}K}{J} + 278.2 = 335.5 \text{ K } (605 \text{ °R})$$

Therefore, T_2 is below 450 K (810 °R) given in Table 5.4. Thus, intercooling is not required, and a single compression stage is adequate.
Now, find the compressibility factor at the compressor outlet, z_2, first calculate the reduced temperature and pressure.

$$T_{R2} = \frac{T_2}{T_C} = \frac{335.5}{370.1} = 0.9065$$

$$P_{R2} = \frac{P_2}{P_C} = \frac{7.0}{42.55} = 0.1645$$

From Equation 5.7.12, $z_2 = 0.93$, and the average value of the compressibility factor,

$z = (0.97 + 0.93)/2 = 0.95$

which is not significantly different than the inlet value. It is not necessary to recalculate the specific volume and volumetric flow rate.

From, Equation 5.7.5, the shaft work,

$$W_{CN} = \frac{4.114 \times 10^6}{0.73\,(0.98)\,(0.95)\,(0.95)} = 6.372 \times 10^6 \text{ J/kgmol } (2.740 \times 10^3 \text{ Btu/lbmol})$$

where conservative values for the seal, bearing and gear efficiencies were taken from Table 5.6.

From Equation 5.7.10, the total shaft power,

$$P_{CP} = \frac{6.372 \times 10^6}{1} \frac{J}{kgmol} \frac{1\ W}{J/s} \frac{1090.0\ kgmol}{1} \frac{1}{h} \frac{h}{3600\ s} \frac{1}{1000} \frac{kW}{W}$$

$$P_{CP} = 1929 \text{ kW } (2590 \text{ hp})$$

The total power required by the electric-motor drive is,

$$P_E = \frac{P_{CP}}{\eta_E} = \frac{1.929 \times 10^6\ W}{0.94} \frac{1}{745.7} \frac{hp}{W} = 2752 \text{ hp}$$

Because electric motors are available in standard sizes from Table 5.1a, select a standard 3000 hp (2.24×10^3 kW) motor. This choice results in a safety factor of 9%.

COMPRESSOR AND PUMP DRIVERS

After calculating the work of compression, a suitable driver must be selected. A compressor driver accounts for about half the cost of a compressor installation [22]. The possible drivers are electric motors, engines, and turbines. Among the electric motors are the synchronous, squirrel cage induction, and wound-rotor induction. The engines include reciprocating steam engines, gas engines, and the oil engines, and turbines consist of steam and gas turbines [24]. The reciprocating steam engine was one of the first drivers, but it is seldom used today [36] and thus will not be given further consideration. The electric motor and steam turbine are the most common, and will be discussed in detail. The gas turbine is used to a lesser extent. Some characteristics of electric motors, steam, and gas turbines are listed in Table 5.9.

Table 5.9: Characteristics of Compressor and Pump Drivers

Driver[a]	Power Range[d] hp	Speed rpm	SpeedControl	Efficiency
Squirrel Cage Induction Motor	1 to 5,000	3600/N (less 2%) N^e = 1 to 8	Constant Speed	10 hp – 86 % 100 hp – 91 % 1,000 hp – 94%
Wound Rotor	1 to 1,500	3,550 - 1,750 - 1,150	100 to 60 %	10 hp – 86 %
Induction Motor		870 - 700 - 580		100 hp – 91 %
	1,500 to 2,500	1,750 - 1,150 - 870	100 to 60%	1,000 hp – 94%
Synchronous Motor	100 to 20,000	3600/N	Constant Speed	90 to 97
Steam Turbine (all)	10 to 20,000	2,000 to 15,000	100 to 35 %	η_A = 50 to 76%
Single Stage[b]	up to 1,000	1,000 to 7,000		
High Back Pressure[b] Single or Multistage	150 to 3,000	5,000 to 10,000		
Multistage[b] Medium Large	750 to 5,000 5,000 to 60,000	up to 10,000 3,000 to 16,000		
Gas Turbine[c]	3,000 to 20,000	10,000		

a) Source: Reference 24 except where indicated.   e) N is the number of poles.
b) Source: Reference 26.
c) Simple cycle.
d) To convert to kW multiply by 0.7457.

Drivers can be grouped according to the type of energy supplied – electrical, expansion of a high pressure gas, and expansion of a high pressure liquid. An important consideration in the selection of a driver is to match the speed of the driver with the speed of the machine. If it is necessary to run both units at different speeds for technical or economic reasons, then gears will be needed to increase or decrease the speed of the driver. Fans for many applications are V-belt driven.

Electric Motors

Most chemical-plant-size compressors are electrically driven [43]. Moore [25] discusses the characteristics of squirrel-cage induction and synchronous electrical motors. Wound rotor induction motors have not been used for compressor drives. For 370 to 4500 kW (500 to 6,000 hp), the induction motors are the first choice. The squirrel-cage induction motor is the most commonly used driver in the process industries from 1/8 to 1,5000 hp (0.0932 to 1,120 kW [25]. From 15,000 hp (149 to 11,200 kw) the synchronous motor could be used [25]. If the compressor is operated at 7,500, 11,000, and 23,000 rad/s (1,200, 1,800 and 3,600 rpm), no step-up gears are required. The least costly speed for an induction motor is 1,000 rad/s (1800 rpm) so that this speed is usually selected. Step-up gears are used to obtain higher speeds.

To calculate the size of an electric motor, divide the compressor shaft power by an electric-motor efficiency. Efficiencies for electric motors are given in Table 5.9. The size of electric motors are standardized according to horsepower, as shown in Table 5.10. If less than the standard horsepower is calculated, then the next standard horsepower is selected.

Table 5.10 Standard Electric-Motor Sizes

Horsepower[a]
1/20, 1/12, 1/8, 1/6, 1/4, 1/3, 1/2, 3/4, 1/2, 1
1-1/2, 2, 3, 5, 7-1/2, 10, 15, 20, 25, 30, 40, 50, 60, 75, 100
125, 150, 200, 250, 300, 350, 400, 450, 500, 600, 700, 800, 900, 1000
1250, 1500, 1750, 2000, 2250, 2500, 3000, 3500, 4000, 4500, 5000
and up to 30,000

a)To convert to kW multiply by 0.7457 .

Expanders

The energy from high-pressure gas streams may be used to drive compressors or pumps. High pressure gases range in temperature from the low-temperature cryogenic fluids to high-temperature combustion gases. The energy source could be the process stream itself or an external working fluid such as steam. Frequently, the energy source is high-pressure steam, but the process engineer should seek opportunities to conserve energy by utilizing the energy from high-pressure process streams whenever possible. In either case, the energy for compression or pumping is obtained by expanding the gas through an expander. Like dynamic compressors, gas expanders are available in either the radial or axial-flow design, where the radial-flow design is used for low flow rates and high-pressure differences and the axial-flow types at high flow rates and low-pressure differences (1 to 40 bar) (0.9869 to 39.5 atm) [28].

The radial-flow expander consists of inflow and outflow types. In the radial-outflow type, the gas flows from the center to periphery of the impeller. The radial-outflow expander is used for very low enthalpy drops, 58 to 70 kJ/kg (25 to 30 Btu/lb) per stage [29]. The radial-inflow expander is similar to a centrifugal compressor used in reverse, i.e., the gas flows radially inward from the periphery of the impeller, exhausting approximately axially. Most radial turbines are of the inflow type. One example of the radial outflow type is the Ljungstrom turbine, which usually uses steam in small in-house generating plants, producing 10 to 35 MW (13,400 to 46,900 hp) of power [30]. Similarly, the axial expander resembles an axial compressor where the gas flows through an annular passage in a direction that is substantially parallel to the axis of the shaft. In both cases, however, the expander blade design differs from the compressor blade design. An expander stage consists of a nozzle followed by a rotor. The purpose of a nozzle is to accelerate a fluid, converting pressure into kinetic energy, and then guide the gas into the rotor where kinetic energy is converted into work. The gas velocity varies from above to below the speed of sound. For a radial flow expander, the nozzle may be a fixed set of vanes, a variable set of vanes, or no vanes at all [27]. A radial-flow expander is shown in Figure 5.18.

Steam Turbines

If the working fluid is steam, then the expander is called a steam turbine. Steam turbines are available as single and multistage units having several blade designs and arrangements [31]. If the power generated is too large for a single stage turbine, or if it is necessary to expand the steam more than once to improve the turbine efficiency, then use a multistage turbine. Inlet steam is limited to about 42 bar (615 psia) and 440 °C (750 °F) [31].

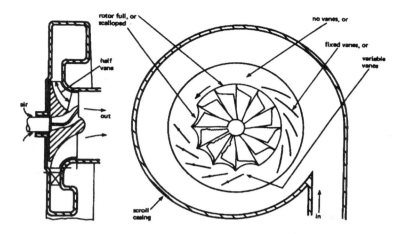

Figure 5.18 A radial-inflow turbine. (Source Ref. 27 with permission).

If the steam is expanded to atmospheric pressure or above, the turbine is called noncondensing. Noncondensing turbines are used when the exhaust steam is needed for process heating. On the other hand, if the steam is expanded to below atmospheric pressure, the turbine is called condensing. Usually, the exhaust pressure is between 0.0040 to 0.0053 bar (3 to 4 mm Hg, 0.058 to 0.0769 psia), but can be anywhere from 0.0013 to 0.020 bar (1 to 15 mm Hg, 0.0189 to 0.29 psia). In condensing turbines, the exhaust steam may contain as much as 15 % moisture by mass, but 10 % is common practice [32].

Because centrifugal and axial compressors are high-speed machines, they could be driven by steam turbines, which are designed for the same high speeds and thus may be directly coupled. To improve efficiency, however, recent developments in steam-turbine technology are in the direction of achieving higher speeds, which will require gears to match the speed of the driven machine [33]. About 2 to 3% of the shaft power is lost by gear friction [26].

To size a steam turbine requires calculating the steam flow rate, which will eventually be needed to size a steam boiler. A summary of equations for sizing a steam turbine are given in Table 5.11 and the calculation procedure in Table 5.12. In this case, the mass balance is simple in that the steam flow rate into the turbine is equal to the steam and the condensate flow rate out of the turbine.

$$m_1 = m_{2L} + m_{2V} \tag{5.36}$$

If Equation 5.36 is divided by m_1 we find that

$$\frac{m_{2L}}{m_1} + \frac{m_{2V}}{m_1} = 1 \tag{5.37}$$

but $x_L = m_{2L}/m_1$ the mass fraction of condensate and $x_V = m_{2V}/m_1$ mass friction of steam. Thus,

$$x_L + x_V = 1 \tag{5.38}$$

This obvious relationship is used in Table 5.11 to obtain mass-fraction averages of thermodynamic properties of steam-condensate mixtures. The macroscopic energy balance, is used to obtain the steam flow rate. Like compressors, the kinetic and potential energy terms are not significant, and the expansion is assumed to be adiabatic.

Table 5.11 Summary of Equations for Sizing Steam Turbines

Subscripts: Isentropic process, s
First subscript: Entering steam, 1 — Exit steam, 2
Second subscript: Condensate, L — Steam, V

Energy Balance

$$P_C' = \eta_T \, m \, (h_1 - h_{2S}) \tag{5.11.1}$$

$$\eta_T = \frac{h_1 - h_2}{h_1 - h_{2S}} \quad \text{— definition} \tag{5.11.2}$$

$$s_1 = s_{2S} \quad \text{— isentropic process} \tag{5.11.3}$$

Single Stage

$$\eta_T = (1 - x/2)\,(\eta_B / c_S) \tag{5.11.4}$$

$$\eta_B = f(\omega', P_1', P_C') \quad \text{— Figure 5.19} \tag{5.11.5}$$

$c_S = f\,(\text{degrees of superheat}) \;-\!-\; \text{Figure 5.2}$ \hfill (5.11.6)

<u>Multistage</u>
$$\eta_T = c_S\, c_P\, \eta_B \tag{5.11.7}$$

$$\eta_B = f\,(P_1{}', P_C{}', x) \;-\!-\; \text{Figure 5.21} \tag{5.11.8}$$

$$c_S = f\,(\text{degrees of superheat}') \;-\!-\; \text{Figure 5.21} \tag{5.11.9}$$

$$c_P = f\,(\text{condensing pressure}') \;-\!-\; \text{Figure 5.21} \tag{5.11.10}$$

Thermodynamic Properties

$$s_{2S} = x_S\, s_{2LS} + (1 - x_S)\, s_{2VS} \tag{5.11.11}$$

$$h_{2S} = x_S\, h_{2LS} + (1 - x_S)\, h_{2VS} \tag{5.11.12}$$

$$h_2 = x\, h_{2L} + (1 - x)\, h_{2V} \tag{5.11.13}$$

$$s_1 = f\,(T_1{}', P_1{}') \tag{5.11.14}$$

$$s_{2LS} = f\,(P_2{}') \tag{5.11.15}$$

$$s_{2VS} = f\,(P_2{}') \tag{5.11.16}$$

$$h_1 = f\,(T_1{}', P_2{}') \tag{5.11.17}$$

$$h_{2LS} = f\,(P_2{}') \tag{5.11.18}$$

$$h_{2VS} = f\,(P_2{}') \tag{5.11.19}$$

$$h_{2L} = f\,(P_2{}') \tag{5.11.20}$$

$$h_{2V} = f\,(P_2{}') \tag{5.11.21}$$

Variables

<u>Single Stage</u>

$\eta_T - \eta_B - c_S - m - h_1 - h_2 - h_{2S} - h_{2LS} - h_{2VS} - h_{2L} - h_{2V} - s_1 - s_{2S} - s_{2LS} - s_{2VS} - x - x_S$

Table 5.11 Continued

Multistage

η_T - η_B - c_S - c_P - m - h_1 - h_2 - h_{2S} - h_{2LS} - h_{2VS} - h_{2L} - h_{2V} - s_1 - s_{2S} - s_{2LS} - s_{2VS} - x - x_S

Degrees of Freedom

Single stage

F = 17 – 17 = 0

Multistage:

F = 18 – 18 = 0

Table 5.12 Calculating Procedure for Sizing Steam Turbines

1. Obtain the thermodynamic properties (Equations 5.11.14 to 5.11.21) at the inlet and discharge of the turbine from the steam tables (44).

2. Calculate the mass fraction of water in the turbine exit stream, x_S, assuming an isentropic expansion of the steam from P_1 to P_2 (Equations 5.11.3 and 5.11.11).

3. Obtain the turbine efficiency, η_B, for a single-stage turbine from Equation 5.11.5 or Equation 5.11.8 for a multistage turbine.

4. Obtain the correction factor for superheated steam, c_S, for a single-stage turbine from Equation 5.11.6. For a multistage turbine, obtain the correction factors c_S and c_P from Equation 5.11.9 and 5.11.10

5. Calculate the exit enthalpy for an isentropic expansion, h_{2S}, from Equation 5.11.12.

6. Calculate the actual mass fraction of water in the exit steam, x, for a single-stage turbine from Equation 5.11.2, 5.11.4, and 5.11.13. For a multistage turbine calculate x, from Equation 5.11.2, 5.11.7, and 5.11.13.

7. Calculate the steam flow rate, m, from Equations 5.11.1.

Thus,

$$W_T = - \Delta h \tag{5.39}$$

The turbine efficiency,

$$\eta_T = \frac{\Delta h}{\Delta h_S} \tag{5.40}$$

where Δh_S is the change in enthalpy for an isentropic expansion.
Therefore,

$$W_T = - \eta_T \Delta h_S = - \eta_T (h_{2S} - h_1) = \eta_T (h_1 - h_{2S}) \tag{5.41}$$

After multiplying Equation 5.41 by the steam flow rate, m, we obtain

$$m W_T = \eta_T m (h_1 - h_{2S}) \tag{5.42}$$

Because power is the rate of doing work, $P_T = m W_T$, Equation 5.42 becomes

$$P_T = \eta_T m (h_1 - h_{2S}) \tag{5.43}$$

When sizing steam turbines, Molich [34] recommends a safety factor of 10%.

Efficiencies for single-stage turbines are given in Figure 5.19 for noncondensing, dry, saturated steam. As it can be seen, the turbine efficiency, which includes mechanical as well as hydraulic losses, depends on brake or shaft power, steam pressure, and turbine speed. To take into account the reduction in efficiency caused by condensation, an arbitrary method, quoted in Reference 14, is to multiply the turbine efficiency by the average of the vapor mass fraction entering and leaving the turbine. Also, the effect of superheated steam on the turbine efficiency is taken into account by dividing by a correction factor, c_S, given in Figure 5.20. Thus, the turbine efficiency of a single-stage turbine, given by Neerkin [31], is

$$\eta_T = \left(1 - \frac{x}{2}\right) \frac{\eta_B}{c_S} \tag{5.44}$$

where x is the mass fraction of water, and η_B is the single-stage isentropic efficiency from Figure 5.19.

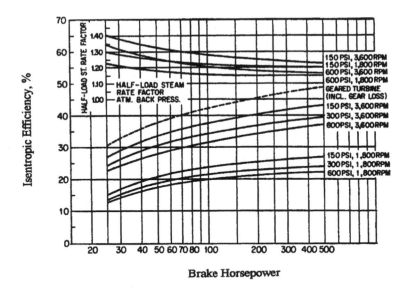

Figure 5.19 Isentropic efficiencies for single-stage noncondensing turbines — dry saturated steam. From Ref. 31 with permission.

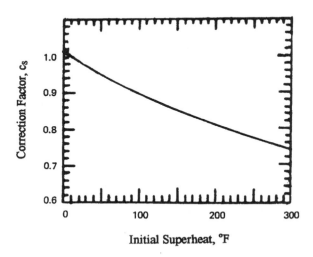

Figure 5.20 Efficiency correction factors for single-stage turbines — superheated steam. From Ref. 31 with permission.

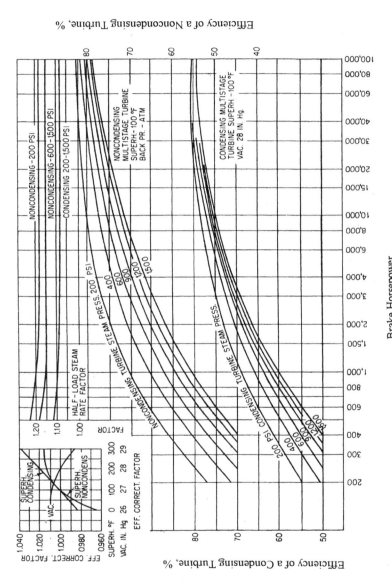

Figure 5.21 Isentropic efficiencies for multistage steam turbines. From Ref. 34 with permission.

A single-stage turbine is limited to about 2,500 hp (1,490 kW) [31]. The efficiencies plotted in Figures 5.19 and 5.20 are used for estimating steam flow rates. Methods for determining more accurate efficiencies and steam flow rates are given in Reference 5.31.

When higher power than a single-stage turbine can provide is needed, then use a multistage turbine for greater efficiency and hence steam economy. Turbine efficiencies for both condensing and noncondensing, multistage steam turbines are given in Figure 5.21. These efficiencies must be corrected for the effect of using superheated steam and the discharge pressure, if it is in the vacuum region. Thus,

$$\eta = c_S \, c_P \, \eta_B \qquad\qquad (5.45)$$

where η_B is the efficiency of dry saturated steam, obtained from Figure 5.19. The superheat correction factor, c_S, and the pressure correction factor, c_P, are also obtained from Figure 5.21. In the upper part of Figures 5.19 and 5.21, a half-load, steam-rate factor is plotted. When the turbine is delivering half its rated power, the steam flow rate will be equal to this factor times one half the full-steam flow rate.

The ideal final state, designated with a subscript s, is reached by conducting an isentropic process from state one to state two. This process is given by Equation 5.11.3 in Table 5.11.

If the steam leaves the turbine part liquid and vapor, the properties of the exit stream are determined by a mass fraction average of the properties of pure liquid and vapor as given by Equation 5.11.11 to 5.11.13. According to the phase rule, these properties are a function of one thermodynamic variable. Because the inlet steam is superheated, the properties depend on two variables as given by Equation 5.11.14 and 5.11.17. Problem 5.3 illustrates the calculation procedure given in Table 5.12.

Example 5.3 Sizing a Steam-Turbine Drive for a Centrifugal Compressor

Superheated steam at 13.0 bar (189 psi) and 260.0 °C (500 °F) is being considered to drive a compressor. The shaft power required by the compressor, P_C, is 100 hp (74.6kW). If a steam turbine rotates at 3,600 rpm and exhausts at 0.15 bar (2.18 psi), what is the power output, steam rate, and steam condensed.

Follow the calculation procedure outline in Table 5.12. First, obtain the thermodynamic properties (Equations 5.8.14 to 5.8.21) at the inlet and discharge of the turbine from the steam tables [44]. These are:

at $P_1 = 1.30$ MPa and $T_1 = 260.0$ °C, $h_1 = 2954.0$ kJ/kg, $s_1 = 6.8301$ kJ/kg-K, and at saturation T = 191.6 °C

at $P_2 = 0.015$ MPa, $T_2 = 45.81$ °C, $h_{2L} = 191.83$ kJ/kg, $h_{2v} = 2584.7$, $s_{2L} = 0.6493$ kJ/kg-K, $s_{2v} = 8.1502$ kJ/kg-K

For an isentropic process (Equation 5.11.3), $s_1 = s_{2S}$. Therefore, from Equation 5.11.11,

$$s_{2S} = 6.8301 = 0.6493 \, x_S + 8.1502 \, (1 - x_S)$$

The mass fraction of water in the exit stream for an isentropic process, x_S, is equal to 0.1760.

From Equation 5.11.12 for an isentropic process, the exit enthalpy for the part-water, part-vapor stream,

$$h_{2S} = 0.1760 \, (191.83) + 0.824 \, (2584.7) = 2164.0 \text{ kJ/kg } (930 \text{ Btu/lb})$$

The compressor power is within the range of single-stage turbines. If it is assumed that the compressor will be directly coupled to the steam turbine, the compressor shaft power must be matched by the steam-turbine shaft power. Allowing for a 10% safety factor, the power delivered to the compressor will be 110 hp (82.0 kW). From Equation 5.11.5, the turbine efficiency, η_B, at 36,000 rpm, 110 hp (82.0 kW), and 1.30 MPa (188.5 psi), is 36%.

The correction factor for superheated steam, c_S, is obtained from Equation 5.11.6 (Figure 5.20). The correction factor depends on the degrees of superheat at the turbine inlet, and it is defined as the difference between the steam temperature and the saturation temperature.

$$\text{superheat} = (260.0 - 191.6) \,^\circ\text{C} \, (9\,^\circ\text{F} / 5\,^\circ\text{C}) = 123.1\,^\circ\text{F}$$

From Figure 5.20, $c_S \approx 0.87$.

After solving Equation 5.11.2, 5.11.4, and 5.11.13 for x by eliminating h_2 and η_T, we obtain

$$x = \frac{h_{2V} - h_1 + (\eta_B / c_S)(h_1 - h_{2S})}{h_{2V} - h_{2L} + (\eta_B / 2 \, c_S) \, h_1 - h_{2S}}$$

$$x = \frac{2584.7 - 2954.0 + (0.36 / 0.87)(2954.0 - 2164.0)}{584.7 - 191.83 + [0.36 / 2 \, (0.87)](2954.0 - 2164.0)} = -0.0762$$

A negative sign means that no condensation occurs. This can also be shown by calculating the actual enthalpy of the exit steam from Equation 5.11.2. If $x = 0$, $\eta_T = \eta_B / c_S$, as can be seen from Equation 5.11.4. Thus, from Equation 5.11.2, the actual enthalpy,

$$h_2 = h_1 - (\eta_B / c_S)(h_1 - h_{2S}) = 2954.0 - (0.36 / 0.87)(2954.0 - 2164.0)$$

= 2627 kJ/kg (1129 Btu/lb)

Therefore, $h_2 > h_{2v}$, which means that the steam leaves the turbine superheated. Although in the isentropic process condensation occurs, friction in the turbine increases the steam temperature and therefore the enthalpy of the steam, preventing condensation.

The steam flow rate can now be calculated from Equation 5.11.1. The adjusted shaft power is 110 hp (82.0 kW). The steam flow rate,

$$
m = \frac{0.87}{0.36} \quad \frac{110.0\ \text{hp}}{1} \quad \frac{745.7\ \text{J/s}}{1\ \text{hp}} \quad \frac{1}{(2954.0 - 2164.0)} \quad \frac{\text{kg}}{\text{J}} = 250.9\ \text{kg/s (553 lb/s)}
$$

Gas Turbines

If the working fluid is a combustion gas, formed by burning a gaseous or liquid fuel, the expander is called a gas turbine. The gas turbine is a relatively recent developed driver for process plants. Figure 5.22 shows a flow diagram for a simple-cycle gas turbine. Fuel is burned with excess compressed air in a combustor at constant pressure. The gas entering the turbine is limited to 760 to 1,000 °C (1,400 to 1,830 °F) because of temperature limits on the materials of construction [37]. The gases are maintained in this temperature range by using excess air. After combustion, the pressurized gas expands through a turbine to about 0.025 bar (10 in H_2O) above atmospheric pressure to allow for the exit-duct losses [38]. The gas turbine drives the air compressor and provides excess power for other process machinery. Inlet-duct pressure losses are about 0.0075 bar (3 in H_2O, 0.109 psia) [36]. The combustion gas typically contains 14 to 19 % oxygen [37]. An efficiently-operated system requires recovering the enthalpy of the hot exhaust gas. The ratio of the output power to the total power generated varies from 0.33 to 0.50 [34].

The gas turbine requires an electric starting motor or steam turbine for starting until the gas-turbine speed reaches 55 % of its final speed and becomes self-supporting. For most applications gears are required to match the speed of the driven equipment [34]. Molich [34] recommends a safety factor of 10 % when justifying sizing gas turbines. The gas turbine for process applications ranges from 1,000 (746 kW) to greater than 100,000 hp (74,600 kW) [34].

Turboexpanders

When the source of high-pressure gas is a process stream, the expander is referred to as a turboexpander. Some process applications of turboexpanders are: the separation of air into oxygen and nitrogen, recovery of condensable hydrocarbons from natural gas, liquefaction of gases, and energy recovery from high pressure gas streams. After conducting chemical reactions at high pressures, the pressure of the effluent stream must be eventually reduced. For example, in the process for synthesizing methanol, the purge gas from the synthesis loop is used as a fuel at 3 to 4

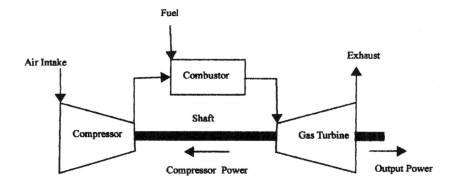

Figure 5.22 A simple cycle gas turbine. From Ref. 34.

bar (43.5 to 58 psia), but the synthesis loop is at 100 to 300 bar (1,450 to 4,350 psia). Thus, the pressure could be dropped through a turbine, partially recovering the energy of the high-pressure stream [28]. Turboexpanders operate at pressures up to 3,000 psia (207 bar) with isentropic efficencies of 75 to 88% [39]. To conserve energy, the turboexpander is frequently used in expanding gas streams in cryogenic processes. For half of these applications, the stream condenses producing, in some cases, more than 50 % by mass of liquid or better [39].

Hydraulic Turbines

Hydraulic turbines are used for recovering energy from high-pressure liquid streams. A common process application is an absorber-stripper combination. In this application, a gas is absorbed in a solvent at a high pressure, where absorption is favored. Then, the solvent is stripped of the absorbed components at a low pressure, where stripping is favored, to recover the solvent. Thus, the energy of the high-pressure solvent stream from an absorber can be partially recovered by a hydraulic turbine. There are three types of hydraulic turbines, the Pelton-wheel turbine, the Francis turbine, and the propeller reaction turbine, an axial type turbine. The propeller reaction turbine is used in hydroelectric applications and will not be considered further. The Pelton-wheel and Francis turbines are shown in Figure 5.23.

In the Pelton-wheel turbine, used for low flow rates, a high-pressure liquid flows through a nozzle to convert the pressure to a high velocity jet which impinges on an impulse wheel or runner. This turbine may contain one to four nozzles. Above approximately 800 m^3/h (28,200 ft^3/h) the Francis turbine becomes

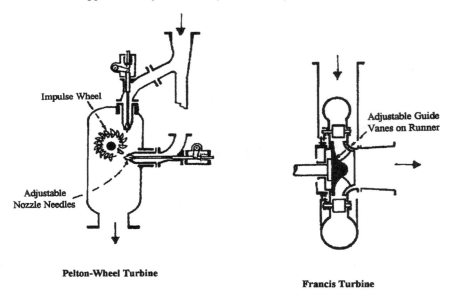

Figure 5.23 Hydraulic turbines. From Ref. 40 with permission.

more economical [28]. The Francis turbine contains a stationary guide case where pressure is partially converted into kinetic energy. In the runner, pressure is further converted into kinetic energy.

Radial-flow centrifugal pumps running backwards can also be used in place of a hydraulic turbine. Although pumps are less expensive, the power recovered by a hydraulic turbine can exceed that of reverse-running pump by 10% or more [28]. Buse [41] has outlined a method for selecting a centrifugal pump that will give the best efficiency when operating as a turbine. The hydraulic efficiency of pumps used as turbines are usually 5 to 10 % below the value given for the pump [39].

The turboexpander is also a hydraulic turbine used for flashing liquids and liquids releasing dissolved gases as discussed by Swearingen [42]. Capacities range from 50 to 1,000 hp (39.3 to 746 kW), suction pressures from 1,000 to 1,500 psia (69 to 103 bar) and discharge pressures from 50 to 200 psia (3.45 to 13.79 bar). In an illustrative example, Swearingen cites an isentropic efficiency of 67 % at a rotational speed of 31,000 rpm. The recovered power can be used to drive a pump or compressor.

The energy available in a high-pressure liquid or gas process stream must be balanced by the energy required by a compressor or pump. Thus, the power delivered by an energy-recovery turbine must be absorbed at the same rate by the compressor or pump. Also, a gas turbine, turboexpander, and hydraulic turbine require an electric motor or steam turbine for starting the driven machinery. Figure 5.22 illustrates a system for starting an axial compressor which is driven by an expander. At startup, the energy from the expander is not available so that the electric motor drives the compressor. When the processes approaches steady operation, the expander supplies some of the energy to operate the compressor. Eventually, the process reaches steady state, and the motor may continue to supply power to the compressor if there is insufficient power delivered by the expander. If the power delivered by the expander exceeds the power needed by the compressor, then the excess power will be absorbed by the generator and delivered to the plant's electrical-distribution system.

PUMPS

Like compressors, pumps are divided into two main categories according to their principle of operation, positive-displacement or dynamic. In positive-displacement pumps, pressure is developed by trapping a quantity of liquid in a chamber and then compressing it to the discharge pressure. In a dynamic pump, the fluid first acquires kinetic energy which is then converted to pressure. The classification of pumps according to this scheme is shown in Figure 5.24, and some characteristics of selected pumps are given in Table 5.13. Examples of these pumps are shown in Figure 5.25 For a more detailed discussion of pumps than will be given here, the reader should refer to Holland and Chapman [45].

POSITIVE-DISPLACEMENT PUMPS

The characteristic feature of positive-displacement pumps is that ideally they will deliver the same volume of liquid at every stroke regardless of the discharge pressure. In practice, the flow rate will decrease with increasing pressure because of increasing leakage pass the seals. This is shown by the characteristic curve in Figure 5.26. The characteristic curve, which is supplied by the pump manufacturer, is a plot of pressure or head against the flow rate of water. Head is the height of a column of liquid that exerts a pressure equal to a given pressure. The difference between the ideal and actual flow rate is called slip. A very high pressure will be developed if the discharge line of a positive-displacement pump becomes blocked. Thus, in order to prevent damage to the pump and piping, a pressure relief valve must be installed across the pump. As soon as the design pressure is exceeded, the relief valve automatically opens and discharges liquid into the pump inlet. Because of this characteristic of positive-displacement pumps, if a valve, located on the discharge side of the pump, is used to vary the flow rate, then, the discharged pressure must be controlled. A variable-speed drive could also be used to vary the flow rate. Generally, positive-displacement pumps are employed where it is required to deliver low flow rates at high pressures. If high flow rates at high pressures are required, then the pumps are installed in parallel. To develop high pressures requires close clearances between the moving parts to minimize leakage, but close clearances means that the pump must move at slower speeds to avoid excessive wear. Thus, pumps designed to develop high pressures are forced to deliver low flow rates. On the other hand, pumps designed to deliver high flow rates usually cannot develop high pressures.

Positive-displacement pumps are self priming, which is the ability of a pump to lift liquids from a level below the center line of the pump. This characteristic of positive-displacement pumps is attributed to the tight seal between the discharge and suction sides of the pump. Thus, at startup air is compressed and

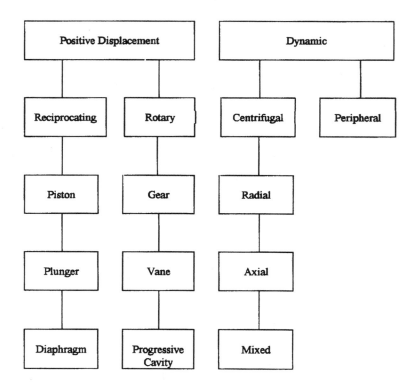

Figure 5.24 Pump-classification chart.

Table 5.13 Pump Characteristics

Pump Type[d]	Flow Range[a] gal/min	Pressure Range[b] Head, ft	Pump Efficiency %
Positive Displacement			
Reciprocating	10 to 10,000	1.0×10^6 max	70 at 10 hp
			85 at 50 hp
			90 at 500 hp
Rotary	1 to 5,000	50,000 max	50 at 80 hp
Dynamic			
Centrifugal			
Single Stage	15 to 5,000	500 max	45 at 100 gal/min
Multistage	20 to 11,000	5,500 max	70 at 500 gal/min
			80 at 10,000 gal/min
Axial	20 to 100,000	40	65 to 85

a) To convert to m^3/min multiply by 0.003785. d) Source of data: Ref 4.
b) To convert to meters multiply by 0.3048.
c) To convert to kW multiply by 0.7457.
  Source Ref. 46.

discharged, creating a vacuum in the suction line allowing liquid to fill the line. As can be seen in Figure 5.24, positive-displacement pumps are classified into two main groups: reciprocating and rotary pumps. These two classes of pumps are discussed in the next two sections.

Reciprocating Pumps

In a reciprocating pump, a piston, plunger or diaphragm moves back-and-forth resulting in an alternating increase and decrease in the volume of the chamber. Examples of common reciprocating pumps are shown in Figure 5.25. As the volume of the chamber is increased by withdrawal of the plunger or diaphragm, a low suction pressure draws liquid into the pump. Then, as the plunger returns, it displaces the liquid forcing it out the discharge. The pump contains check valves to prevent backflow. Reciprocating pumps have a pulsating discharge as contrasted to rotary or centrifugal pumps which produce steady flow. The pulsation causes piping to flex and vibrate. This, in turn, may cause piping connections to leak and piping to fail in fatigue. To minimize pulsation, the designer could select

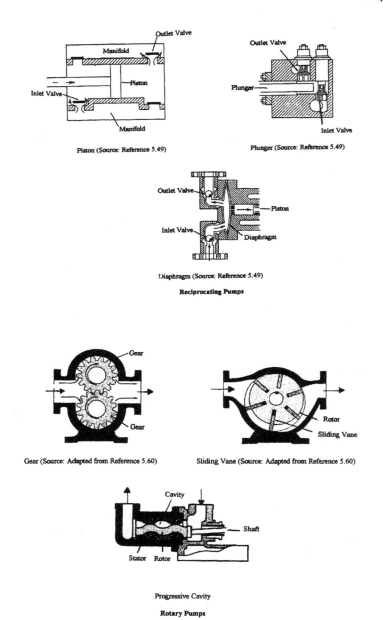

Figure 5.25 Common positive-displacement pumps. Adapted from Ref-ererence 49, 60 with permission.

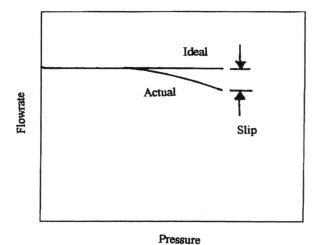

Figure 5.26 Characteristic curve for a rotary positive-displacement pump.

reciprocating pumps with multiple cylinders arranged in parallel so that one discharge stroke begins before another has ended. Alternatively, or in addition, the designer can use pulsation dampers, discussed by Reynolds [47]. The piston pump uses a piston as the displacement element, the plunger pump, a rod, and the diaphragm pump a flexible diaphragm. Also, because the piston and plunger pumps have close clearances between parts, they cannot pump liquids containing any solids. The diaphragm pump is used where leakage or contamination cannot be tolerated, and it is suitable for pumping liquids containing solids, shear-sensitive liquids, and viscous liquids.

Rotary Pumps

Figure 5.25 shows some common rotary pumps. Rotary pumps, as contrasted to reciprocating pumps, produce a smooth-flowing discharge and do not require check valves at the inlet and discharge sides of the pump. Rotary pumps rotate at higher speeds than reciprocating pumps, and thus they can deliver a higher flow rate but at the expense of delivering lower pressures than reciprocating pumps. A gear pump is shown in Figure 5.25, where a drive gear and driven gear are contained in a casing. Liquid flows around the periphery of the revolving gears from the suction to the discharge sides of the pump. Between the gears and side plates and between the gear tips and the housing requires a certain amount of clearance. Clearance is necessary to prevent seizure, but it also results in leakage, called slippage. A sliding-vane pump, shown in Figure 5.25, is similar to a sliding-vane compressor or vacuum pump. Rectangular vanes that are free to move in a radial slot are placed at regular intervals around the rotor. As the rotor revolves, the vanes are thrown outwards against the casing to form a seal. In the suction side of the pump, the space between the vanes fills with liquid, which is then compressed and discharged. Both the gear and sliding-vane pumps are not suitable for pumping liquids containing solids.

A progressive-cavity pump is shown in Figure 5.25. A helical screw revolves in a fixed casing which is shaped to produce cavities. At the suction side of the pump, the liquid flows into a partial vacuum created in a cavity, which moves the liquid to the discharge side of the pump as the helical screw rotates. Toward the discharge side, the shape of the casing causes the cavity to close. This action generates an increase in pressure forcing the liquid into the outlet line. The discharge pressure determines the length and pitch of the helical-screw rotor. Unlike the other types of rotary pumps, the progressive-cavity pump, can pump liquids containing large amounts of nonabrasive suspended solids.

DYNAMIC PUMPS

Dynamic pumps are divided into two main classes as shown in Figure 5.24, centrifugal and peripheral. Dynamic pumps are characterized by their ability to deliver high flow rates at low pressures. To achieve high flow rates requires that the impeller rotate at high speeds. Thus, the clearances between the impeller and the pump housing are larger than those between moving and stationary parts of positive-displacement pumps. This, in turn, means that the pressures developed by dynamic pumps cannot be as large as the pressures developed by positive-displacement pumps. Dynamic pumps, with the exception of peripheral pumps, are not self priming. The large clearances between the impeller and casing does not facilitate the removal of air from the pump at startup. Thus, dynamic pumps must be filled with the liquid being pumped before starting, which is called priming. The flow rate from dynamic pumps is smooth and is easily be controlled by installing a control valve on the discharge side of the pump.

Centrifugal Pumps

About 95 % of the pumps used in the chemical industry are centrifugal pumps. The centrifugal pump contains an impeller, usually having curved blades that are mounted on a shaft. The blades rotate inside a volute casing, as shown in Figure 5.27. A liquid enters axially into the eye of the impeller and velocity is imparted to the liquid by rotating blades. An appreciable amount of the kinetic energy of the liquid is then converted into pressure in the casing. Centrifugal pumps are employed where it is required to deliver a high flow rate at a medium pressure. To achieve higher discharge pressures with centrifugal pumps, several stages are in-

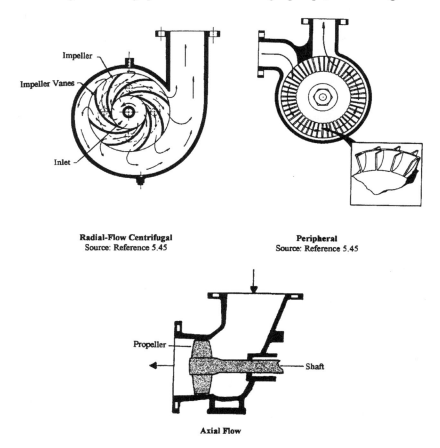

Radial-Flow Centrifugal
Source: Reference 5.45

Peripheral
Source: Reference 5.45

Axial Flow

Figure 5.27 Dynamic pumps.

stalled on one shaft or the pumps are installed in series. Centrifugal pumps are not self priming. Methods of priming centrifugal pumps are discussed by Kern [48]. A centrifugal pump operating at a constant speed will develop the same head in feet regardless of the specific gravity of the fluid being pumped, provided that the viscosity of the fluids do not differ significantly. For this reason, it is usual to plot the pump characteristic curve as head against volume flow rate for a given rotational speed. Viscosities of less than 50 cp (0.05 Pa-s) will not affect the head appreciably. The effect of viscosity is to change the internal friction of the pump and thus the head developed by the pump. Although the head developed by different fluids are the same, the pressure will differ. The denser fluid will exert the greater pressure. The power consumed will also be greater for the denser fluid.

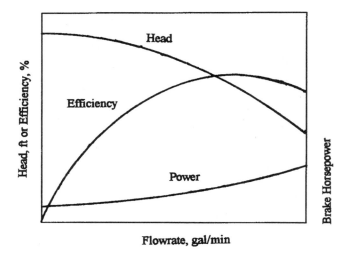

Figure 5.28 Characteristic curve for a centrifugal pump.

Figure 5.28 shows typical performance curves for a centrifugal pump. The pump manufacturer supplies these curves for water. When a control valve in the pump discharge opens or closes, the pump will follow these performance curves.

Pumps wear and the curve will change with time. In addition, friction factors will generally increase with time because of corrosion and deposits. For these reasons, pumps are usually oversized and thus will initially deliver larger flow rates than required. A control valve installed on the discharge side of the pump will bring the pump to the desired operating point on the curve.

Peripheral Pumps

A peripheral pump, shown in Figure 5.27, is sometimes referred to as a regenerative pump or a turbine pump because of the shape of the impeller. This pump employs a combination of mechanical impulse and centrifugal force to produce heads of several hundred feet at low flow rates. The impeller, which rotates at high speed with small clearances, has many short radial passages milled on each side at the periphery of the impeller. Similar passages are milled in the mating surfaces of the casing. Upon entering, the liquid flows into the impeller passages and proceeds in a spiral pattern around the periphery, passing alternately passing alternately from impeller to the casing, receiving successive impulses. In effect, this pump may be considered a multi-stage pump with the stages built into the periphery of the impeller. A characteristic curve is shown in Figure 5.29.

Peripheral pumps are particularly useful for pumping low-flow-rate, low-viscosity liquids at high pressures than are normally available with centrifugal pumps. Close clearances limit their use to clean liquids. Also, because of the close clearances between the impeller and casing, a peripheral pump has excellent suction lift – up to 8.5 m (128 ft) of head.

Axial-Flow Pumps

At very high flow rates and low heads, axial and mixed-flow pumps provide more efficient pumping in smaller casings than centrifugal pumps. The higher flow rates are achieved with higher pumping speeds than centrifugal pumps. The impeller of an axial-flow pump resembles a boat propeller as shown in Figure 5.27. A typical characteristic curve is shown in Figure 5.30. Because the suction lift of an axial-flow pump is not good, the intake must be located below or only slightly above the liquid surface.

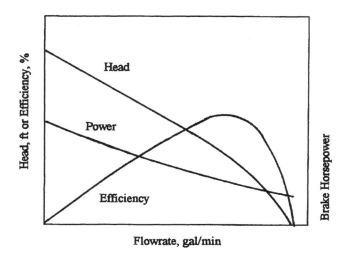

Figure 5.29 Characteristic curve for a peripheral pump.

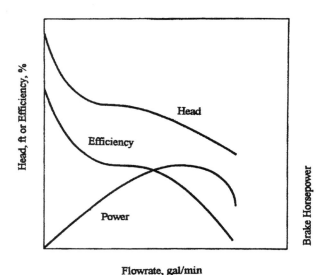

Figure 5.30 Characteristic curve for an axial-flow pump.

PUMP SELECTION

Figure 5.31 shows the operating range of the pumps discussed. As you would expect there is overlap in the operating range of the various pumps. In the overlaping region, selecting the right pump requires experience, but as a general rule a centrifugal pump should be considered first. Figure 5.31 shows that centrifugal pumps can be used to produce high pressures by staging.

PUMP SIZING

Because liquids are incompressible, Equation 5.2 may be used to calculate the work required to pump a liquid. The kinetic energy term is small compared to the other terms and is neglected. Therefore, Equation 5.2 reduces to

$$(g/g_C) \, \Delta z \; + \; \Delta p/\rho + \; W + \; E = 0 \tag{5.46}$$

where the units of each term is in $ft\text{-}lb_F/lb_M$.

Next, define the flow system as point 1 for the inlet and point 2 for the outlet. After expanding Equation 5.46, we obtain.

$$(g/g_C)(z_2 - z_1) + (p_2 - p_1)/\rho + W + E = 0 \tag{5.47}$$

The friction pressure loss term is split into two parts, one for the suction side of the pump, E_S, and, the other for the discharge side of the pump, E_D. Thus, Equation 5.47 becomes, after solving for W.

$$W = (g/g_C)(z_1 - z_2) + (p_1 - p_2)/\rho - (E_S + E_D) \tag{5.48}$$

Frequently, we must make a preliminary estimate of the pump work. Manufacturers do not stock all pumps and other expensive machinery because of the cost of carrying an inventory. The machinery is manufactured on receipt of an order from a customer. Manufacturing some process machinery, may take six months or longer. To save time in implementing a project will require ordering equipment having long delivery times before completing a detailed design. Also, the management of a firm will require an estimate of the cost of a project to prepare a budget or a proposal for a customer.

To estimate the size of a pump at the preliminary stages of a process design before the flow system is completely defined, requires experience with similar designs. From Equation 5.48, we see that the elevation, pressure difference, and frictional pressure losses of the system have to be estimated. Before using Equation 5.48, clearly define the system. Points 1 and 2 are usually selected, when the pressures are known at these points. The elevation, Δz, of the flow system can be made from a rough estimate of the size and the location of equipment. Table 5.14 gives rules-of-thumb for locating equipment. The pressure at both ends of the system will be known. Because the exact length of piping and the kind and number of fittings will not be known until all equipment is exactly located and the piping designed, a rule-of-thumb must be used to estimate the frictional pressure losses, E. Valle-Riestra [50] recommends using a very liberal pipeline frictional pressure drop of 0.345 bar (5.0 psi) and a 0.345 bar pressure drop across control valves. Walas [46], however, states that a 0.69 bar (10.0 psi) pressure drop across a control valve is required for adequate control. Table 5.16 contains rules-of-thumb for frictional pressure losses for some equipment.

Once the work is estimated, the pump shaft power,

$$P_P = m W / \eta_P \tag{5.49}$$

where η_P, the pump efficiency, includes both the hydraulic and mechanical frictional losses. Pump efficiencies are given in Table 5.13 for several pumps. Table 5.17 outlines a calculation procedure for calculating an approximate pump size. Example 5.4 illustrates the procedure.

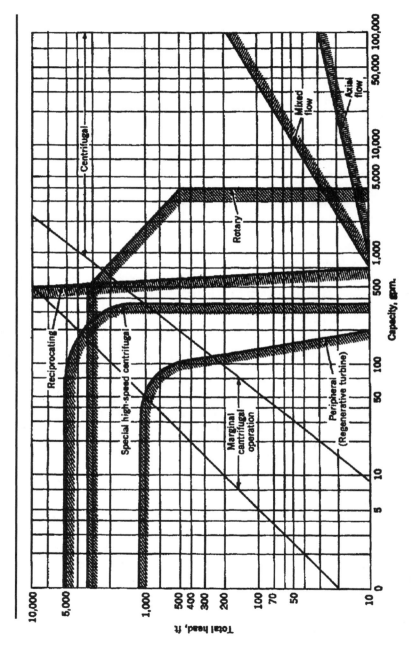

Figure 5.31 Pump-selection chart. From Ref. 49 with permission).

Table 5.14 Rules-of-Thumb for Locating Process Equipment

Process Equipment	Location Above Ground Level[c], ft
Pumps	0
Condensers	20
Reflux Drums	10
Phase Separators	3 to 5
Skirt[a] Height for Columns[b] (2 to 12 ft in diameter)	3 to 6
Heat Exchangers	1 to 4

a) A "skirt" supports the column. The skirt diameter equals the column diameter.
b) Source: Reference 57
c) To convert to meters multiply by 0.3048.

Table 5.15 Approximate Frictional Pressure Drop Across Process Equipment

Flow System Component	Pressure Drop[b], bar	Reference
Pipeline	0.35	5.50
Control Valve	0.70	5.46
Interchanger	0.35[a]	5.55
Air Cooler	0.60	5.23
Surge Vessel	Small	

a) Pressure drop for a fluid with a viscosity less than 1 cp (0.001 Pa-s)
b) To convert to psi multply by 14.5.

Table 5.16 Approximate Pump-Sizing Calculation Procedure

1. Define the flow system, i.e., locate points 1 and 2. The pressures p_1 and p_2 will be known at these points.
2. Locate the process equipment according to the rules-of-thumb listed in Table 5.14.
3. Estimate z_1 and z_2.
4. Estimate the frictional pressure losses E_S and E_D using the rules-of-thumb given in Table 5.15.
5. Calculate the pump work from Equation 5.48.
6. Calculate the pump shaft horsepower using Equation 5.49 and the pump efficiencies given in Table 5.13.
7. Calculate the electric-motor horsepower using the motor efficiency given in Table 5.9.
8. Select a standard electric-motor horsepower using Table 5.10 to obtain approximately a 10% safety factor.

Example 5.4: Approximate Pump Sizing
It is required to estimate the amine-circulating-pump size in the separation process shown in Figure 5.4.1.   The pump, which is a centrifugal pump, will be delivered six months after placing an order. In order to put the process on stream as soon as possible, a process engineer must place the order within three days.

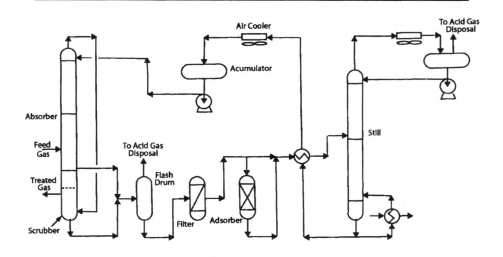

Figure 5.4.1 An acid-gas-removal process. From Ref. 51.

Process Description

In this process, acid gas, i.e., a gas containing CO_2 and H_2S is removed from a natural-gas stream by absorption in a solution containing 0.15 mass fraction of monoethanolamine (MEA) dissolved in water. Removal of CO_2 and H_2S, from gas streams is a common processing problem. These gases react with the mono-ethanolamine at a high pressure in the absorber, shown in Figure 5.4.1, and are removed from the gas stream. The exit gases are then recycled to the bottom of the absorber to scrub out any entrained liquid drops. After absorption, the liquid stream is flashed across the valve where some CO_2, H_2S, and other dissolved gases are desorbed from the solution in the gas-liquid separator. Solids are frequently present in the liquid stream because of corrosion and degradation of the MEA. The solids are removed by the filter. Also, soluble degradation products of MEA are removed in the purge stream by the carbon adsorber to reduce foaming and corrosion.

In the amine stripper, the MEA solution is regenerated by stripping the solution of CO_2 and H_2S using hot vapors from the reboiler. The hot liquid from the stripper is cooled before returning to the absorber by first preheating the feed stream to the still in an interchanger and then by air cooling. An accumulator in the line dampens the solution flow rate to the absorber.

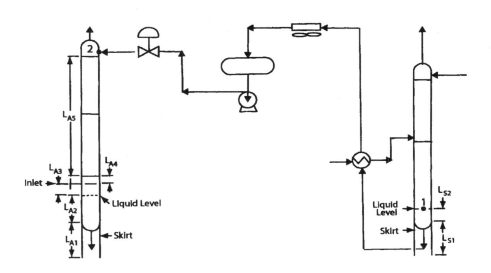

Figure 5.4.2 A simplified acid-gas-removal process.

Analysis

A simplified flow diagram for the process is shown in Figure 5.4.2. Table 5.4.1 list specifications for the process, obtained from Maddox and Burns [52, 53]. The shaft work for the pump will be calculated from Equation 5.48 and the shaft power from Equation 5.49. Because the process at this point is not well defined, all approximations must be made to maximize the estimated power so that the pump will not be undersized.

First, designate the terminal points of the flow system. Point 1 is located at the liquid surface at the bottom of the stripper, as shown in Figure 5.4.2. Point 2 is located at the top of the absorber. These points are selected because the pressures, given in Table 5.4.1, are known.

Calculate the elevation on the discharge side of the pump – the first term in Equation 5.48. The elevation consists of:

1. height of the column support, "the skirt"
2. the liquid level at the bottom of the absorber
3. distance between the liquid level and the gas inlet
4. the distance between the gas inlet and the bottom tray
5. the number of trays
6. distance between trays.

Table 5.4.1 Specifications for an Acid-Gas-Removal Process

Mass Fraction of MEA[a]	0.15
Liquid Flow Rate[a]	2.08 m^3/min
Average Density	974 kg/m^3

Specification	Stripper[a]	Absorber[b]
No. of Trays	20	26
Tray Spacing, m	0.61 (2 ft)	0.61 (2 ft)
Top Temperature, °C	93.0 (199 °F)	38.0 (100 °F)
Top Pressure, bar	1.35 (20.8 psia)	34.8 (505 psia)
Bottom Temperature, °C	116.0 (241 °F)	57.0 (135 °F)
Bottom Pressure, bar	1.65 (23.9 psia)	35.5 (515 psia)
Column Diameter, m	2.41 (7.81 ft)	1.75 (5.74 ft)

a) Source: Reference 52
b) Source: Reference 53

The skirt diameter is equal to the diameter of the column, and its height varies from 2 to 12 ft (0.61 to 3.66 m). The height of the skirt is determine by maintenance requirements. Maintenance workers need space to repair the bottom section of a column. Columns could also be supported on a structure. Assume that the skirt height for the absorber, L_{A1}, in Figure 5.4.2, is 2.0 m (6.56 ft).

To dampen flow-rate fluctuations, requires liquid holdup at the bottom of the column. Ludwig [54] recommends 5 to 20 min for the surge time, i.e., the liquid residence time at the bottom of a column. On the other hand, it is desirable to minimize the solvent inventory in a process to minimize cost and to minimize the amount of flammable liquids. Also, if the liquid contains heat-sensitive organic compounds, it is necessary to reduce the residence time, particularly in strippers, where the temperature is high. For this problem, select a surge time of 5.0 min to keep the residence time low for the stated reasons. Therefore, the liquid height in the absorber,

$$L_{A2} = \frac{2.08 \text{ m}^3}{1 \text{ min}} \quad \frac{5.0 \text{ min}}{1} \quad \frac{4}{\pi (1.75)^2 \text{ m}^2} = 4.323 \text{ m} (14.2 \text{ ft})$$

The distance between the liquid level and the gas inlet in the absorber, L_{A3}, is 1.5 m (4.92 ft), and the distance from the gas inlet to the bottom tray, L_{A4}, is 0.9 m (2.95 ft). From Table 5.4.1, it is seen that there are 26 trays with a spacing of 0.61 m (2.0 ft) Thus, the trays for the absorber will occupy a height of

$$L_{A5} = 0.61 (26 - 1) = 15.25 \text{ m} (50 \text{ ft})$$

The elevation of the pump discharge line,

$$z_1 = L_{A1} + L_{A2} + L_{A3} + L_{A4} + L_{A5}$$

$$z_1 = 2.0 + 4.323 + 1.5 + 0.9 + 15.25 = 24.0 \text{ m} (78.7 \text{ ft})$$

The elevation for the suction side of the pump consists of the sum of the stripper-skirt height and the liquid level at bottom of the stripper. Again, assume that the skirt height is 2.0 m (6.56 ft).

To calculate the liquid height, L_{S2}, we again assume 5.0 min for the surge time.

$$L_{S2} = \frac{2.08 \text{ m}^3}{1 \text{ min}} \quad \frac{5.0 \text{ min}}{1} \quad \frac{4}{\pi (2.412)^2 \text{ m}^2} = 2.280 \text{ m} (7.48 \text{ ft})$$

Therefore, the elevation of the pump suction line,

$z_2 = 2.0 + 2.280 = 4.28$ m (14.0 ft)

From Table 5.4.1, the stripper-bottom pressure, $p_1 = 1.65$ bar (23.9 psia), and the absorber-top pressure $p_2 = 34.8$ bar (505 psia). Thus, we can calculate the second term in Equation 5.48.

The total-frictional pressure drop in the system is the sum of the pressure drops caused by the piping and fittings, control valve, interchanger, and air cooler. Estimates of these loses are listed in Table 5.15. Therefore, the total frictional pressure drop in the system,

$E_S + E_D = 0.35 + 0.70 + 0.35 + 0.60 = 2.0$ bar (29.0 psi)

Now, substitute numerical values into Equation 5.48. In the SI system of units g_C, is not needed. Because one bar equals 1×10^5 Pascals (N/m²), we have, after multiplying and dividing the first term in Equation 5.48 by kg,

$$W = \frac{9.8 \text{ m}}{1 \text{ s}^2} \cdot \frac{(4.28 - 24.0) \text{ m-kg}}{1} + \frac{(1.65 - 34.8) \text{ bar}}{1} \cdot \frac{1 \times 10^5 \text{ N}}{1 \text{ m}^2\text{-bar}} \cdot \frac{1 \text{ m}^3}{974.0 \text{ kg}}$$

$$- \frac{2.0 \text{ bar}}{1} \cdot \frac{1 \times 10^5 \text{ N}}{1 \text{ m}^2\text{-bar}} \cdot \frac{1 \text{ m}^3}{974.0 \text{ kg}} = -3.391 \times 10^3 \text{ N-m/kg} \ (-3.217 \text{ Btu/lb})$$

$(-3.391 \times 10^3$ J/kg$) \ (-3.217$ Btu/lb$)$

Because kg-m/s² equals one Newton, the first term has units of N-m/kg, as does the other terms. The negative sign means that the work is done on the system. Work done by a system is positive.

According to Table 5.13, the centrifugal pump efficiency depends on the volumetric flow rate. From Table 5.13, the pump efficiency is 45 %, Therefore, from Equation 5.49 the pump shaft power,

$$P_P = \frac{m \text{ W}}{\eta_P} = \frac{1}{0.45} \cdot \frac{2.08 \text{ m}^3}{1 \text{ min}} \cdot \frac{974 \text{ kg}}{1 \text{ m}^3} \cdot \frac{3391 \text{ J}}{1 \text{ kg}} \cdot \frac{1 \text{ min}}{60 \text{ s}}$$

$= 2.544 \times 10^5$ J/s (254 kW) (341.2 hp)

Assuming that a squirrel-cage electric-motor drive for the pump is selected, the electric-motor efficiency is determined by interpolating between 100 and 1,000 hp (in Table 5.9). An acccurate determination of the motor efficiency requires knowing the final motor size. With some loss in accuracy, we will use the power calculated above. Thus, the motor efficiency is 91.9 %, and electric-motor power,

$P_P = 341.2 / 0.919 = 371.3$ hp (276.9 kw)

From Table 5.10, select a standard 400 hp (298 kW) motor, which results in a safety factor of 7.73 %. The next size motor is 450 hp (336 kW) resulting in a safety factor of 21.2 %. Based on his past experience, the process engineer would have to decide what safety factor to chose.

NOMENCLATURE

English

c_P heat capacity at constant pressure
or condensing-pressure efficiency correction factor for a multistage steam turbine

c_S superheat efficiency correction factor for a single stage
or multistage steam-turbine

E friction losses

E_D friction losses on the discharge side of a pump

E_S friction losses on the suction side of a pump

g acceleration of gravity

g_C conversion factor

h enthalpy

k ratio of the heat capacity at constant pressure to the heat capacity at constant volume

L length

m mass flow rate

M molecular weight

n number of moles or polytropic exponent

N number of compression stages for cooling

p pressure

P   power or pressure

P_C compressor power

P_{CP} compressor power for a polytropic compression

P_E electric motor power

P_F fan power

P_P pump power

P_T turbine power

R   gas constant

s   entropy

T   absolute temperature

v   average velocity or specific volume

V   gas volumetric flow rate

W   work

W_A actual work of compressing a gas

W_C compressor work

W_{CN} compressor work for N cooling stages

W_F fan work

W_P polytropic or pump work

W_{PN} polytropic work for N cooling stages

W_S isentropic work

W_N turbine work

x   mass fraction of moisture in steam

y   mole fraction

z   elevation

Greek

α kinetic energy correction factor

η_A isentropic efficiency

η_B bearing efficiency or uncorrected steam-turbine efficiency

η_C compressor efficiency

η_F fan efficiency

η_G gear efficiency

η_H hydraulic efficiency

η_M mechanical efficiency

η_P polytropic efficiency

η_S seal efficiency

η_T steam turbine efficiency

μ viscosity

ρ density

ω rotational speed

Subscripts

C   compressor or critical conditions

L   liquid phase

R   reduced state

s   constant entropy

V   vapor phase

z   elevation or compressibility factor

REFERENCES

1. Pollak, R., Selecting Fans & Blowers, Chem. Eng., 80, 2, 86, 1973.
2. Kusay,R.G.P., Vacuum Equipment for Chemical Processes, Brit. Chem.Eng., 16, 1, 29, 1971.
3. Ryans, J.L., Croll, S., Selecting Vacuum Systems, Chem. Eng., 88, 25, 72, 1981.
4. Patton, P.W., Joyce, C. F., Lowest Cost Vacuum System, Chem. Eng., 83, 3, 84, 1976.
5. Dobrowolski, Z., High Vacuum Pumps, Chem. Eng., 63, 9, 181, 1956.
6. Neerken, R.F., Compressor Selection for the Chemical Process Industries, Chem. Eng., 82, 2, 78, 1975.

7. Summerell, H.M., Consider Axial-Flow Fans When Choosing a Gas Mover, Chem. Eng., 88, 11, 59, 1981.
8. Thompson, J.E., Tickler, C.J., Fans and Fan Systems, Chem. Eng., 90, 6, 48,1983.
9. Bird, R.B., Stewart, W.E., Lightfoot, E.N., Transport Phenomena, John Wiley & Sons, New York, NY 1960.
10. Booth, R.G., Epstein N., Kinetic Energy and Momentum Factors for Rough Pipes, Can. J. of Chem. Eng., 47, 10, 515, 1969.
11. Fischer, J., Practical Pneumati Conveyor Design, Chem. Eng., 65, 11, 114,1958.
12. Dimoplon, W., What Process Engineers Need to Know About Compressors, Hydrocarbon Proc., 57, 5, 221, 1978.
13. James, R., Positive Displacement Compressors, Encyclopedia of Chemical Processing and Design, McKetta, J. J., Cunningham, W. A., eds., Marcel Dekker, Inc., New York, NY, 1979.
14. Ludwig, E.E., Applied Process Design for Chemical and Petrochemical Plants, Vol. 3, 3rd ed., Gulf Professional Publishing, Boston, MA, 2001.
15. Brochure, Gardner-Denver, Quincy, IL, No Date.
16. Advertisement, Atlas Copco Comptec, Voorheesville, NY, Chem. Eng., 93, 23, 10, 1986.
17. Severns, W.H., Degler, W.H., Steam Air and Gas Power, John Wiley & Sons, 4th ed.. New York , NY, 1948.
18. Esplund, D.E. Schildwachter, J. C., How to Size and Price Axial Compressors, Hydrocarbon Proc., 42, 1, 141, 1963.
19. Shemeld, D.E., Turbocompressors, Applications, Selections, Limitations, Dresser Industries, Olean, NY, No Date.
20. Ryans, J., Bays, J., Run Clean with Dry Vacuum Pumps, Chem. Eng. Prog., 97, 10, 32,
21. Brochure, Clark Centrifugal Compressors, Bulletin 336, Dresser Industries, Cranford, NJ, 1983.
22. Moens, J.P.C., Adapt Process to Compressor, Hydrocarbon Proc., 50, 12, 96, 1971.
23. Ulrich, G.P., A Guide to Chemical Engineering Process Design and Eco nomics, John Wiley & Sons, New York, NY, 1984.
24. Hancock, R., Drivers, Controls and Accessories, Chem. Eng., 63, 6, 227, 1956.
25. Moore, J.C., Electric Motor Drivers for Centrifugal Compressors, Hydrocarbon Proc., 54, 6, 133, 1975.
26. Willoughby, W.W., Steam Rate: Key to Turbine Selection, Chem. Eng., 85, 20, 146, 1978.
27. Harman, R.T.C., Gas Turbine Engineering, John Wiley & Sons, New York City, NY, 1981.
28. Rex, M.J., Choosing Equipment for Process Energy Recovery, Chem. Eng., 82, 16, 98, 1975.

29. Rossheim, D.B., Peterson, F. W., Vogrin, C. M., Mechanical, Plant, and Project Engineering, Perry's Chemical Engineers' Handbook, 4th ed., p. 24-1, McGraw-Hill Book Co., New York, NY, 1950.
30. Turton, R.K., Principles of Turbomachinery, E & F, N. Spon, London, Eng land, 1984.
31. Neerken, R.F., Use Steam Turbines as Process Drivers, Chem. Eng., 87, 17, 63, 1980.
32. Scheel, L.F., What You Need to Know About Gas Expanders, Hydrocarbon Proc., 49, 2, 105, 1970.
33. Makansi, J.M., Advances in Steam Turbine Technology Focus on Efficiency, Power, 126, 7, 19.
34. Molich, K., Consider Gas Turbines for Heavy Loads, Chem. Eng., 87, 17, 79, 1980.
35. Gatmann, H., DeLaval Engineering Handbook, McGraw-Hill, New York, NY, 1970.
36. Bloch, H.P., Driver Selection, Encylopedia of Chemical Processing and Design, McKetta, J. J., and Cunningham, W.A., eds., Marcel Dekker, New York, NY, 1979.
37. Campagne, W.V.L., Select HPI Gas Turbines, Hydrocarbon Proc. 64, 3, 77, 1985.
38. Campagne, W.V.L., Gas Turbine Selection for the Chemical Industry. Summer National Meeting, Philadelphia, PA, American Institute of Chemical Engineers, New York, NY, 1984.
39. Process Machinery Drives, Bloch, H.P., Daugherty, R.H., Geiter, F.K., Boyce, M.P., Sweringen, J.S., Jennet, E., Calistrat, M.M., Perry's Chemical Engineers Handbook, Perry, R.H., Green, T.D.W., Maloney, J.O., eds., 7th ed., p. 24-1, McGraw- Hill, New York , NY, 1997.
40. Franzke, A., Benefits of Energy Recovery Turbines, Chem. Eng., 77, 109, 1970.
41. Buse, F., Using Centrifugal Pumps as Hydraulic Turbines, Chem. Eng., 88, 2, 113, 1981.
42. Swearingen, J.S., Flashing Liquid Runs Turboexpander, Oil and Gas J., 14, 27, 70, 1976.
43. Frank, O., Personal Communication, Consulting Engineer, Convent Station, Jan. 2002.
44. Keenan, J.H., Keyes, F. G. Hill, P. G., Moore, J. G., Steam Tables, John Wiley & Sons, New York, NY, 1978.
45. Holland, F.A., Chapman, F.S., Pumping of Liquids, Rheinhold Publishing, New York, NY, 1966.
46. Walas, S.M., Rules of Thumb, Chem. Eng., 94, 4, 75, 1987.
47. Reynolds, J.A., Pump Installation and Maintenance, Chem. Eng., 78, 23, 67, 1971.
48. Kern, R., How to Design Piping for Pump Suction Conditions, Chem. Eng., 82, 9, 119, 1975.

49. Stindt, W.H., Pump Selection, Chem. Eng., <u>78</u>, 23, 43, 1971.
50. Valle-Riestra, J.F., Project Evaluation in the Chemical Process Industries, McGraw-Hill., New York, NY, 1983.
51. Brochure, Gas Treating, Pro-Quip Corp., Tulsa, OK, No Date.
52. Maddox, R.N., Burns, M.D., Here are Principal Problems in Designing Stripping Towers, Oil and Gas J., <u>65</u>, 40, 110, 1967.
53. Maddox, R.N., Burns, M. D., How to Design Amine Absorbers, Oil and Gas J., <u>65</u>, 38, 114, 1967.
54. Ludwig, E.E., Applied Process Design for Chemical and Petrochemical Plants, Vol. 2, 3rd ed., Gulf Publishing, Houston, TX, 1997.
55. Frank, O., Simplified Design Procedures for Tubular Heat Exchangers, Practical Aspects of Heat Transfer, Chem. Eng. Prog. Tech. Manual, Am. Inst. of Chem. Eng., New York, NY, 1978.
56. Elliott Multistage Compressors, Bulletin P-25A, Carrier Corp., Jeannette, PA, 1975.
57. Kern, R., Pipe Systems for Process Plants, Chem. Eng., <u>82</u>, 24, 209, 1975.
58. Cole, P., Personal Communication, Dresser Industries, Olean, NY, 1985.
59. Brochure, Liquid Ring Vacuum Pumps and Compressors, GMCI/1, Graham Manufacturing Co., Batavia, NY, 1978.
60. Cody, D.J., Graig, A.V., Stratt, D.K Selecting Positive Displacement Pumps, Chem. Eng., <u>92,</u> 15, 38, 1985.

6

Separator Design

Chapter 1 discussed two major types of separation processes, component and phase separation. In component separation, the components are separated from a single phase by mass transfer. An example is gas absorption where one or more components are removed from a gas by dissolving in a solvent. In phase separation, two or more phases can be separated because a force acting on one phase differs from a force acting on another phase or because one of the phases impacts on a solid barrier. The forces are usually gravity, centrifugal, and electromotive. Examples are removal of a solid from a liquid by impaction (filtration), gravity (settling), centrifugal force, and the attraction of charged particles in an electrostatic precipitator. One exception to these mechanisms is drying by evaporating unbonded water from a solid. In this case, separation of a liquid from a solid occurs by mass transfer. For example, the water mixed with sand can be removed by evaporating the water. Because many component separations require contacting two phases, like liquid-liquid extraction, component separation is frequently followed by phase separation. Phase separators can be classified according to the phases in contact: liquid-gas, liquid-liquid, liquid-solid, solid-gas, and solid-solid. Some of the more common phase separators will be discussed.

In addition to discussing phase separators, it is also appropriate to consider the application of accumulators in processes. Accumulators or surge vessels are necessary to reduce fluctuations in flow rate, pressure and composition and thereby improve process control. Although accumulators are not phase separators, they are discussed here because they are sometimes contained in the same vessel as a phase separator. For example, in a gas-liquid separator, the volume of liquid at the bottom of the separator is determined by the need to dampen fluctuations in flow rate.

VESSEL DESIGN

Although a mechanical or civil engineer normally designs vessels, the process engineer should have some knowledge of the mechanical design of vessels. For example, the process engineer may have to make a preliminary design of vessels for a cost estimate. Reactors, fractionators, absorbers, heat exchangers, and some phase separators are classified as vessels. What makes an absorber an absorber, for example, is its internal design. A vessel consists of a cylindrical shell and end caps, called heads. For safety, vessel design is governed by codes. An example is the ASME (American Society of Mechanical Engineers) Boiler and Pressure Vessel Code. Engineers who agreed on what is a safe procedure for designing vessels formulated this code.

Most vessels in the process industries are thin-walled vessels, which have a wall thickness of less than about 5% of the inside diameter of a vessel. Internal pressure acting on the walls of a cylindrical vessel produces a longitudinal and radial stress, also called hoop stress. For thin-wall vessels, it may be assumed that the radial stress is approximately uniform across the wall. Rase and Borrow [1], for example, showed that the radial stress, produced by an internal pressure, P, is given by Equation 6.1.

$$S = \frac{P\,D}{4\,t_S} \tag{6.1}$$

where the diameter of the vessel is D. The radial stress is larger than the longitudinal stress, and thus it must be used to calculate the wall thickness, t_S. If a cylindrical vessel fails, it will split longitudinally.

Vessels larger in diameter than about 30 in (0.672 m) and above are fabricated from plates, which are formed into cylinders, called shells, and welded longitudinally. Shells smaller than 30 in (0.672) may be extruded and thus will not contain a longitudinal weld. Shells may then be joined by welding circumferentially to form longer shells. After fabricating the shell, end caps, called heads, are welded to the shell to form the vessel. Because the weld may have imperfections, the radial stress will be less than its maximum value. Thus, S is multiplied by a joint or weld efficiency, ε, which depends on the type of x-ray inspection of the weld. Thus,

$$\varepsilon\,S = \frac{P\,D_M}{4\,t_S} \tag{6.2}$$

where the mean diameter, D_M, is the average of the outside and inside diameters.

$$D_M = \frac{D + (D + 2\,t_S)}{2} \qquad (6.3)$$

If Equation 6.3 is substituted into Equation 6.2, the wall thickness,

$$t_S = \frac{P\,D}{\varepsilon\,S - P} \qquad (6.4)$$

Sivals [10] summarizes values of the weld efficiency in Table 6.1. Radiographic examination locates imperfections in the weld using x-rays or gamma rays. This technique is described by Gumm and Turner [2]. Shells are either seamless or contain a longitudinal weld. As Table 6.1 shows, the weld efficiency depends on whether the shell is seamless or not. To use Table 6.1, first decide if the shell will be seamless or contain a longitudinal weld. Next select the type of x-ray required to inspect weld.

Even in a thin-walled vessel the radial stress is not exactly uniform over the vessel thickness. To correct for this, the internal pressure in the denominator of Equation 6.4 is multiplied by 1.2 to obtain a more accurate formula. Thus,

$$t_S = \frac{P\,D}{2\,\varepsilon\,S - 1.2\,P} \qquad (6.5)$$

To account for corrosion, the vessel thickness is increased by adding a corrosion allowance, t_C, to assure that the vessel operates safely during the lifetime of a process. Therefore, Equation 6.5 becomes

$$t_S = \frac{P\,D}{2\,\varepsilon\,S - 1.2\,P} + t_C \qquad (6.6)$$

The minimum corrosion allowance frequently selected is 1/8 in (3.18 mm). Wallace and Webb [3], however, point out that arbitrarily selecting 1/8 in can be unnecessarily costly. There may be situations where there is no corrosion at all. The corrosion allowance should be determined by past experience, laboratory tests, or data taken from the literature.

Table 6.1 Weld Joint Efficiencies for Ellipsoidal and Torispherical Heads (Source: Adapted from Ref. 10).

Weld Efficiency – Head/Shell				
Seamless Shell[a] with a Circumferential Weld		Shell with a Longitudinal Weld		
		Full X-Ray	Spot X-Ray	No X-Ray
Full X-Ray	1.0/1.0	1.0/1.0	1.0/0.85	1.0/0.85
Partial X-Ray	1.0/1.0	1.0/1.0	1.0/0.85	1.0/0.85
Spot X-Ray	0.85/0.85	0.85/0.85	0.85/0.85	0.85/0.85
No X-Ray	0.80/0.80	0.85/0.85	0.85/0.85	0.80/0.70

a) Two or more shells joined with a circumferential weld to make a longer shell

Several head designs are shown by Walas [6], but not all of these are common designs. The most common head designs are shown in Figure 6.1. According to Markovitz [7] ellipsoidal heads, where the ratio of the semi-major to semi-minor axis is 2:1, are commonly used when the pressure is greater than 150 psig (10.3 barg). Below 150 psig, a torispherical head (dished head) is used.

The wall thickness for a 2:1 ellipsoidal head is given by

$$t_H = \frac{P D}{2 \varepsilon_H S - 0.2 P} + t_C \qquad (6.7)$$

where H/D = ¼ in Figure 6.1.

For a torispherical head the wall thickness,

$$t_H = \frac{1.104 \ P D}{2 \varepsilon_H S - 0.2 P} + t_C \qquad (6.8)$$

where R/L = 0.06 and L = D in Figure 6.1.

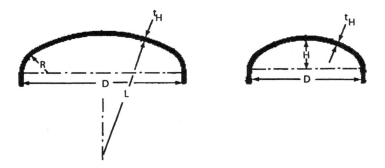

Figure 6.1 Ellipsoidal and torispherical vessel heads.

Table 6.2 Wall-Thickness Rounding Increments for Pressure Vessels

Wall Thickness, in	Rounding[a] Increment[b], in
≤ 1.0	1/32
> 1.0 ≤ 2.0	1/16
> 2.0 ≤ 3.0	1/8
> 3.0	1/4

a) Because alloy steels and nonferrous metals are more costly than carbon steel, the rounding increment should be smaller than the above.
b) To convert to mm multiply by 25.4.

Minimum Vessel Thickness for Pressure Vessels		
Metal	Service	Minimum Thickness, in
Carbon and low-alloy steels	noncorrosive	3/32
High-alloy steels and nonferrous metals	noncorrosive	1/16
High-alloy steels and nonferrous metals	corrosive	3/32

Because the operating pressure in a vessel may fluctuate, for safety, process engineers will use a design pressure in Equations 6.6 to 6.8 to calculate the wall thickness. The design pressure is 1.10 times the expected operating pressure or the expected operating pressure plus 25 psi, whichever is greater. For carbon steel, the calculated vessel thickness is rounded off according to the rules listed in Table 6.2. For high columns, the thickness at the bottom of the column may have to be increased further because of wind load. Mulet et al. [8] describe a calculation procedure to determine the effect of wind load on wall thickness. Other factors will also affect the strength of vessels, such as nozzles and manholes. To take these factors into account, an engineer must follow the ASME pressure vessel code. Table 6.3 summarizes the equations for calculating the vessel wall thickness, and Table 6.4 outlines the calculation procedure.

Table 6.3 Summary of Equations for Calculating Vessel Wall Thickness

$$P = 1.1\, P_o' \quad\text{— or} \tag{6.3.1}$$
$$P = P_o' + 25\text{ psi} \quad\text{— whichever is larger}$$

For a shell

$$\alpha_S = \frac{P}{2\, \varepsilon_S'\, S' - 1.2\, P} \tag{6.3.2}$$

For a torispherical head:

$$\alpha_H = \frac{1.104\ P}{2\, \varepsilon_H'\, S' - 0.2\, P} \quad\text{— or} \tag{6.3.3}$$

For a 2:1 ellipsoidal head:

$$\alpha_H = \frac{P}{2\, \varepsilon_H'\, S' - 0.2\, P}$$

$$t_S = \alpha_S\, D' + t_C' \tag{6.3.4}$$

$$t_H = \alpha_H\, D' + t_C' \tag{6.3.5}$$

Table 6.3 Continued

Variables

P, t_S, t_H, α_S, α_H

Table 6.4 Calculation Procedure for Calculating Vessel Wall Thickness

1. Calculate the design pressure, P (psig), from Equation 6.3.1 where P_o is the expected operating pressure.

2. Select the shell and head weld efficiencies, ε_S and ε_H, from Table 6.1.

3. Calculate the shell factor, α_S, in the hoop stress formulas from Equation 6.3.2.

4. Calculate the head factor, α_H, from Equation 6.3.3. If $P \leq 150$ psig, select a torrispherical head. Above 150 psig select an ellipsoidal head.

5. Calculate the shell thickness, t_S, from Equation 6.3.4.

6. Calculate head thickness, t_H, from Equation 6.3.5.

7. Select a standard thickness from a vessel manufacturer.

VORTEX FORMATION IN VESSELS

Vortex formation in separators must be prevented to reduce gas entrainment in the liquid, which can result in the following: loss of valuable vapor, pump damage, loss of flow, erroneous liquid level readings resulting in poor control, and vibrations caused by unsteady two-phase flow. Vortexes appear frequently in nature such as in hurricanes, tornados, and whirlpools. The mechanism of atmospheric generated of vortices is an active area of research. Even the more common bathtub vortex is of scientific interest. Sibulkin [15] describes experiments to determine the effect of the earth's rotation on the rotation of a bathtub vortex. Although the earth's rotation induces a small angular velocity when draining water, the direction of rotation of a bathtub vortex is usually accidental. It is determined mainly by residual motion caused by the method of filling the tub. If, however, care is taken to reduce residual motions, then the direction of vortex rotation will consistently be counterclockwise in the Northern Hemisphere and clockwise in the Southern Hemisphere.

Vortexes will form in process vessels for bottom, side, and top outlets as illustrated in Figure 6.2. The development of a vortex starts with a dimple on the liquid surface. Below the dimple rotational flow of the liquid reaches to the outlet. As the dimple deepens, a surface vortex develops that resembles an inverted cone penetrating into the liquid. In a fully developed vortex, the cone funnel extends to the vessel outlet. As long as the liquid level is above a minimum value, a vortex will not form. As discussed by Patterson [16], the minimum liquid level depends on the following factors: vessel outlet size and position, tangential velocity components in the liquid induced by the inlet flow, whether the vessel is draining or the level is constant, outlet liquid velocity, and viscosity. For a draining tank, with no inflow of liquid, the outlet velocity only affects the minimum level up to a velocity of 2.6 ft/s [16]. When designing a vessel, considering the above factors may reduce the minimum level at which a vortex forms. Tangential velocity components will induce a vortex so that a tangential entrance pipe should be avoided. When the outlet line is at the top of the vessel, locate the line

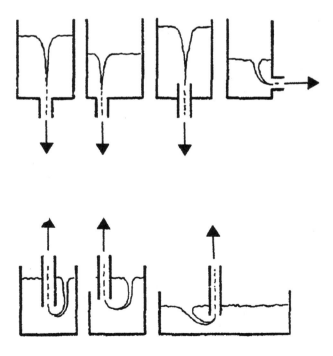

Figure 6.2 Vortex formation in vessels. From Ref. 16 with permission.

close to the side of the vessel as recommended by Patterson [16]. A vortex will occur at a higher liquid level when a tank is draining with no inflow of liquid than when a tank has both equal inflow and outflow. Finally, to minimize the formation of a vortex at low liquid levels, a vortex breaker is installed at a vessel outlet. Vortex breakers may be flat plates, crosses, radial vanes or gratings. Although Patterson [16] gives dimensions for a flat-plate design he recommends radial vanes, as shown in Figure 6.4, or a grating. Another design, recommended by Frank [75], is four vanes at right angles with a flat circular plate welded at the top.

ACCUMULATORS

Accumulators are not separators. In one application, an accumulator placed after a total condenser provides reflux to a fractionator and prevents column fluctuations in flow rate from affecting downstream equipment. In this application the accumulator is called a reflux drum. A reflux drum is shown in Figure 6.3. Liquid from a condenser accumulates in the drum before being split into reflux and product streams. At the top of the drum is a vent to exhaust noncondensable gases that may enter the distillation column. The liquid flows out of the drum into a pump. To prevent gases from entering the pump, the drum is designed with a vortex breaker at the exit line.

The total volume of an accumulator is calculated using a residence time, also called surge time, which is obtained from experience, according to the type and degree of the process control required. After examining 18 accumulators in service, Younger [11] recommended a residence time of 5 to 10 min. Once a residence time is selected, size the accumulator for half-full operation to accommodate either an increase or decrease in liquid level. Thus, the accumulator volume is calculated from Equation 6.5.1 in Table 6.5, where equations for sizing an accumulator are listed. The volumetric liquid flow rate, V_L, is obtained from a mass balance on the system. After calculating the total accumulator volume, calculate the accumulator diameter and length by solving Equations 6.5.3 and 6.5.4. Equation 6.5.4 is a rule-of-thumb for L/D. According to Younger [11], for an L/D ratio of 2.5 to 6 the cost varies by only 2%. After surveying several accumulators in use, Younger [11] found that fifteen were horizontally placed and three were vertically placed.

Table 6.6 outlines a calculation procedure for sizing an accumulator. According to Gerunda [4], the calculated diameter for a vessel is rounded off in six-inch increments, starting with a 30 in (0.762 m) diameter vessel. Six-inch increments are required to match standard-diameter heads for the ends of a vessel (Aerstin, 6.5). The maximum vessel diameter is limited to about 13.5 ft (4.11 m), because of shipping limitations by rail or truck. If a larger diameter than 13.5 ft (4.11 m) is required, then the process engineer must consider either specifying two or more vessels in parallel or fabricating a larger diameter vessel at the construction site. If a vessel is less than 30 in (0.762 m) in diameter, use standard pipe. After calculating the vessel length, round it off in three-inch increments.

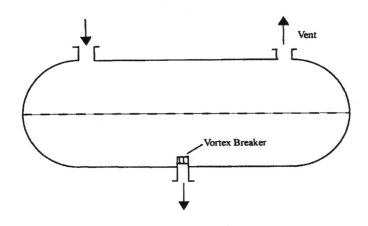

Figure 6.3 An accumulator.

Table 6.5 Summary of Equations for Sizing Accumulators

Subscripts: L = liquid — HV = vessel head

$$V = 2 V_L' t_S \tag{6.5.1}$$

$$t_S = 5 \text{ to } 10 \text{ min} \tag{6.5.2}$$

$$V = \frac{\pi D^2 L}{4} + 2 f_{HV} D^3 \tag{6.5.3}$$

$$L/D = 2.5 \text{ to } 6 \tag{6.5.4}$$

$$f_{HV} = 0.1309 \text{ — for a 2:1 ellipsoidal head, or} \tag{6.5.5}$$
$$f_{HV} = 0.0778 \text{ — for a torispherical head}$$

Unknowns
t_S, D, L, P, V, f_{HV}

Table 6.6 Calculation Procedure for Sizing Accumulators

1. Select a residence or surge time, t_S, from Equation 6.5.2.

2. Calculate the accumulator volume, V, from Equation 6.5.1

3. Select a vessel head. If the internal pressure is 150 psig (10.3 barg) or less, select a torispherical head. If the internal pressure is above 150 psig (10.3 barg), select a 2:1 ellipsoidal head.

4. Select the geometrical factor for the volume of the head, f_{HV}, from Equation 6.5.5.

5. Substitute Equation 6.5.4 into Equation 6.5.3 and solve for the vessel diameter, D.

6. Round off D in 6 in (.152 m) increments, starting with 30 in (0.762 m). If the diameter is less than 30 in (0.762 m), use standard pipe.

7. Calculate the length, L, of the accumulator using Equation 6.5.4.

8. Round off L in 3 in (7.62 cm) increments, for example, 5.0, 5.25, 5.5, 5.75 ft, etc.

Example 6.1 Sizing a Reflux Drum

A fractionator separates dimethylformamide from water and acetic acid. The distillate contains a trace amount of acetic acid. Assuming that the fractionator uses a total condenser, estimate the diameter, length, and wall thickness of the reflux drum. Because the mixture contains acetic acid, use stainless steel (SS 316) for the drum.

Data
Distillate flow rate	16,000 lb/h (7,260 kg/h}
Acetic acid	20 ppm
Temperature	212 °F (100 °C)
Pressure	14.7 psia (11.013 bar)
Density	62.38 lb/ft³ (9993 kg/m³}

Follow the calculation procedure outlined in Table 6.6. First, calculate the reflux-drum volume from Equations 6.5.1 and 6.5.2 in Table 6.5. From Equation 6.5.2, take the average of the surge times.

$$V = 2 \ \frac{16000 \text{ lb}}{1 \text{ h}} \ \frac{7.5 \text{ min}}{1} \ \frac{1}{62.38 \text{ lb}} \ \frac{\text{ft}^3}{60 \text{ min}} \ \frac{1 \text{ h}}{} = 64.12 \text{ ft}^3 \ (1.816 \text{ m}^3)$$

From Equation 6.5.4, select an average L/D ratio.

L/D = 4.25

Substitute this ratio into Equation 6.5.3, and solve for D^3 to obtain

$$D^3 = \frac{V}{1.063 \ \pi + 2 \ f_{HV}}$$

Calculate the design pressure from Equation 6.3.1. Because the pressure is atmospheric, the gage pressure $P_o = 0$, and therefore the design pressure is 25 psig (1.72 barg) According to step 3 in Table 6.6, select a torispherical head because the design pressure is less than 150 psig (10.3 barg). Thus, from Equation 6.5.5, $f_{HV} = 0.0778$. From the above equation for D^3, we obtain.

$$D^3 = \frac{64.12}{1.063 \ (3.142) + 2 \ (0.0778)} = 18.34 \text{ ft}^3 \ (0.5194 \text{ m}^3)$$

D = 2.637 ft (31.64 in, 0.803 m)

Because the drum diameter is greater than 30 in (0.762 m) but less than 36 in (0.914 m), round off D to the highest 6 in increment, which is 36 in (0.914 m). From Equation 6.5.4, L = 4.25 (3.0) = 12.75 ft (3.89 m). This length requires no rounding.

Now, calculate the head thickness following the procedure outlined in Table 6.4. From Table 6.1, with no X-ray inspection, the weld efficiency for the weld joining the head to the shell is 0.80. Because of the acetic acid present in the distillate, we select SS 316, which has an allowable stress of 15,200 psi $(1.04 \times 10^5$ kPa). For the moment, neglect the corrosion allowance. From Equations 6.3.1 and 6.3.3 for a torispherical head, the head thickness

$$t_H = \frac{1.104 \ (25) \ (36)}{2 \ (0.80) \ (15200) - 0.2 \ (25)} = 0.04086 \text{ in } (1.04 \text{ mm})$$

Next, calculate the shell thickness from Equations 6.3.2 and 6.3.4. From Table 6.1, with no x-ray inspection of the longtudinal weld, $\varepsilon = 0.7$. Again, if we neglect the corrosion allowance,

$$t_S = \frac{25\,(36)}{2\,(0.7)\,(15200) - 1.2\,(25)} = 0.04235 \text{ in}\,(1.08 \text{ mm})$$

Thus, the shell wall thickness is essentially the same as the head thickness. According to Table 6.2, the minimum wall thickness is 3/32 in (2.38 mm) for high-alloy steels. The application of this rule-of-thumb more than doubles the wall thickness, which should be an adequate corrosion allowance. The selection of a corrosion allowance in the final design must be based on past experience or from laboratory and pilot plant tests.

PHASE SEPARATORS

Gas-Liquid Separators

As stated by Holmes and Chen [12], the reasons for using gas-liquid or vapor-liquid separators are to recover valuable products, improve product purity, reduce emissions, and protect downstream equipment. Gas-liquid separators are used after flashing a hot liquid across a valve. In this case the separator is called a flash drum.

A vertical gas-liquid separator is shown in Figure 6.4. The gas-liquid mixture is separated by gravity and impaction. The mixture enters the separator about midway where a splash plate deflects the stream downward. Most of the liquid flows downward, and the vapor, containing liquid drops, flow upward. As the vapor rises, large drops settle to the bottom of the separator by gravity. According to Watkins [14], 95 % separation of liquid from vapor is normal. If greater than 95 % liquid separation is required, then use a wire-mesh mist eliminator, installed near the vapor outlet. Very small drops are separated by impaction using a wire-mesh pad located at the top of the separator. The mesh usually consists of 0.011 in (0.279 mm) diameter wires interlocked by a knitting machine to form a pad from 4 to 6 in (0.102 to 0.152 m) thick [12]. Entrained liquid drops in the vapor impact on the wires and coalesce until the drops become heavy enough to break away from the wire and fall to the bottom of the separator. Because of the large free volume of the pad – 97 to 99 % – the pressure drop across the pad is usually less than 1.0 in of water [13]. The separation efficiency of a pad is about 99.9% or greater.

The major objective in sizing a gas-liquid separator is to lower the gas velocity sufficiently to reduce the number of liquid droplets from being entrained in the gas. Thus, the separator diameter must be determined. The separator is also designed as an accumulator for the liquid portion of the stream. Thus, the liquid

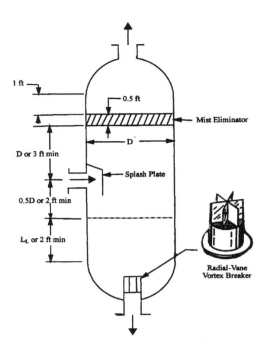

Figure 6.4 A vertical gas-liquid separator.

height is calculated by allowing sufficient surge time to dampen flow-rate variations of the liquid stream, as was discussed earlier for accumulators. Presumably,
this liquid height will also be sufficient to allow vapor bubbles to rise to the top of
the liquid before being trapped in the outlet stream at the bottom of the vessel.
This can be achieved by reducing the outlet liquid velocity by increasing the diameter of the outlet nozzle.

In a separator, there is not a single drop size but a distribution of drop sizes.
To prevent all drops from being carried out by the gas stream would require an
uneconomically large separator. Thus, a maximum gas velocity is specified so
that all but the very small drops are recovered. An empirical expression for the
maximum gas velocity is derived by considering the forces acting on a small drop
suspended in a gas stream. These forces are gravity acting downward and the
buoyant and drag forces acting upward. Thus,

$$F_G = F_B + F_D \tag{6.9}$$

From Newton's law for the gravitation force, Archimedes principle for the buoyant force, and the definition of the drag force, the force balance on a drop becomes

$$m_L \, g = m_L \, \frac{\rho_V}{\rho_L} \, g + C_D \, A_L \, \frac{\rho_L \, v_V^2}{2 \, g} \qquad (6.10)$$

where: m_L is the mass of a drop, g the acceleration of gravity, ρ the density of either liquid or vapor, C_D the drag coefficient, A_L the projected area of a drop, and v_V, the maximum vapor velocity.

Solving for the maximum vapor velocity, we find that

$$v_V = \left(\frac{2 \, m_L \, g^2}{C_D \, A_L \, \rho_L} \right)^{1/2} \left(\frac{\rho_L - \rho_V}{\rho_V} \right)^{1/2} \qquad (6.11)$$

Equation 6.11 does not accurately describe the physical situation. In practice, what is done is to set the coefficient of Equation 6.11 equal to k_V so that

$$v_V = k_V \left(\frac{\rho_L - \rho_V}{\rho_V} \right)^{1/2} \qquad (6.12)$$

where k_V is an empirical constant that depends on the properties of the fluids, the design of the separator, the size of the drops, the vapor velocity, and the degree of separation required.

Knock-Out Drums

Knock-out drums, used when the liquid content of the incoming stream is low, is a special case of a gas-liquid separator. The drum is placed before a compressor inlet to prevent liquid drops from entering and damaging the compressor. In this case, allowing a sufficient residence time for the liquid is not a consideration.

To determine the length and diameter of knock-out drums, Younger [11] recommends using a value of k_V of 0.2 ft/s (0.01 m/s) without a mist eliminator or 35 ft/s (0.107 m/s) with a mist eliminator, and an L/D ratio of 2. A calculation procedure for solving the equations listed in Table 6.7, is given in Table 6.8. The volume of the dished heads is not considered in the procedure. Knock-out drums are mostly installed in a vertical position. Younger [11] found that out of eleven drums installed in several plants, nine were vertical.

Table 6.7 Summary of Equations for Sizing Knock-Out Drums

Subscripts: L = liquid — V = vapor

$$v_V = k_V \left(\frac{\rho_L' - \rho_V'}{\rho_V'} \right)^{1/2}$$ (6.7.1)

$k_V = 0.2$ ft/s (0.061 m/s) — with no mist eliminator (6.7.2)
$k_V = 0.35$ ft/s (0.107 m/s) — with a mist eliminator

$V_V' = v_V A$ (6.7.3)

$A = \pi D^2/4$ (6.7.4)

$L/D = 2$ (6.7.5)

Variables
v_V - k_V - A - D – L

Table 6.8 Calculation Procedure for Sizing Knock-Out Drums

1. Select a value of k_V from Equation 6.7.2.

2. Calculate a maximum gas velocity, v_V, from Equations 6.7.1.

3. Calculate the cross-sectional area of the separator, A, from Equation 6.7.3.

4. Calculate the diameter, D, of the separator from Equation 6.7.4.

5. Round off D in 6 in (0.152 m) increments, starting at 30 in (0.762 m). If D is less then 30 in (0.762 m), use standard pipe.

6. Calculate the length of the separator from Equation 6.7.5. Round off L in 3 in (76.2 mm) increments, for example, 5.0, 5.25, 5.5, 5.75 ft, etc.

Example 6.2 Sizing a Compressor Knock-Out Drum

A gas stream having the composition given in Table 6.2.1 flows into a compressor suction. Size the knockout drum to prevent liquid from entering the compressor. The gas enters the drum at 105 °F (40.6 °C) and 150 psig (10.3 bar).

Table 6.2.1 Gas Composition

Gas	Flow Rate[a] lbmol/h
H_2	2312.8
CH_4	277.5
C_2H_6	246.7
C_3H_8	185.0
i-Butane	61.7

a) To covert to kgmol/h multiply by 0.4536.

Follow the calculation procedure outlined in Table 6.8. Assume that the drum will have a mist eliminator. From Equation 6.7.2, $k_V = 0.35$ ft/s (0.107 m/s). The effect of the mist eliminator is to increase the maximum allowable velocity and therefore to reduce the drum diameter. The densities obtained from ASPEN [57] are: $\rho_V = 0.2493$ lb/ft^3 (3.99 kg/m^3) and $\rho_L = 33.19$ lb/ft^3 (532 kg/m^3). The volumetric flow rate, also obtained from ASPEN, is 1.134×10^5 ft^3/h (3210 m^3/h).

From Equation 6.7.1, the maximum-allowable gas velocity,

$$v_V = 0.35 \left(\frac{33.19 - 0.2493}{0.2493} \right)^{1/2} = 4.023 \text{ ft/s (1.23 m/s)}$$

From Equation 6.7.3, the cross-sectional area,

$$A = \frac{1.134 \times 10^5 \text{ ft}^3/\text{h}}{1} \frac{1}{3600 \text{ s/h}} \frac{1}{4.023 \text{ ft/s}} = 7.830 \text{ ft}^2$$

The drum diameter from Equation 6.7.4 is

$$D = \left(\frac{7.83\ (4)}{3.142} \right)^{1/2} = 3.157 \text{ ft } (0.9677 \text{ m})$$

According to step five in Table 6.8, round off the diameter to 3.5 ft (1.07 m).

Finally, from Equation 6.7.5, the length of the drum is

$$L = 2\ (3.5) = 7.0 \text{ ft } (2.13 \text{ m})$$

Vertical Gas-Liquid Separators

There are several design procedures reported in the literature – not all of them are in agreement. A schematic diagram of a vertical gas-liquid separator is shown in Figure 6.4. Gas-liquid separators may be designed for horizontal or vertical operation, but Younger [11] found that for seven separators in use, with L/D varying from 1.7 to 3.6, all were installed vertically. This is consistent with the rule given by Branan [49] that if L/D > 5, a horizontal separator should be used. Equations for sizing vertical gas-liquid separators are summarized in Table 6.9, and a calculation procedure is outlined in Table 6.10. The volume of the dished heads is not included in the calculation procedure. As for sizing knockout drums, first calculate the drum diameter by solving Equations 6.9.1 to 6.9.4.

Next calculate the droplet settling length. This is the length from the center line of the inlet nozzle to the bottom of the mist eliminator. Scheiman [72] recommends that the settling length should be to 0.75 D or a minimum of 12 in (0.305 m) whereas Gerunda [4] specifies a length equal to the diameter or a minimum of 3 ft (0.914 m). Gerunda's recommendation is used in Figure 6.4.

Also, to prevent flooding the inlet nozzle, Scheiman allows a minimum of 6 in (0.152 m) from the bottom of the nozzle to the liquid surface or a minimum of 12 in (0.305 m) from the center line of the nozzle to the liquid surface. Branan [49] recommends using 12 in (0.305 m) plus ½ of the inlet nozzle outside diameter or 18 in (0.4570 m) minimum. Gerunda specifies a length equal to 0.5 D or 2 ft (0.610 m) minimum, which is used in Figure 6.4.

Now calculate the liquid height. The separator is also sized as an accumulator to dampen variations in the liquid flow rate by allowing sufficient liquid residence time or surge time in the separator. Scheiman [72] recommends a surge time in the range of 2 to 5 min, whereas Younger [11] recommends 3 to 5 min. In Table 6.9, 3 to 5 min is selected. There is a minimum liquid height required to prevent a vortex from forming. The design of the separator will have to include a vortex breaker. The minimum liquid level should cover the vortex breaker plus an additional liquid height. Experiments conducted by Patterson [16] showed that the

Table 6.9 Summary of Equations for Sizing Vertical Gas-Liquid Separators

---

Subscripts: L = liquid — V = vapor

$$V_V' = v_V \, A \tag{6.9.1}$$

$$v_V = k_V \left(\frac{\rho_L' - \rho_V'}{\rho_V'} \right)^{1/2} \tag{6.9.2}$$

$$
\begin{aligned}
&k_V = 0.1 \text{ ft/s } (0.03045 \text{ m/s}) \text{ — with no mist eliminator} \\
&k_V = 0.35 \text{ ft/s } (0.0107 \text{ m/s}) \text{ — with a mist eliminator}
\end{aligned} \tag{6.9.3}
$$

$$A = \pi \, D^2/4 \tag{6.9.4}$$

$$L_L \, A = V_L' \, t_S \text{ — where the minimum value of } L_L \text{ is 2 ft (0.610 m)} \tag{6.9.5}$$

$$3 \le t_S \le 5 \text{ min} \tag{6.9.6}$$

$$
\begin{aligned}
&L = L_L + 1.5 \, D + 1.5 \text{ ft or} \\
&L = 8.5 \text{ ft (2.59 m) — whichever is larger}
\end{aligned} \tag{6.9.7}
$$

Variables
v_V - A - k_V - D - L - L_L - t_S

---

lower liquid level varies slightly with the liquid velocity in the outlet nozzle. For a velocity of 7 ft/s (2.13 m/s) in the outlet piping of a tank, with no vortex breaker, a vortex forms at a liquid level of about 5 in (0.127 m). The flow should be turbulent to break up any vortex. Thus, Gerunda's recommendation, allowing a 2 ft (0.610 m) minimum liquid level, should suffice.

To complete calculating the length of the separator, specify the thickness of the mist eliminator, which must be thick enough to trap most of the liquid droplets rising with the vapor. The thickness of the eliminator is usually 6 in (0.152 m). Finally, an additional 12 in (0.305 m) above the eliminator is added to obtain uniform flow distribution across the eliminator. If the eliminator is too close to the outlet nozzle, a large part of the flow will be directed to the center of the eliminator, reducing its efficiency. The total length of the separator can now be calculated by summing up the dimensions given in Figure 6.4. According to Branan [49], if L/D is greater than 5, use a horizontal separator. Also, Branan states that if L/D < 3,

Table 6.10 Calculation Procedure for Sizing Vertical Gas-Liquid Separators

---

1. Select k_V from Equation 6.9.3.

2. Calculate the maximum gas velocity, v_V, from Equations 6.9.2.

3. Calculate the cross-sectional area, A, from Equation 6.9.1.

4. Calculate D from Equation 6.9.4.

5. Round off D in 6 in (0.152 m) increments, starting at 30 in (0.762 m). If D is less than 30 in (0.762 m), use standard pipe.

6. Select a liquid-phase surge time, t_S, from Equation 6.9.6.

7. Calculate the liquid-level height from Equation 6.9.5.

8. Calculate the total separator height from Equation 6.9.7. Round off L in 3 in (0.0762 m) increments, for example, 5.0, 5.25, 5.5, 5.75 ft etc.

9. If L/D < 3.0, then recalculate L so that L/D > 3.0 by letting L/D = 3.2. If L/D > 5 use a horizontal separator.

---

increase L in order that L/D > 3, even if the liquid surge volume is increased. Increasing the surge volume is in the right direction.

Horizontal Gas-Liquid Separators

Like vertical gas-liquid separators, there are several design procedures reported in the literature – not all of them are in agreement. A schematic diagram of a horizontal gas-liquid separator is shown in Figure 6.5. For horizontal separators, the calculation procedure for sizing is essentially the same as vertical separators except increase k_V by 25 % [49]. Also, the minimum value of the cross-sectional area for gas flow should be at least 20 % of the total cross-sectional area of the separator [49]. Use a 6 in (0.152 m) mist eliminator and a distance of 12 in (0.3048m) above the eliminator. According to Gerunda [4], the distance from the bottom of the mist eliminator to the liquid level should be at least 2 ft (0.610 m) and should not be below the center of the separator. Scheinman [72] recommends 6 in (0.152 m). Use an average of 1.25 ft. The main consideration is to prevent

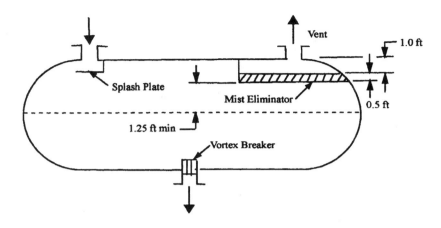

Figure 6.5 A horizontal gas-liquid separator.

Table 6.11 Summary of Equations – Sizing Horizontal Gas-Liquid
Separators

Subscripts: L = liquid —— V = vapor

$$V_V' = 0.5 \, v_V \, A \tag{6.11.1}$$

$$v_V = 1.25 \, k_V \left(\frac{\rho_L' - \rho_V'}{\rho_V'} \right)^{1/2} \tag{6.11.2}$$

$k_V = 0.10$ ft/s (0.0305 m/s) —— with no mist eliminator \qquad (6.11.3)
$k_V = 0.35$ ft/s (0.107 m/s) —— with a mist eliminator

$$A = \pi \, D^2/4 \quad \text{—— minimum D} = 5.5 \text{ ft (1.67 m)} \tag{6.11.4}$$

$$0.5 \, L \, A = V_L' \, t_S \tag{6.11.5}$$

$$7.5 \leq t_S \leq 10 \text{ min} \tag{6.11.6}$$

Variables

v_V - A - k_V - D - L - t_S

the mist eliminator from flooding because of a rising liquid level. We will design
for a liquid level at the center of the separator. These rules result in a minimum
diameter of 5.5 ft if the liquid level is at the center of the separator, as shown in

Figure 6.5. This diameter might result in a short separator length if the liquid flow rate is small. If this occurs it may be necessary to increase the separator length, or employ other designs for reducing the diameter as given by Sigales [73]. The equations are listed in Table 6.11, and the calculation procedure for calculating L and D is given in Table 6.12. As was the case for vertical gas-liquid separators, if L/D < 3, increase L so that L/D > 3, even if the liquid surge volume is increased. Similarly, if L/D > 5 increase D so that L/D < 5. Increasing D will reduce the gas velocity and increase the liquid surge volume, which is in the right direction. The volume of the dished heads is not included in the design procedure. Example 6.3 illustrates the calculation procedure for sizing horizontal gas-liquid separators.

Table 6.12 Calculation Procedure for Sizing Horizontal Gas-Liquid Separators

1. Select k_V from Equation 6.11.3.

2. Calculate the maximum vapor velocity, v_V, from Equation 6.11.2.

3. Calculate the cross-sectional area, A, from Equation 6.11.1.

4. Calculate D using Equation 6.11.4. Round off D in 6 in (0.152 m) intervals, starting at 30 in (0.762 m). If D is less then 30 in (0.762), use standard pipe.

5. Select a liquid phase surge time, t_S, from Equation 6.11.6.

6. Calculate the separator length from Equation 6.11.5. Round off L in 3 in (0.0762 m) intervals (for example, in feet, 5.0, 5.25, 5.5, 5.75 etc.)

7. If L/D < 3.0, then recalculate L so that L/D > 3.0 by setting L/D = 3.2. If L/D > 5.0, then recalculate D so that L/D < 5.0 by setting L/D = 4.8.

Example 6.3 Sizing a Gas-Liquid Separator

Calculate the length and diameter of a gas-liquid separator to separate 200.7 ft^3/min (5.68 m^3/min) of vapor from 5.0 gal/min (0.0189 m^3/min) of a liquid.

Data
vapor density	1.372 lb/ft^3 (21.98 kg/m^3)
liquid density	31.15 lb/ft^3 (499.0 kg/m^3)
design pressure	50 psig (3.45 barg)
design temperature	200 °F (93.3 °C)
material	carbon steel

noncorrosive service

Follow the procedure outlined in Table 6.10. Assume that a vertical separator with a mist eliminator will be used. From Equation 6.9.3, $k_V = 0.35$ ft/s (0.107 m/s).

From Equation 6.9.2, the maximum vapor velocity,

$$v_V = 0.35 \left(\frac{31.15 - 1.372}{1.372} \right)^{1/2} = 1.631 \text{ ft/s } (0.497 \text{ m/s})$$

From Equation 6.9.1, the cross-sectional area of the separator,

$$A = 200.7 / 60 \, (1.631) = 2.051 \text{ ft}^2 \, (0.191 \text{ m}^2)$$

From Equation 6.9.4, the separator diameter,

$$D = [\, (4 / 3.142) \, (2.051) \,]^{1/2} = 1.616 \text{ ft } (0.493 \text{ m})$$

Because the separator diameter is below 30 in (0.762 m), select standard pipe. From the chemical engineering handbook (6.66), the closest pipe size is 20 in (0.508 m), Schedule 10 pipe, which has an inside diameter of 19.50 in (1.625 ft, 0.495 m), an inside cross-sectional area of 2.074 ft^2 (0.193 m^2), and a wall thickness of 0.25 in (6.35 mm). From piping tables, the allowable pressure for carbon steel at 200 °F (93.3 °C) is 186 psig (12.8 barg), which is above the design pressure of 50 psig (3.45 barg).

Now, calculate the length of the separator. First, calculate the height of the liquid from Equation 6.9.5. Use an average of the residence times given by Equation 6.9.6.

$$L_L = \frac{5.0 \text{ gal/min}}{7.481 \text{ gal/ft}^3} \quad \frac{4 \text{ min}}{1} \quad \frac{1}{2.074 \text{ ft}^2} = 1.289 \text{ ft } (0.393 \text{ m})$$

The minimum liquid level is 2.0 ft (0.610 m).

From Equation 6.9.7,

$$L = 2.0 + 1.5 \, (1.625) + 1.5 = 5.938 \text{ ft } (1.81 \text{ m})$$

Round off the length in three-inch intervals. Therefore, $L = 6.0$ ft (1.83 m), but according to Figure 6.4, the minimum length is 8.5 ft (1.83 m).

L/D = 8.5 / (1.625) = 5.23. Because L/D is greater than 5.0, size a horizontal separator. We could stop here, however, because five is not a precise number and 5.23 is close to 5.0.

Now, size a horizontal separator using the procedure outlined in Table 6.12 and the equations listed in Table 6.11.

Select a mist eliminator. From Equation 6.11.3, k_V = 0.35 ft/s (0.0107 m/s).

From Equation 6.11.2, the maximum vapor velocity,

$$v_V = 1.25\,(0.35)\left(\frac{31.15 - 1.372}{1.372}\right)^{1/2} = 2.038 \text{ ft/s } (0.621 \text{ m/s})$$

From Equation 6.11.1,

$$A = 200.7 / 60\,(0.5)\,(2.038) = 3.283 \text{ ft}^2 \,(0.0283 \text{ m}^2)$$

From Equation 6.11.4, the separator diameter,

$$D = [\,(4 / 3.142)\,(3.283)\,]^{1/2} = 2.044 \text{ ft } (0.623 \text{ m})$$

From Figure 6.5, the minimum vapor-phase height is 2.75 ft (0.838 m). Because the liquid level is at the middle of the separator, the minimum D = 5.5 ft (1.68 m).

From Equation 6.11.,

$$A = (3.142 / 4)\,(5.5)^2 = 23.76 \text{ ft}^2 \,(2.21 \text{ m}^2)$$

Now, from Equation 6.11.5 for a separator that is half filled with liquid,

$$L = \frac{5.0 \quad \text{gal/min}}{7.481 \quad \text{gal/ft}^3}\;\frac{8.75 \text{ min}}{1}\;\frac{1}{0.5\,(23.76) \text{ ft}^2} = 0.4923 \text{ ft } (0.150 \text{ m})$$

which, clearly, is not satisfactory.

The L/D ratio should be in the range of 3.0 < L/D < 5.0. If we select 3.2, then,

$$L = 3.2\,(5.5) = 17.60 \text{ ft } (5.36 \text{ m})$$

Round off the length to 17.75 ft (5.41 m). This separator is larger than the vertical separator.

Let us try to reduce the size of the horizontal separator. If we move the mist eliminator to outside of the separator shell, as shown in Figure 6.3.1, the diameter

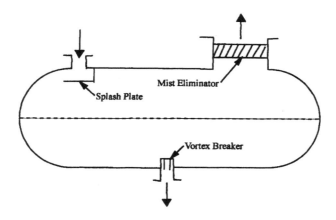

Figure 6.3.1 A horizontal gas-liquid separator with an external mist eliminator.

will be reduced. From the calculation of separator diameter given above, D =2.044 ft (24.53 in, 0.623 m). Because D < 30 in (0.762 m), we can use pipe. From the chemical engineering hand book [66], select a 30 in (0.762 m) Schedule ST pipe, which has an inside diameter of 29.25 in (2.438 ft, 0.743 m), an inside cross-sectional area of 4.666 ft² (0.4335 m²), and a wall thickness of 0.375 in (9.53 mm). The allowable pressure for carbon steel at 200 °F (93.3 °C) is 266 psig (18.3 barg), which is above the design pressure of 50 psig (3.45 barg).

$$ L = \frac{5.0 \ \text{gal/min}}{7.481 \ \text{gal/ft}^3} \ \frac{8.75 \ \text{min}}{1} \ \frac{1}{0.5 \ (4.707) \ \text{ft}^2} = 2.507 \ \text{ft} \ (0.764 \ \text{m}) $$

Round off the length to 2.5 ft (0.762 m).
    Check the L/D ratio.

L/D = 2.5 / 2.438 = 1.03

which is not within the limits of 3.0 < L/D < 5.0.

If we select L/D = 3.2,

L = 3.2 (2.448) = 6.541 ft (1.994 m)

Round off the length to 6.75 ft (2.06 m). The effect of increasing the separator length is to increase the surge time above the original 8.75 min.

The vertical separator is 1.625 ft (0.494 m) in diameter and 8.5 ft (2.59 m) long. At this point, it appears that the vertical separator is the best choice because of its smaller diameter and wall thickness. Also, locating the mist eliminator outside of the separator shell will add to the cost of the horizontal separator.

Liquid-Liquid Separators

Liquid-liquid separators are also called decanters or settlers. The flow to the settler consists of a dispersed phase and a continuous phase, and the function of a settler is to coalesce and separate the dispersed phase from the continuous phase. The separator volume must be sufficiently large to allow sufficient time for the dispersed-phase drops to reach the liquid-liquid interface and coalescence. Thus, the residence time has two components. These are: the time required for the droplets to reach the interface and the time required for the droplets to coalesce.

Figure 6.6 shows a design for a decanter. After the two-phase mixture enters the decanter at the feed nozzle, the liquid jet must be diffused to prevent mixing of the two phases and promote settling of the dispersed phase. One way to accomplish this is to insert two closely spaced, perforated parallel plates across the jet, as shown in Figure 6.6. The first plate drops the pressure of the jet, and the second plate decreases its velocity. Jacobs and Penny [17] recommend that the flow area of the first plate be 3 to 10% of the decanter flow area, and the second plate 20 to 50% of the decanter flow area. Another way to disperse the entering liquid jet, and at the same time enhance coalescence of the dispersed phase, is to use a wire-mesh pad in front of the feed nozzle.

After flowing past the plates, the liquid-liquid mixture flows down the length of the decanter. Either the light or heavy phase could be dispersed, depending on the properties of both phases. The dispersed-phase drops will either

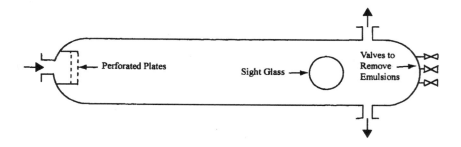

Figure 6.6 A liquid-liquid separator

move downward or upward toward the interface, depending on the specific gravity of the two liquids. Then, at the interface, drops will accumulate before coalescing with one of the phases.

To prevent entraining either the light or heavy phase in the outlet streams, the liquid velocity in both outlet nozzles should be low. According to Jacobs and Penny [17], the liquid velocity in each outlet nozzle should not be any more than 10 times the average velocity of each phase in the decanter. This rule allows sizing the outlet nozzles.

If either a surface-active agent or a dispersion of fine solids is present, a stable emulsion could form, which is analogous to foam in a gas-liquid system. The emulsion or "rag" accumulates and will eventually have to be removed from the decanter using valves located at the end of the vessel. After removal, the emulsion can be de-emulsified by filtration, heating, adding chemical de-emulsifying agents, or reversing the phase that is dispersed.

There appears to be no satisfactory sizing procedure for decanters. Drown and Thomson [18] compared three sizing procedures and found that all were unsatisfactory. We will develop a simple method here to illustrate some of the factors involved and to obtain a preliminary estimate of the decanter size. Accurate sizing must be supplemented by testing. Even though settling and coalescing of drops occur simultaneously, it will be assumed that first the drops flow to the interface, and then the drops coalesce with the appropriate phase. This simple model is illustrated in Figure 6.7.

The first step in developing a sizing procedure is to determine which phase is dispersed. Selker and Sleicher [19] found that the value of the parameter θ, defined by Equation 6.15.1 in Table 6.15, could be used as a guide to determine the dispersed phase. After calculating θ, then use Table 6.13 to identify the dispersed phase.

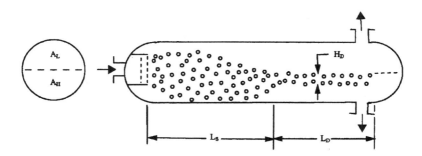

Figure 6.7 An idealized liquid-liquid-separator model.

Table 6.13 Dispersed-Phase Parameter in Liquid-Liquid Separation

θ	Result
< 0.3	light phase always dispersed
0.3 - 0.5	light phase probably dispersed
0.5 - 2.0	phase inversion probable, design for worst case
2.0 - 3.3	heavy phase probably dispersed
>3.3	heavy phase always dispersed

Source: Ref. 19.

Table 6.14 Effect of Turbulence on Liquid-Liquid Separation

Reynolds Number	Effect
<5000	little problem
5000 - 20,000	some hindrance
20,000 - 50,000	major problem may exist
>50,000	expect poor separation

Source Ref. 21.

Liquid-liquid separation is hindered by turbulence. Bailes et al. [21] determined the effect of turbulence on the separation, which is given in Table 6.14. The separator diameter is calculated to minimize turbulence. Increasing the separator diameter reduces the Reynolds number and therefore turbulence. Thus, use Table 6.14 as a guide in calculating the diameter. Because there are two phases, calculate the Reynolds numbers for both the light and heavy phases. Because the flow area for both phases is not circular, the diameter in the Reynolds number must be replaced by an equivalent diameter for the noncircular flow area. The equivalent diameter is equal to four times the hydraulic radius, which is defined as the cross-sectional area of the stream (flow area) divided by the wetted perimeter. The definition of hydraulic radius is only valid for turbulent flow, as discussed by Bird et al. [68]. For a liquid-liquid interface located at the center of the decanter, the flow area is equal to ½ the cross-sectional area of the separator, and the wetted perimeter is equal to the separator diameter plus ½ its circumference.

Table 6.15 Summary of Equations for Sizing Liquid-Liquid Separators

Subscripts: L = light phase – H = heavy phase
D = dispersed phase – C = continuous phase

Transport Relations

$$\theta = \frac{V_L'}{V_H'} \left(\frac{\rho_L' \, \mu_H'}{\rho_H' \, \mu_L'} \right)^{0.30} \tag{6.15.1}$$

$$V_D = V_L' \text{ or } V_D = V_H' \quad \text{---} \quad \text{from Table 6.13} \quad \text{---} \quad V_D, \text{ light or heavy phase} \tag{6.15.2}$$

$$v_D = v_L \text{ or } v_D = v_H \quad \text{---} \quad \text{from Table 6.13} \quad \text{---} \quad v_D, \text{ light or heavy phase} \tag{6.15.3}$$

$$\rho_C = \rho_L' \text{ or } \rho_C = \rho_H' \quad \text{---} \quad \text{from Table 6.13} \quad \text{---} \quad \rho_C, \text{ light or heavy phase} \tag{6.15.4}$$

$$\rho_D = \rho_L' \text{ or } \rho_D = \rho_H' \quad \text{---} \quad \text{from Table 6.13} \quad \text{---} \quad \rho_D, \text{ light or heavy phase} \tag{6.15.5}$$

$$\mu_C = \mu_L' \text{ or } \mu_C = \mu_H' \quad \text{---} \quad \text{from Table 6.13} \quad \text{---} \quad \mu_C, \text{ light or heavy phase} \tag{6.15.6}$$

$$v_d = \frac{g \, (d')^2 \, (\rho_D' - \rho_C')}{18 \, \mu_C} \tag{6.15.7}$$

$$t_D = D \, / \, 2 \, v_d \tag{6.15.8}$$

$$L_S = v_D \, t_D \tag{6.15.9}$$

$$H_D = 0.1 \, D \tag{6.15.10}$$

$$t_R' = \frac{(1/2) \, H_D \, A_I}{V_D} \tag{6.15.11}$$

Table 6.15 Continued

$$A_I = L_D D \tag{6.15.12}$$

$$L = L_S + L_D \tag{6.15.13}$$

$$A_L = \pi D_L^2 / 8 \tag{6.15.14}$$

$$A_H = \pi D_H^2 / 8 \tag{6.15.15}$$

$$v_L = V_L' / A_L \tag{6.15.16}$$

$$v_H = V_H' / A_H \tag{6.15.17}$$

$$Re_L = \frac{4 R_{h\,L}\, \rho_L'\, v_L}{\mu_L'} \tag{6.15.18}$$

$$Re_H = \frac{4 R_{h\,H}\, \rho_H'\, v_H}{\mu_H'} \tag{6.15.19}$$

$$R_{h\,L} = \frac{\pi D_L / 4}{2 + \pi} \tag{6.15.20}$$

$$R_{h\,H} = \frac{\pi D_H / 4}{} \tag{6.15.21}$$

$$Re_L \le 10{,}000 \tag{6.15.22}$$

$$Re_H \le 10{,}000 \tag{6.15.23}$$

$$D = D_L \text{ or } D = D_H \;\; -\!- \;\; \text{whichever is greater} \tag{6.15.24}$$

Variables

v_d - v_D - v_L - v_H - V_D - ρ_D - ρ_C - μ_C - Re_L - Re_H - R_{hH} - $R_{h\,L}$ - D - D_L - D_H - L - L_S - L_D - A_L - A_H - A_I - H_D - t_D - θ

Therefore, the flow area,

$$A_F = \frac{1}{2} \frac{\pi D^2}{4} \qquad (6.13)$$

and the wetted perimeter,

$$P = D + \frac{\pi D}{2} \qquad (6.14)$$

For both phases, the hydraulic radius,

$$R_h = \frac{A_F}{P} = \frac{\pi D/4}{2 + \pi} \qquad (6.15)$$

provided the interface is in the center of the decanter.

In a horizontal decanter, dispersed phase drops are being carried along the decanter by the flow of the continuous phase. If the velocity of the two separated layers is more than a few centimeters per second, the shape of the dispersion zone will be distorted by drag, and there will be entrainment of drops [21]. Therefore, the Reynolds number for both phases must be limited. The effect of Reynolds number on liquid-liquid separation is shown in Table 6.14. This limitation on the Reynolds number will also be used for the dispersed phase to determine the decanter diameter. The minimum diameter is 10.0 cm (0.328 ft) because of wall effects [19].

Stokes' Law is usually used to estimate the settling time of liquid drops in decanters, and hence the length of the settling zone, even though the assumptions used to derive Stokes' Law are not strictly met. These assumptions are:

1. the continuous phase is a quiescent fluid.
2. the drop is a sphere with no internal circulation.
3. the drop moves in laminar flow.
4. the drop is large enough to ignore Brownian motion.
5. the drop movement is not hindered by other droplets or by the wall of the separator.

Stokes' Law, which gives the terminal velocity of a drop in a stationary, continuous-phase liquid is given by

$$v_d = \frac{g \, d^2 \, (\rho_H - \rho_L)}{18 \, \mu_C} \qquad (6.16)$$

where the subscripts refer to heavy (H), light (L), and continuous phase (C).

The drop diameter, d, for use in Equation 6.16 is difficult to determine. There is not a single drop size but a distribution of drop sizes. Jacobs and Penny [17] recommend a drop diameter of 150 micrometers, which is conservative and compensates somewhat for the other assumptions in Equation 6.16.

Once the drop terminal velocity is found, the time taken for the dispersed phase to reach the interface is given by Equation 6.15.8 in Table 6.15, and the decanter length required for the droplets to settle is given by Equation 6.15.9. The maximum distance that the disperse phase droplets have to travel to reach the interface, which is located at the center of the separator, is D/2. The distance varies from zero to D/2. Also, the path of the droplets is not straight down or up but will curve because of the motion of the phases.

The length of the coalescing zone of the decanter is determined by the time required for the dispersed phase to coalesce. Coalescence could occur by drop to drop coalescence and drop to interface coalescence. There is no relationship that can predict the time required for coalescence, which according to Drown and Thomson [18] could vary from seconds to many hours. Coalescence is enhanced when the continuous phase viscosity is small, the density difference between phases large, the interfacial tension large, and the temperature high. Because of the time it takes for coalescence, the dispersed phase drops accumulate near the interface to form a dispersion zone. Jacobs and Penny [17] recommend that the dispersion zone thickness be kept to less than or equal to 10% of the decanter diameter as given by Equation 6.15.10. Also, the drops occupy about half of the volume of the dispersion zone volume. Neglecting the curvature of the separator, the dispersion zone volume is equal to $H_D A_I$, where H_D is the thickness of the dispersion zone, and A_I is the area of the interface. Therefore, the residence time, t_R, of the drops in the dispersion zone is given by Equation 6.15.11. The residence time is specified by experience, and the interfacial area required for coalescence is calculated. If it is assumed that the interface will be located at the center of the decanter, then the length of the coalescing zone, L_D, is calculated from Equation 6.15.12. The total length of the decanter is the sum of the lengths required for settling and coalescence. The procedure for calculating the dimensions of a decanter is given in Table 6.16, and Example 6.4 illustrates the procedure.

Table 6.16 Calculation Procedure for Sizing Liquid-Liquid Separators

1. Calculate θ to determine the dispersed phase from Equation 6.15.1 using Table 6.13.

2. Solve Equations 6.15.14, 6.15.16, 6.15.18, 6.15.20 and 6.15.22 for D_L, the inside diameter of the decanter, assuming that the light phase determines the diameter.

3. Also, solve Equations 6.15.15, 6.15.17, 6.15.19, 6.15.21, and 6.15.23 for D_H, the inside diameter of the decanter, assuming that the heavy phase flow determines the diameter.

4. The decanter diameter is the larger of the diameters calculated in Steps 2 and 3.

5. Round off D in six-inch (0.152 m) increments starting with 30 in (0.762 m). Below 30 in (0.762 m) use standard pipe.

6. Calculate v_d, the droplet velocity, from Equations 6.15.7 and 6.15.4 to 6.15.6

7. Calculate t_D, the dispersed-phase settling time, from Equation 6.15.8.

8. Calculate L_S, the decanter length required for settling of the dispersed phase from Equation 6.15.9.

9. Calculate H_D, the dispersion-zone height, from Equation 6.15.10.

10. Calculate A_I, the interfacial area required for coalescing the dispersed phase from Equation 6.15.11.

12. Calculate L_D, the decanter length required for coalescing the dispersed phase from Equation 6.15.12.

13. Calculate L, the total length of the decanter, from Equation 6.15.13. Round off L in 3 in (0.0762 m) increments, for example, 5.0, 5.25, 5.5, 5.75 ft, etc.

Example 6.4 Sizing a Liquid-Liquid Separator

An oil-water mixture is separated in a decanter. The properties of oil and water from an example by Hooper and Jacobs [22] are summarized in Table 6.4.1. If the residence time required for coalescence is 5.0 min, obtained from experiments, find the dimensions of the decanter.

The volumetric flow rates of both phases are

$$V_L = \frac{m_L}{\rho_L} = \frac{1.26 \text{ kg}}{1} \frac{1}{s} \frac{m^3}{897 \text{ kg}} = 1.405 \times 10^{-3} \text{ m}^3/\text{s } (0.0356 \text{ ft}^3/\text{s})$$

and

$$V_H = \frac{5.04 \text{ kg}}{1 \text{ s}} \quad \frac{1}{1000} \quad \frac{\text{m}^3}{\text{kg}} = 5.040 \times 10^{-3} \text{ m}^3/\text{s} \ (0.178 \text{ ft}^3/\text{s})$$

Table 6.4.1 Properties of Water-Oil Mixtures

Property	Oil	Water
ρ (kg/m^3)	897	1000
μ (Pa-s)	0.01	7.0×10^{-4}
m (kg/s)	1.26	5.04
d (m)	150×10^{-6}	
t_R (s)	300	

Follow the procedure given in Table 6.16. Step 1, requires determining the dispersed phase. From Equation 6.15.1,

$$\theta = \frac{1.405 \times 10^{-3}}{5.040 \times 10^{-3}} \left(\frac{897}{1000} \quad \frac{7 \times 10^{-4}}{0.01} \right)^{0.3} = 0.1215$$

Therefore, according to Table 6.13 the light phase or oil is dispersed, and the heavy phase or water is continuous. Therefore, from Equations 6.15.2 to 6.15.6,

$V_D = V_L$, $v_D = v_L$, $\rho_D = \rho_L$, $\rho_C = \rho_H$, and $\mu_C = \mu_H$.

Now, calculate the decanter diameter. After substituting A_L from Equation 6.15.14 into Equation 6.15.16, the superficial velocity for the light phase,

$$v_L = 8 \, V_L / \pi \, D^2$$

Next, substitute this equation and Equation 6.15.20 into Equation 6.15.18. Thus, the Reynolds number for light phase,

$$Re_L = \frac{8 \, \rho_L \, V_L}{(\pi + 2) \, \mu_L \, D_L}$$

Similarly, for the heavy phase substitute Equation 6.15.15 into Equation 6.15.17. Thus, the superficial velocity for the heavy phase,

$$v_H = 8 \, V_H / \pi \, D^2$$

Substituting this equation and Equation 6.15.21 into Equation 6.15.19, the Reynolds number for the heavy phase,

$$Re_H = \frac{8 \, \rho_H \, V_H}{(\pi + 2) \, \mu_H \, D_H}$$

From Equation 6.15.22 and the Reynolds number for the light phase given above, the decanter diameter,

$$D_L = \frac{8 \, \rho_L \, V_L}{(\pi + 2) \, \mu_L \, Re_L} = \frac{8}{(3.142 + 2)} \quad \frac{897 \text{ kg}}{1 \text{ m}^3} \quad \frac{1 \text{ m-s}}{0.01 \text{ kg}} \quad \frac{1.405 \times 10^{-3} \text{ m}^3}{1} \quad \frac{1}{\text{s } 1 \times 10^4}$$

$$= 0.01961 \text{ m} \, (0.06434 \text{ ft})$$

From Equation 6.15.23 and the Reynolds number for the heavy phase given above, the decanter diameter,

$$D_H = \frac{8 \, \rho_H \, V_H}{(\pi + 2) \, \mu_H \, Re_H} = \frac{8}{(3.124 + 2)} \quad \frac{1000 \text{ kg}}{1 \text{ m}^3} \quad \frac{1}{7.0 \times 10^{-4}} \quad \frac{\text{m-s}}{\text{kg}} \quad \frac{5.04 \times 10^{-3} \text{ m}^3}{1} \quad \frac{1}{\text{s } 1 \times 10^4}$$

$$= 1.124 \text{ m} \, (3.688 \text{ ft})$$

Therefore, the decanter diameter is 3.688 ft (1.124 m), which is rounded off to 4.0 ft (1.219 m). For the same conditions, but with the interface located above the center of the decanter, Hooper and Jacobs [22] obtained a diameter of 3.0 ft (0.914 m). Hooper and Jacobs located the interface above the center of the decanter, which lowers the heavy-phase velocity and hence the diameter.

The next step is to calculate the length of the decanter. The length is equal to the sum of the length required for the oil drops to reach the interface and the length required for the oil drops to coalesce with the oil phase at the interface.

From Equations 6.15.4 to 6.15.6, $\rho_C = \rho_H$, $\rho_D = \rho_L$, and $\mu_C = \mu_H$. The drop diameter used by Hooper and Jacobs [22] is 150 mm. According to Walas [6], 150 mm is a common drop diameter for the design of decanters. Then, from Equation 6.15.7, the settling velocity of a drop of oil,

$$v_d = \frac{1}{18} \quad \frac{9.807\ m}{1\ s^2} \quad \frac{(150 \times 10^{-6})^2\ m^2}{1} \quad \frac{(897-1000)\ kg}{1} \quad \frac{1}{m^3} \quad \frac{m\text{-}s}{7 \times 10^{-4}\ kg}$$

$$= -1.804 \times 10^{-3}\ m/s\ (-5.91 \times 10^{-3}\ ft/s)$$

The negative sign means that the oil drops move upward instead of downward.
From Equation 6.15.24, $D = D_H$. Substituting into the equation for v_L, given above,

$$v_L = 8\ V_L\ /\pi\ D_H{}^2$$

but from Equation 6.15.2, $V_D = V_L$ and from Equation 6.15.3, $v_D = v_L$. Therefore,

$$v_D = \frac{8\ (1.405 \times 10^{-3})\ m^3}{1} \quad \frac{1}{s} \quad \frac{1}{3.124\ (1.219)^2\ m^2} = 2.421 \times 10^{-3}\ m/s\ (7.94 \times 10^{-3}\ ft/s)$$

From Equations 6.15.8 and 6.15.9, the settling length,

$$L_S = \frac{v_D\ D}{2\ v_d} = \frac{2.421 \times 10^{-3}\ m}{2} \quad \frac{1.219\ m}{s} \quad \frac{1}{1} \quad \frac{s}{1.804 \times 10^{-3}\ m} = 0.8180\ m\ (2.68\ ft/s)$$

From Equation 6.15.10, the dispersion layer thickness is

$$H_D = 0.1\ D = 0.1\ (1.219) = 0.1219\ m\ (0.400\ ft)$$

From Equations 6.15.11, the interfacial area required for coalescence,

$$A_I = \frac{2\ V_D\ t_R}{H_D}$$

The residence time for the oil drops in the dispersion layer is 5 min and from Equation 6.15.2 $V_D = V_L$. Therefore, the interfacial area is

$$A_I = \frac{2\ (1.405 \times 10^{-3})\ m^3}{1} \quad \frac{300\ s}{s} \quad \frac{1}{0.1219\ m} = 6.916\ m^2\ (0.642\ ft^2)$$

and from Equation 6.15.12, the dispersion length is

$$L_D = \frac{A_I}{D} = \frac{6.916 \text{ m}^2}{1.219 \text{ m}} = 5.674 \text{ m (18.5 ft)}$$

Thus, the total decanter length is

$$L = L_S + L_D = 0.8180 + 5.624 = 6.442 \text{ m} \quad (21.14 \text{ ft})$$

Rounding the length off in 3 in increments, L = 21.25 ft (6.447 m). We should increase the decanter length to account for the diffuser plates at the entrance of the decanter. There appears to be no rule on the needed length except that the plates are closely spaced. We will assume six inches will be needed. Thus, L = 21.75 ft (6.63 m).

The length to diameter ratio is

$$\frac{L}{D} = \frac{6.63}{1.219} = 5.44$$

The ratio recommended by Barton [20] is five for settlers without considering the coalescence time for the droplets.

Solid-Liquid Separators

An example of a solid-liquid phase separation – often referred to as a mechanical separation – is filtration. Filters are also used in gas-solid separation. Filtration may be used to recover liquid or solid or both. Also, it can be used in waste-treatment processes. Walas [6] describes many solid-liquid separators, but we will only consider the rotary-drum filter. Reliable sizing of rotary-drum filters requires bench and pilot-scale testing with the slurry. Nevertheless, a model of the filtering process will show some of the physical factors that influence filtration and will give a preliminary estimate of the filter size in those cases where data are available.

Rotary-Drum Filters

As shown in Figure 6.8, a rotary-drum filter consists of three parts: a drum with an automatic filter valve, a filter tank with a slurry agitator, and a scraper for removing the cake. The drum rotates from 0.1 to 2 rpm about its horizontal axis [23]. Other characteristics are: drum diameters from 4 to 14 ft (1.22 to 4.27 m), drum length from 1.5 to 18 ft (0.427 to 5.49 m), and drum surface area from 18 to 783 ft^2 (1.67 to 72.7 m^2) [24]. A filter cloth is wrapped around the drum, which

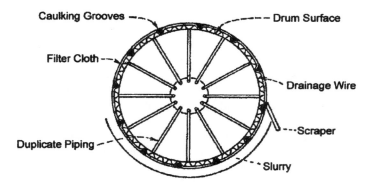

Figure 6.8 A rotary-drum filter. From Ref. 24 with permission.

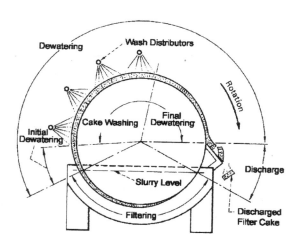

Figure 6.9 Filtration cycle for a rotary-drum filter. From Ref. 28 with permission.

is usually divided into 12 to 24 longitudinal compartments [25], depending on the drum diameter. Each compartment contains channels for collecting liquid that flows into filtrate piping, which leads to the filter valve at one end of the drum. A vacuum can be applied separately to each compartment. The drum is partially submerged in a slurry tank, which contains an agitator to prevent solids from settling. Usually, the slurry tank is designed to submerge about 40% of the drum area, but the maximum effective submerged filter area that can be subjected to vacuum is about 37.5%. As the drum rotates, each compartment is connected to an external system by the filter valve to apply vacuum, to collect filtrate, to collect wash water, or to apply air pressure to assist in removing solids from the drum.

The operation of a rotary-drum filter can be followed by examining Figure 6.9. In the cake-forming zone, slurry is drawn from the slurry tank onto the drum by a vacuum, depositing solids on the drum. After leaving this zone, the cake is dewatered, washed, if it is necessary, and then dewatered again before being discharged. In one method of cake removal, compressed air pushes the filter cloth against a knife that scrapes the cake from the cloth. The cake could also be removed by a roll, string or belt, depending on the cake thickness.

A simple rotary-filter system consists of a rotary filter and auxiliary equipment such as a compressor, a filtrate receiver, a filtrate pump, a vacuum pump, and a separator-silencer, as shown in Figure 6.10. Auxiliary equipment usually runs 25 to 40% of the filter cost [25]. When solids deposit on the drum, air and filtrate are drawn into the filtrate receiver, which is a gas-liquid separator. After

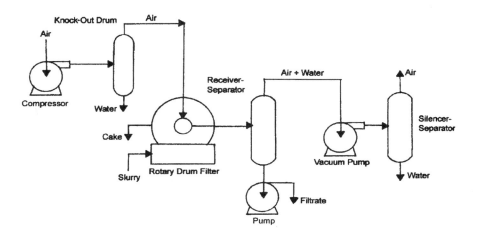

Figure 6.10 A rotary-drum filtration system.

separating the air and filtrate, the filtrate is pumped out for further processing, and the air is removed by the vacuum pump. The air then flows into the separator-silencer. The separator-silencer is another gas-liquid separator or knock-out drum. These drums are for small amounts of liquid entrained in the entering gas. In addition, the silencer attenuates the noise produced by the vacuum pump. An air compressor provides air to push the filter cloth against the scraper for cake removal. After the compressor is a knock-out drum for removing water drops produced by cooling of the compressed air. Other auxiliary equipment may be added to the filtration system, depending on the composition of the slurry. For example, if the liquid is an organic solvent, a component separator, such as an absorber, will be necessary to remove the solvent from the exhaust air. Also, if it is necessary to keep the filtrate and wash water separate, two receivers are used.

To obtain a formula for sizing a rotary-drum filter, the mechanism of liquid flow through a porous medium must be considered. As the filter drum rotates through the slurry tank, a porous solid deposits on the surface of the drum, increasing the resistance to liquid flow. The surface of the filter cake is at atmospheric pressure. If it is assumed that the pressure downstream of the filter medium is constant (created by a vacuum pump), then the pressure drop across the filter cake and medium is constant. As the filter cake thickens, the liquid flow rate decreases because of the increasing resistance to flow.

The starting point for deriving a formula to calculate the filtration area is the Kozeny-Carmen equation for flow through porous media. The flow, which is laminar, follows a tortuous path through the cake. The Kozeny-Carmen equation, for a differential cake thickness, is

$$\frac{dP}{dx} = \frac{4.17\, s^2\, \mu\, v_S\, (1 - \varepsilon)^2}{\varepsilon^2} \tag{6.17}$$

In Equation 6.17, P is the pressure at any point in the cake shown schematically in Figure 6.11, s, the specific surface (surface area per unit volume of particle), μ, the liquid viscosity, v_S, the superficial liquid velocity, and ε, the porosity of the cake. The Kozeny-Carmen equation is derived in a number of texts. See, for example, Bird et al. [26], who have called the equation the Blake-Kozeny equation.

Replace 4.17 in Equation 6.17 with k, because the coefficient varies with the type of material. In most cases, k is assigned a value of 5.0 for an isentropic cake having a porosity of $0.3 < \varepsilon < 0.6$ [67].

The differential mass of dry cake, dm, shown in Figure 6.11, is given by

$$dm = (1 - \varepsilon)\, \rho_S\, A_F\, dx \tag{6.18}$$

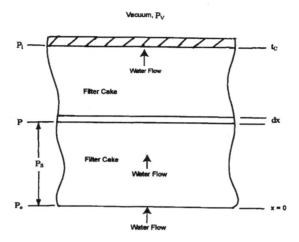

Figure 6.11 Section of a filter cake.

where dm is the differential amount of dry cake in a layer of thickness dx and the volume fraction of solids in the wet cake is $(1 - \varepsilon)$.

After substituting Equation 6.18 into Equation 6.17 to eliminate dx, we obtain

$$-\frac{dP}{dm} = \frac{k \, s^2 \, \mu \, v_S \, (1 - \varepsilon)}{\rho_S \, \varepsilon^2 \, A_F} \qquad (6.19)$$

Define a specific resistance, α.

$$\alpha = \frac{k \, s^2 \, (1 - \varepsilon)}{\rho_S \, \varepsilon^2} \qquad (6.20)$$

The specific resistance, which has units of m/kg, depends on the characteristics of the cake. As the pressure across the cake increases, the porosity of the cake decreases because the cake becomes compressed. Consequently, the specific resistance increases. The specific resistance at any point in a compressible cake can be expressed as

$$\alpha = \alpha_o \, P_S{}^n \qquad (6.21)$$

Table 6.17 Filter-Cake Specific-Resistance Parameters (Source Ref. 27).

Substance	$\alpha_o\ (10^{10})$ m/kg[a]	Exponent, n
Asbestos	---	---
Calcium carbonate	---	0.19
Celite	---	0.14
Crushed limestone	---	---
Gairome clay	282	0.60
Ignition plug clay	---	0.56
Kaolin	---	---
Kaolin, Hong Kong pink	101	0.33
Solkofloc	0.0024	1.01
Talc	8.66	0.51
Titanium dioxide	32	0.32
Zinc sulfide	14	0.69
General range	1×10^8 to 1×10^3	0 to 1.2

a) To convert to ft/lb multiply bv 1.488.

where $P_S = P_o - P$ is equal to the pressure drop across the cake at any point. The exponent, n, usually varies from 0.2 to 0.8. If n = 0, the cake is incompressible. Values of the specific resistance and n are given in Table 6.17.

After substituting Equation 6.20, 6.21, and $dP_S = -dP$ into Equation 6.19, and after separating variables, we obtain,

$$\frac{dP_S}{P_S^n} = \frac{\alpha_o\, \mu\, v_S\, dm}{A_F} \qquad (6.22)$$

The limits of integration for Equation 6.22 are: at $x_1 = 0$, $P_S = 0$, m = 0 and at $x = x_2$, $P_S = P_o - P_i$, and $m = m_S$, where P_i is the pressure at the interface of the cake and the filter medium, as shown in Figure 6.11. Thus, after integrating Equation 6.22 across the cake, we obtain

$$\frac{(P_o - P_i)^{(1-n)}}{(1-n)} = \frac{\alpha_o\, \mu\, v_S\, m_S}{A_F} \qquad (6.23)$$

In many cases, the pressure drop across the filter medium is low, and P_i is approximately equal to the pressure at the downstream side of the filter medium. If the pressure, P_V, is produced by the vacuum pump, then $P_i \approx P_V$, and

$$\alpha = \alpha_o (P_o - P_V)^n \qquad (6.24)$$

After substituting Equation 6.24 into Equation 6.23, we obtain

$$\frac{(P_o - P_V)}{(1 - n)} = \frac{\alpha \mu v_S m_S}{A_F} \qquad (6.25)$$

The superficial velocity,

$$v_S = (dV/dt) / A_F \qquad (6.26)$$

The dry solids, $m_S = c_{1,2} V_1$, where the first subscript (1) in the solids concentration refers to the incoming stream and the second subscript (2) to the solids. The filtrate volume, V, is the total volume of filtrate collected up to time t. Now, substitute Equation 6.26 and the expression for the mass of dry solids into Equation 6.25.

$$\frac{(P_o - P_V)}{(1 - n)} = \frac{\alpha \mu c_{1,2} V}{A_F^2} \frac{dV}{dt} \qquad (6.27)$$

Assume that the filtration is conducted at constant pressure. Then, after separating variables and integrating from 0 to t_F and 0 and V_F, we obtain

$$\frac{(P_o - P_V)}{(1 - n)} t_F = \frac{\alpha \mu c_{1,2} V_F^2}{2 A_F^2} \qquad (6.28)$$

Next, solve for A_F^2.

$$A_F^2 = \frac{(1 - n) \alpha \mu c_{1,2} V^2}{2 t_F (P_o - P_V)} \qquad (6.29)$$

The total drum area, A_T, is greater than the filtering area because of the need to wash, dewater, and dry the cake. Thus,

$$A_T = A_F / f \qquad (6.30)$$

The equations for sizing rotary-drum filters are summarized in Table 6.18. Equation 6.18.1 is the liquid mass balance. In this procedure, y is a mass fraction. Because the cake is wet, the liquid entering the filter will be less then the liquid leaving. Equation 6.18.2 is the solids mass balance, assuming that all the solids in the slurry are removed. Solve Equation 6.18.2 for the cake formation rate, m_C. Then, solve Equation 6.18.1 for the filtrate volumetric flow rate, V_2. Next, calculate the filtration area from Equation 6.18.5 and the drum area from Equation 7.18.6. Finally, select a standard rotary filter from Table 6.20. The calculation procedure for sizing a rotary filter is outlined in Table 6.19. Example 6.5 illustrates the sizing procedure.

Operating data for filtering slurries could also be used to estimate rotary-filter areas. Some filtration rates are given by Walas [6]. Thus, by dividing the feed rate of solids onto the filter by the filtration rate, expressed as kg of solids/h m^2, the filter area can be estimated.

Table 6.18 Summary of Equations for Sizing Rotary-Drum Filters

---

First Subscript: Entering Stream = 1 — Leaving Stream = 2
Second Subscript: Liquid = 1 — Solids = 2 — c = wet cake

Mass Balance

$$y_{1,1}\, \rho_1\, V_1' = y_{C,1}'\, m_C + \rho_2'\, V_2 \tag{6.18.1}$$

$$y_{1,2}'\, \rho_1\, V_1' = y_{C,2}\, m_C \tag{6.18.2}$$

$$y_{1,1} + y_{1,2}' = 1 \tag{6.18.3}$$

$$y_{C,1}' + y_{C,2} = 1 \tag{6.18.4}$$

Rate Equation

$$A_F^2 = \frac{(1 - n')\, \alpha\, \mu'\, c_{1,2}\, V_F^2}{2\, t_F'\, (P_o' - P_V')} \tag{6.18.5}$$

$$A_F = f'\, A_T \tag{6.18.6}$$

System Properties

$$\rho_1 = y_{1,1}\, \rho_{1,1}' + y_{1,2}'\, \rho_{1,2}' \tag{6.18.7}$$

$$c_{1,2} = y_{1,2}'\, \rho_1 \tag{6.18.8}$$

$$\alpha = \alpha_O'\, (P_o' - P_V')^n \tag{6.18.9}$$

Variables

$y_{1,1}$ - $y_{C,2}$ - m_S - ρ_1 - V_2 - $c_{1,2}$ - α - A_F - A_T

Table 6.19 Calculation Procedure for Sizing Rotary-Drum Filters

1. Calculate the mass fraction of liquid in the entering stream, $y_{1,1}$, from Equations 6.18.3

2. Calculate the density of the entering stream, ρ_1, from Equation 6.18.7.

3. Calculate the mass fraction of solids in the cake, $y_{C,2}$, from Equation 6.18.4.

4. Calculate the concentration of solids in the entering stream, $c_{1,2}$, from Equation 6.18.8.

5. Calculate the rate of wet cake formation, m_C, from Equation 6.18.2.

6. Calculate the volumetric flow rate of liquid in the exit stream, V_2, from Equation 6.18.1.

7. Calculate the specific cake resistance, α, from Equations 6.18.9.

8. Calculate the filter area, A_F, from Equation 6.18.5.

9. Calculate the drum area, A_T, from Equation 6.18.6.

10. Select a standard rotary-drum filter from Table 6.20.

Table 6.20 Standard Rotary-Drum Filters

37 ½ % Drum Submergence

Filter Size Diameter[a] ft	Nominal Length[a] ft	Drum Area[b] ft^2	Drum Drive[c] hp	Agitator Drive[c] hp
8	6	150	1	1 ½
10	6 1/3	200	1 ½	3
10	8	250	1 ½	3
10	10	300	1 ½	3
11.5	10	360	1 ½	3
11.5	12	430	1 ½	3
11.5	14	500	2	5
11.5	16	575	2	5
12	20	750	5	5

a) To convert to meters multiply by 0.3048.
b) To convert to square meters multiply by 10.76.
c) To convert to kilowatts multiply by 0.7457.

Example 6.5 Sizing a Rotary-Drum Filter

A rotary-drum filter filters 20 m³/h (706 ft³/h) of a calcium carbonate slurry at 20 °C (68 °F). The pressure drop across the cake is 0.658 bar (9.541 psi). If the slurry contains 0.15 mass fraction of calcium carbonate, and the filter cake contains 0.40 mass fraction of water, estimate the surface area of the rotary-drum filter.

Data
water density   998.3 kg/m³ (62.3 lb/ft³)
water viscosity (20 °C)   0.001 Pa-s (1 cp)
$CaCO_3$ density   2709 kg/m³ (169 lb/ft³)

From Equation 6.18.3, $y_{1,1} = 0.85$, and from Equation 6.18.7 the average density of the slurry,

$\rho_1 = 0.85\,(998.3) + 0.15\,(2709) = 1255$ kg/m³ (78.35 lb/ft³)

Equation 6.18.4 gives $y_{C2} = 0.60$. The formation rate of wet cake, m_C, can now be calculated from Equation 6.18.2. Thus,

$$m_C = \frac{0.15}{0.60}\ \frac{1255\ \text{kg}}{1\ \text{m}^3}\ \frac{20\ \text{m}^3}{1\ \text{h}} = 6.275\times10^3\ \text{kg/h}\ (1.3\times10^4\ \text{lb/h})$$

The volumetric flow rate of the filtrate, V_2, obtained from Equation 6.18.1, is

$$V_2 = \frac{y_{1,1}\,\rho_1\,V_1 - y_{C,1}\,m_C}{\rho_{2,1}}$$

$$y_{1,1}\,\rho_1\,V_1 = \frac{0.85}{1}\ \frac{1255\ \text{kg}}{1\ \text{m}^3}\ \frac{20\ \text{m}^3}{1\ \text{h}} = 2.134\times10^4\ \text{kg/h}\ (4.71\times10^4\ \text{lb/h})$$

$y_{C,1}\,m_C = 0.40\,(6275\ \text{kg/h}) = 2.510\times10^3$ kg/h (5.535×10³ lb/h)

$$V_2 = \frac{21340 - 2510}{998.3} = 18.86\ \text{m}^3/\text{h}\ (666\ \text{ft}^3/\text{h})$$

Table 6.17 does not contain α_o for calcium carbonate. According to Walas [6], the specific cake resistance for filtering $CaCO_3$, given by Equation 6.18.9, is

$\alpha = 1.604 \times 10^{10} (P_o - P_i)^{0.2664}$

where $P_o - P_i$ is in bars and α in m/kg. The pressure at the interface of the filter cake and filter medium, P_i, is assumed to be equal to the pressure downstream of the filter medium, P_V. Therefore, $P_o - P_V = 0.658$ bar or 6.58×10^4 Pa according to Walas [6.6] for filtering a $CaCO_3$ slurry. Therefore,

$\alpha = 1.604 \times 10^{10} (0.658)^{0.2664} = 1.435 \times 10^{10}$ m/kg $(2.14 \times 10^{10}$ ft/lb)

McCabe and Smith [23] state that the cycle time for filtering $CaCO_3$ is 5 min. According to Table 6.20, 37.5 % of the drum is submerged during filtration. Because the drum is only partially submerged, the filtering time,

$t_F = 0.375 (5) (60) = 112.5$ s

The volume of filtrate collected,

$V_F = V_2 t_F = [(18.86 / 3600)] (112.5) = 0.5894$ m^3 (20.8 ft^3)

From Equation 6.18.8, the concentration of solids in the entering stream,

$c_{1,2} = 0.15 (1255) = 188.3$ kg/m^3 (11.76 lb/ft^3)

Finally, the filter area can now be calculated from Equation 6.18.5.

$$A_F{}^2 = \frac{(1 - 0.2664)}{2\ (115.5)\ s} \frac{1.435 \times 10^{10}\ m}{1} \frac{0.001\ Pa\text{-}s}{kg} \frac{188.3\ kg}{6.58 \times 10^4\ Pa} \frac{(0.5894)^2\ m^6}{1\ m^3} \frac{}{1}$$

$A_F = 6.731$ m^2 (72.4 ft^2)

According to Equation 6.18.6, the drum area,

$A_T = 6.731 / 0.375 = 17.95$ m^2 (193 ft^2)

From Table 6.20, a standard filter has 250 ft^2 (23.2 m^2) of surface area. This choice will result in a safety factor of 29.5%. The final decision on the filter size, will require laboratory or pilot plant tests. In most cases, the filter manufacturer will provide this service.

COMPONENT SEPARATORS

The most frequently used component separators are absorbers, strippers, fractonators, and extractors. According to Humphrey [74], fractionators are used in 90 to 95% of the separations in the US. The principles of component separators are covered extensively in several texts such as Treybal [29], King [30] and Henley and Seader [31, 65]. We will only consider short cut sizing methods. These methods are useful for preliminary design estimates and for first guesses for more exact calculations, requiring iterative calculation procedures.

A fractionator or absorber consists of a cylindrical shell containing internals, either trays or packing, as shown in Figures 6.12. By creating surface area trays and packing promote mass transfer between liquid and gas. A liquid film forms on the packing and vapor bubbles through the liquids on the trays. Packed separators are usually used for diameters less than 2.5 ft (0.762 m). In both separator types, the liquid enters at the top of the column and at the feed tray for fractionators and flows downward by gravity. Gas enters the separator at the bottom and then flows upward countercurrent to the liquid flow.

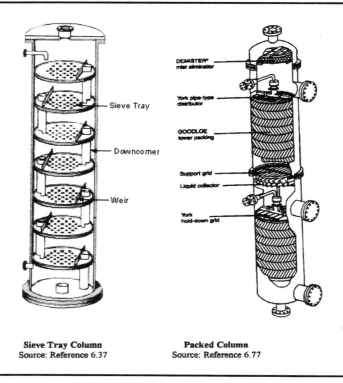

Sieve Tray Column
Source: Reference 6.37

Packed Column
Source: Reference 6.77

Figure 6.12 A fractionator or absorber design, with permission.

Tray Columns

The purpose of a tray is to provide thorough contact between gas and liquid, facilitating mass transfer between the two phases on each tray. Gas bubbles through liquid flowing across the tray. The most common designs are the sieve, valve, and bubble-cap trays shown in Figure 6.13. According to Harrison and France [32], the sieve and valve tray have mostly displaced the bubble-cap tray because they are less expensive and have a higher capacity. The sieve tray is the most widely used and should be considered first because of its lower installed cost, well known design procedures, low fouling tendency, large capacity, and high efficiency [33].

When comparing tray designs the turndown ratio is important because it is a measure of the flexibility of a column in dealing with a change in flow rate. The turndown ratio is defined as the ratio of the maximum to minimum operating flow rate. For bubble cap and valve trays, the turndown ratio is about ten whereas for sieve trays it is only about three.

Engineers realize that all equipment have a maximum operating capacity, and because of uncertainty in system property data, the equipment will be over-designed to insure that adequate capacity will be available. Overdesigning, however, – besides being costly – can cause operating difficulties because all equipment have a turndown ratio. Below the minimum or above the maximum capacity, equipment may become inoperable or very inefficient. This is illustrated in Figure 6.15 which shows that the tray efficiency, expressed as a percentage of the flooding gas velocity, is relatively constant over a range of gas velocities. Close to the flooding point the gas velocity is high so that an excessive amount of liquid drops are carried to the tray above. This form of backmixing causes the tray efficiency to decrease. On the other hand, at low gas velocities, mixing of gas with

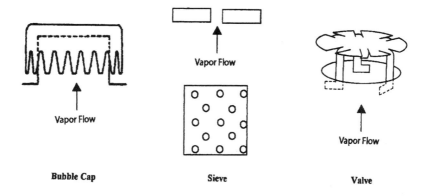

Figure 6.13 Bubble cap, sieve, and valve trays.

Raschig Ring   **Pall Ring**   **Structured**
                                                                   Source: Reference 78

Figure 6.14 Examples of random and structure packings.

liquid is poor, reducing the mass-transfer rate. Also, at low gas velocity, liquid will leak – called weeping – through the openings in the tray to the tray below, reducing the column efficiency. Both of these effects cause the efficiency to drop sharply.

Packed Columns

In packed columns, liquid spreads over the packing and flows downward. The gas flows upward through the void space in the packing countercurrent to the liquid flow. Like trays, the purpose of the packing is to provide surface area to enhance mass transfer between gas and liquid. There are two broad classes of packing, random and structured packing. Random packing is loaded into the separator by first filling the separator with water. Then, the packing is gradually loaded into the separator. After settling the packing will assume random positions within the column. Also, the water prevents breaking fragile packing. For structured packing, the position of the packing is definite. Three types of random packing are shown in Figure 6.14, the oldest being the Raschig ring, which is a hollow cylinder. Later, more efficient packings were developed, like the Pall ring, which is the most widely used packing [6]. An example of structure packing is given in Figure 6.14. Because of low liquid holdup and pressure drop, structured packing is suitable for vacuum separations. There are numerous packing types on the market. For example, see Walas [6].

Similar to tray columns, packed columns operated at high gas velocities causes backmixing, and low gas velocities reduce the mass transfer rate. If the gas velocity is too high, the column will flood. In addition, at low liquid flow rates the packing will not wet completely, resulting in a reduction in mass-transfer. Another problem is the tendency for the liquid to channel. To minimize this effect, redistributors have to be installed every 5 to 10 m (16.4 to 30.5 ft) [23] to even out the liquid flow.

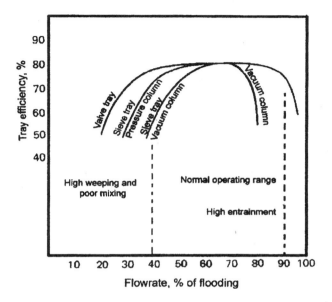

Figure 6.15 The effect of vapor flow rate on tray efficiency. From Ref. 33 with permission.

Absorber and Stripper Sizing

Assuming dilute solutions, Table 6.21 lists the equations for sizing absorbers and strippers in terms of a key component and Table 6.22 outlines the calculation procedure. In numbering the relationships in Table 6.21, A, S, P, and T means absorption, stripping, packed columns and tray columns, respectively. Processing dilute solutions implies that heat effects will be small, and therefore, the separation is essentially isothermal. If the column is both isothermal and isobaric, the equilibrium value will be constant. Also, dilute solution means that the gas and liquid flow rates will essentially be constant. In absorption, the gas flow rate is fixed and the liquid flow rate must be estimated, whereas in stripping the liquid flow rate is fixed and the gas flow rate must be estimated.

The first step in the sizing procedure is to determine the minimum liquid flow rate for an absorber or the minimum gas flow rate for a stripper. For gas absorption, the entering liquid and gas concentrations are known, which is shown in Figure 6.16. The subscript 1 refers to the top of the separator, and the subscript 2 to the bottom of the separator, as shown in Figure 6.16. The fractional absorption and therefore the exit gas concentration is also known, fixing point 1 – at the top of the column. The exit liquid concentration is not known. Therefore, point 2 – at the bottom of the column – is not fixed. The minimum liquid flow rate occurs when the liquid leaving the absorber is in equilibrium with the entering gas. This occurs when the operating line intersects the equilibrium curve, as shown by the dashed line in Figure 6.16. The intersection is given by Equation 6.21.1A, which is derived by the simultaneous solution of a component balance and an equilibrium relation.

Table 6.21 SUMMARY OF EQUATIONS FOR SIZING ISOTHERMAL ABSORBERS AND STRIPPERS – COLUMN HEIGHT

Subscripts: L = liquid — V = vapor
1 = top of column — 2 = bottom of column — see Figure 6.16
m = minimum — k = key component

Minimum Flow Rates

Absorbers

$$\frac{m_{Lm}}{m_V'} = \frac{y_{2k}' - y_{1k}}{y_{2k}'/K_k - x_{1k}'} \tag{6.21.1A}$$

$$y_{1k} = (1 - \varepsilon') y_{2k}' \tag{6.21.2A}$$

Strippers

$$\frac{m_{Vm}}{m_L'} = \frac{x_{1k}' - x_{2k}}{K_k x_{1k}' - y_{2k}'} \tag{6.21.1S}$$

$$x_{2k} = (1 - \varepsilon') x_{1k}' \tag{6.21.2S}$$

Optimum Flow Rates

Absorbers

$$m_L = 1.5 \, m_{Lm} \tag{6.21.3A}$$

Strippers

$$m_V = 1.5 \, m_{Vm} \tag{6.21.3S}$$

Number of Equilibrium Stages

$$A_A = m_L / K_k m_V \tag{6.21.4}$$

Absorbers

$$(A_A)^{Ne} = \frac{y_{2k}' - K_k x_{1k}'}{y_{1k} - K_k x_{1k}'} \left(1 - \frac{1}{A_A}\right) + \frac{1}{A_A} \qquad (6.21.5A)$$

Strippers

$$(1/A_A)^{Ne} = \frac{x_{1k}' - y_{2k}'/K_k}{x_{2k} - y_{2k}'/K_k} (1 - A_A) + A_A \qquad (6.21.5S)$$

Tray Columns	**Packed Columns**
Column Diameter	
D ≥ 2.5 ft	**D < 2.5 ft**
Use Equations 6.23.1 to 6.23.6 in Table 6.23 $\qquad$ (6.21.6T)	Use Equations 6.23.1 to 6.23.3 and Equations 6.23.7 to 6.23.9 in Table 6.23 $\qquad$ (6.21.6P)

Column Height

$Z = N_A Z_T + 3 \text{ ft} + 0.25\, D + L_S$ $L_S = V_L\, t_S/A$ — $t_S = 5$ min or $L_S = 0.06\, N_A + 2.0$ — all terms in meters $\qquad$ (6.21.7T)	$Z = N_e\,(\text{HETS}) + 3 \text{ ft} + 0.25\, D + L_S$ $L_S = V_L\, t_S/A$ — $t_S = 5$ min or $L_S = 0.06\, N_A + 2.0$ — all terms in meters $\qquad$ (6.21.7P)
$N_A = N_e/E_o \qquad (6.21.8T)$	
$Z_T = f(P)$ — Table 6.25 $\qquad (6.21.9T)$	

System Properties

$E_o = f(\mu_L, \alpha)$ — Figure 6.17 $\qquad (6.21.10T)$	$K_k = f(T', P') \qquad (6.21.8P)$
$K_k = f(T', P') \qquad (6.21.11T)$	HETS $= 0.5$ m — for $D < 0.5$ m or HETS $= D^{0.3}$ — for $D > 0.5$ m $\qquad (6.21.9P)$
$\alpha = 10\, K_k \qquad (6.21.12T)$	
$\mu_L = f(T') \qquad (6.21.13T)$	

Table 6.2.1 Continued

Variables

Absorbers

Tray Columns — m_L - m_{Lm} - y_{1k} - K_k - A_A - N_E - N_A - E_o - Z - Z_T - α - μ_L - D

Packed Columns — m_L - m_{Lm} - y_{1k} - K_k - A_A - N_E - Z - D - HETS

Strippers

Tray Columns — m_V - m_{Lm} - x_{2k} - K_k - A_A - N_E - N_A - E_o - Z - Z_T - α - μ_L - D

Packed Columns — m_V - m_{Lm} - x_{2k} - K_k - A_A - N_E - HETS

TABLE 6.22 Calculation Procedure for Sizing Isothermal Absorbers or Strippers–Column Height

---

1. For absorbers, calculate the minimum liquid flow rate, m_{Lm}, from Equations 6.21.1A, 6.21.2A, and 6.21.11T. For strippers, calculate the minimum gas flow rate, m_{Vm}, from Equations 6.21.1S, 6.21.2S and 6.21.8P.

2. For absorbers, calculate the actual liquid flow rate, m_L, from Equation 6.21.3A. For strippers, calculate the actual gas flow rate, m_V, from Equation 6.21.3S.

3. Calculate the column diameter, D, for tray columns, from Equation 1.21.6T and for packed columns from Equation 6.21.6P.

4. Calculate the absorption factor, A_A, from Equation 6.21.4.

5. Calculate the number of equilibrium stages, N_e, from Equation 6.21.5A for absorbers or Equation 6.21.5S for strippers.

6. Calculate the actual number of stages, N_A, from Equations 6.21.8T and 6.21.10T to 6.21.13T.

7. Calculate the tray spacing, Z_T, from Equation 6.21.9T

8. Calculate the column height, Z, from Equation 6.21.7T for tray columns. For packed columns calculate Z from Equations 6.21.7P and 6.21.9P.

For stripping, the entering liquid and gas concentrations are known. The fraction stripped and therefore the exit liquid concentration is also known, but the exit gas concentration is unknown. Therefore point 2 – at the bottom of the column is fixed, but point 1 – at the top of the column – is not fixed. The maximum exit and minimum gas flow rate gas concentration is obtained when the operating line intersects the equilibrium curve, as shown by the dashed line in Figure 6.16. The intersection is given by Equation 6.21.1S, which is also obtained by the simultaneous solution of a component balance and an equilibrium relation.

After finding the minimum flow rate, the optimum or operating flow rate can be calculated by using the rules-of-thumb from Treybal [29], which are given by Equations 6.21.3A or 6.21.3S. The 1.5 given in the equations is within the range of 1.2 to 2.0 for both absorbers and strippers given by McNulty [36].

To minimize channeling of liquid in packed absorbers and strippers require that the packing be sufficiently small when compared to the column diameter. Small packing, however, will result in a high pressure drop.Treybal [29] specifies that the ratio of separator diameter to the packing diameter should be 15/1.

The recovery of the key component is specified to calculate the exit composition of the gas stream for absorbers or the exit composition of the liquid stream for strippers from Equation 6.21.2A or 6.21.2S. For both cases, it is assumed that the operating line intersects the equilibrium curve at one end and not at some intermediate point between the ends of the operating line. This is the case for dilute solutions when both the operating and equilibrium lines are linear. Separation of dilute solutions occurs frequently when purifying waste streams. Because both the equilibrium and operating curves are linear for dilute solutions, the equation derived by Kremser [59] can be used to calculate the number of equilibrium stages.

Next calculate the number of actual stages. For tray columns the efficiency, E_o, is obtained from Equation 6.21.10T. For packed columns the HETS (height equivalent to a theoretical stage) is given by Equation 6.21.9P, as recommended by Ulrich [50]. The column height is the sum of the height occupied by packing or trays, a section above the top tray, room for manholes and handholes, and an additional section below the bottom tray. The manholes and handholes are required for inspection and maintenance. The top section de-entrains liquid from gas (phase separation). For a packed column, Vatavuk and Neveril [60] recommended adding 2 ft (0.610 m) to 3 ft (0.914 m) plus 25% of the column diameter to allow for gas-liquid separation, handholes, and manholes. Ulrich [50]. Both Henley and Seader [31] and Valle-Riestra [53] recommend 4 ft (1.22 m) above the top tray. Valle-Riestra's recommendation is based on a two-foot diameter column. He recommends adjusting the number for a larger or smaller diameter column, but he did not give any recommendations for making the adjustment. In Reference 76, 2.0 m (6.56 ft) is recommended for an ethane column.

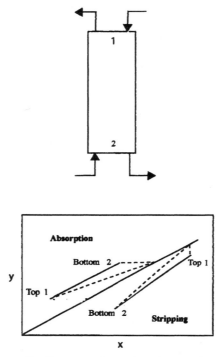

Figure 6.16: Schematic diagram of an absorber, stripper, or extractor.

The height of the bottom section of the column is required for liquid surge capacity and reboiler return for fractionators and a gas inlet nozzle for absorbers and strippers. Henley and Seader [31] recommend 10 ft (3.05 m) section below the bottom tray and Valle-Riestra 6 ft (1.83 m) section for a two-foot diameter column. Again, Valle-Riestra recommends adjusting the number for other diameter columns. Another way is to calculate the height assuming a five-minute surge time. In Reference 76 for an ethane separation column, the height of the lower section is given by: $L_S = 0.06\,N_A + 2.0$ (all terms in meters). The number of actual stages is N_A. All these recommendations should result in a reasonable estimate for the height of the bottom section of the column. In Table 6.21, the height is estimated by using both the 5-min surge time and the above formula.

If one of the components is heat sensitive, the volume of liquid and hence the contact time at the bottom of the column should be a minimum to minimize degradation. Sufficient liquid height, however, is necessary for level control. Then, the bottom section should be designed for adequate control using the minimum liquid height. The height of a tray column is calculated from Equation 6.21.7T and the height of a packed column from Equation 6.21.7P.

The calculation procedure in Table 6.22 could also be used for a multi-component mixture. After calculating the number of stages for separating the key component from the mixture, then the composition of all other components in the exit stream can be calculated using the Kremser equation, Equation 6.21.5A for absorbers or 6.21.5S for strippers.

The tray spacing, Z_T, calculated from Equation 6.21.9T, depends on the pressure [75]. To obtain HETS from Equation 6.21.9P requires the column diameter. To obtain the column diameter, calculate the maximum allowable gas velocity to prevent entrainment of liquid. First, find the maximum value of the parameter k, from Equation 6.23.4 in Table 6.23, which occurs when the column is about to flood. The column is then designed to operate below the flood point. The maximum value of k is found in Figure 6.18 for trays. Fair [52] recommends that the flooding parameter, obtained from Figure 6.18 for tray columns, be multiplied by 0.9 for nonfoaming liquids. Treybal [29] recommends 0.75 for foaming liquids. In Figure 6.18, the flooding parameter requires correcting for surface tension and tray geometry using corrections given by Fair [52]. Figure 6.18 was developed by Fair [52] for fractionators. Henley and Seader [65] used an earlier version of Figure 6.18 for absorbers and strippers. Fair [52] lists restrictions on Equation 6.23.5 for tray columns. He also corrects k for tray geometry, but for preliminary estimates we want to avoid designing trays.

If the column diameter $\geq$ 2.5 ft (0.762 m), use a tray column. Because of maintenance, a tray column has to be internally accessible [75, 31]. The minimum diameter column that is accessible is 2.5 ft. For packed columns, Figure 6.19 only gives factors for a limited number of packings. For other packings use the flooding ratios in Table 6.26. To obtain k for a packing listed in Table 6.26 multiply the flooding ratio by the flooding factor for 50 mm (2 in) Pall rings obtained from Figure 6.19. For packed columns use 0.7 for nonfoaming liquids, a commonly accepted value, and 0.4 for foaming liquids [6]. After obtaining k_V, the column diameter is then calculated from equations listed in Table 6.23.

Table 6.23 Summary of Equations for Sizing Absorbers, Strippers, or Fractionators – Column Diameter

Subscripts: L = liquid — V = vapor

Column Diameter

$$A = V_V' / v_V \tag{6.23.1}$$

$$A = \pi D^2 / 4 \tag{6.23.2}$$

$$v_V = k_V \left[(\rho_L' - \rho_V') / \rho_V' \right]^{1/2} \quad -- \quad v_s = 0.9\, v_v \tag{6.23.3}$$

Table 6.23 Continued

Tray Columns for D ≥ 2.5 ft

$$k = f \left[\frac{m_L' \, M_L'}{m_V' \, M_V'} \left(\frac{\rho_V'}{\rho_L'} \right)^{0.5}, \, Z_T \right] \quad - \text{Figure 6.18} \qquad (6.23.4)$$

$k_V = 0.9 \, k \, (\sigma/20)^{0.2}, \sigma$ (dyne/cm) — for non-foaming liquids $\qquad$ (6.23.5)
or $k_V = 0.75 \, k \, (\sigma/20)^{0.2}, \sigma$ (dyne/cm) — for foaming liquids

$Z_T = f(D)$ — Table 6.25 $\qquad$ (6.23.6)

Packed Columns for D < 2.5 ft

$$k = f \left[\frac{m_L' \, M_L'}{m_V' \, M_V'} \left(\frac{\rho_V'}{\rho_L'} \right)^{0.5}, \, d, \text{ packing type}' \right] \quad - \text{Figure 6.19, Table 6.26} \qquad (6.23.7)$$

$k_V = 0.7 \, k \, (\sigma/20)^{0.2}, \sigma$ (dyne/cm) — for non-foaming liquids $\qquad$ (6.23.8)
or $k_V = 0.4 \, k \, (\sigma/20)^{0.2}, \sigma$ (dyne/cm) — for foaming liquids

$D = 15 \, d$ $\qquad$ (6.23.9)

Variables

Tray Columns

$A - D - Z_T - v_S \, v_V - k - k_V$

Packed Columns
$A - D - d - v_V - k - k_V$

Table 6.24 Calculation Procedure for Sizing Absorbers, Strippers, and Fractionators – Column Diameter

1. Calculate a preliminary column diameter from Equations 6.23.1 and 6.23.2 by assuming a superficial velocity of 2 ft/s. If D < 2.5 ft, select a packed column. Otherwise, select a tray column.

2. Calculate the actual diameter from Equations 6.23.1 to 6.23.6 for tray fractionators or Equations 6.23.1 to 6.23.3 and Equations 6.23.7 to 6.23.9 for packed columns.

Table 6.25 Tray Spacing for Absorbers, Strippers, and Fractionators. (Source Ref. 75).

Pressure	Tray Spacing, ft[a]
Vacuum	2.0 to 2.5
Atmospheric	1.5
High Pressure	1.0

a) To obtain the tray spacing in meters multiply by 0.3048.

Table 6.26 Relative Flooding Factors for Column Packings (Source Adapted from Ref. 52 with permission).

Pall Rings—Metal		Intalox Saddles—Ceramic		Koch Sulzer BX	
50 mm	1.00*	50	0.89	BX	1.00
38	0.91	38	0.75		
25	0.70	25	0.60	Koch Flexipac	
12	0.65	12	0.40	No. 1	0.69
				No. 2	1.08
Raschig Rings—Metal		Berl Saddles—Ceramic		No. 3	1.35
50	0.79	50	0.84		
38	0.71	38	0.70	Tellerettes	
25	0.66	25	0.54		1.0
12	0.55	12	0.37		
				Nor-Pak Plastic	
Raschig Rings—Ceramic		Intalox Saddles—Metal		No. 25	1.0
50	0.78	No. 25	0.88	No. 35	1.20
38	0.65	No. 40	0.98		
25	0.50	No. 50	1.10		
12	0.37	No. 70	1.24		

*The tabulated values are the ratio of k for a packing type and size to k for 50 mm (2 in) Pall rings.

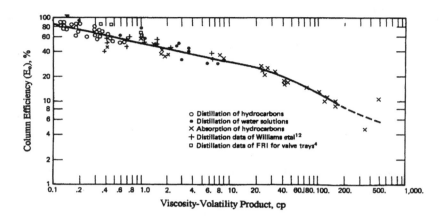

For fractionators the relative volatility and viscosity of the key components are taken at the average of the the top and bottom tray temperature and at the feed composition. For absorbers, separating hydrocarbons, the volatility is taken as ten times K_k for the key component.

Figure 6.17 Column efficiency for fractitionators, absorbers, and strippers. From Ref. 31 with permission.

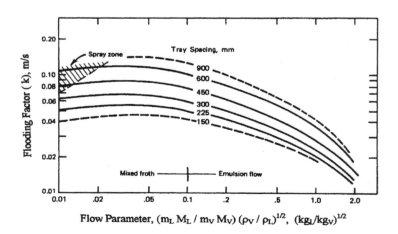

Flow Parameter, $(m_L M_L / m_V M_V)(\rho_V / \rho_L)^{1/2}$, $(kg_L/kg_V)^{1/2}$

Figure 6.18 Flooding factor for sieve, bubble cap, and valve trays. From Ref. 52 with permission.

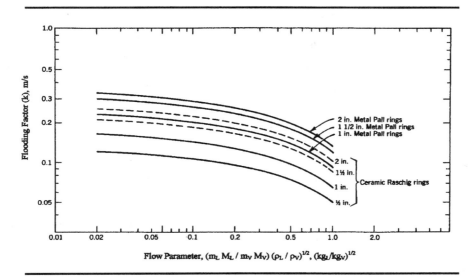

Figure 6.19 Flooding factor for packed columns. From Ref. 52 with permission.

Example 6.6 Stripping Methylene Chloride from Wastewater

A wastewater stream contains 100 ppm of methylene chloride. One million gallons a day of water will be stripped of the methylene chloride using air. If 99% of the methylene chloride is removed, what will be the size of the separator?

<u>Data</u>

MW of air	2.9
wastewater temperature	100 °F (37.78 °C)
water viscosity	0.703 cP (7.03×10^{-4} Pa-s) (1.47×10^{-4} lb/ft-s)
water density	62.00 lb/ft^3 (993 kg/m^3)
air temperature	100 °F and saturated with water
air density (100 °F and 1 atm)	0.07395 lb/ft^3 (1.18 kg/m^3)
surface tension of water	69 dynes/cm (0.069 N/m)

Either steam or air could be used to remove methylene chloride from the wastewater stream. Air will be used in this design. To simplify the problem, assume that the air is saturated with water. If the air is dry, the temperature in the column will vary because water will evaporate into the air, cooling the water.

Follow the procedures outlined in Tables 6.22 and 6.24 for calculating the height and diameter of the stripper. First, calculate the air rate. This calculation requires calculating the minimum air rate, which is determined by the intersection of the operating line and the equilibrium curve as shown in Figure 6.16. Because the concentration of methylene chloride is low, Henry's law applies. Shukla and Hicks [68] have compiled Henry's–law–constant relationships for many organic compounds found in wastewater. For methylene chloride, Henry's law constant, H, is given by

$$\log H \text{ (mm Hg)} = 9.58 - 1139/(t\,^\circ C + 231) = 5.342$$

$$H = 2.200 \times 10^5 \text{ mm Hg}$$

$$y = (H/P)\, x = (2.2 \times 10^5 / 760)\, x = 289.5\, x$$

Thus, the equilibrium value $K_k = 289.5$. This calculation satisfies Equation 6.21.11T or Equation 6.21.8P. We have yet to determine if the column will contain trays or packing.

Convert the mass fraction of methylene chloride in the entering wastewater stream to mole fraction.

$$x_{1k} = 1 \times 10^{-4}\, (18 / 84.94) = 2.119 \times 10^{-5}$$

From Equation 6.21.2S, the exit water concentration,

$$x_{2k} = (1 - 0.99)\, 2.119 \times 10^{-5} = 2.119 \times 10^{-7}$$

Next, convert the wastewater flow rate from gal/day to lbmol/h.

$$m_L = \frac{1 \times 10^6 \text{ gal}}{1 \text{ day}} \cdot \frac{1 \text{ day}}{24 \text{ h}} \cdot \frac{1 \text{ ft}^3}{7.481 \text{ gal}} \cdot \frac{62.00 \text{ lb}}{1 \text{ ft}^3} \cdot \frac{1 \text{ lbmol}}{18 \text{ lb}}$$

$$m_L = 1.919 \times 10^4 \text{ lbmol/h } (8.707 \times 10^4 \text{ kgmol/h})$$

From Equation 6.21.1S, the minimum air flow rate,

$$m_{Vm} = \frac{2.119 \times 10^{-5} - 2.119 \times 10^{-7}}{289.5\,(2.119 \times 10^{-5}) - 0}\ 1.919 \times 10^4 = 65.59 \text{ lbmol/h } (29.7 \text{ kgmol/h})$$

According to Equation 6.21.3S, the actual air flow rate,

m_V = 1.5 (65.59) = 98.39 lbmol/h (44.6 kgmol/h)

Now, calculate the number of equilibrium stages, N_e, from Equation 6.21.5S, and then divide the result by the column efficiency to obtain the actual number of trays, N_A.

From Equation 6.21.4, the absorption factor,

$A_A = 1.919 \times 10^4 / 289.5 (98.39) = 0.6734$

From Equation 6.21.5S, we obtain.

$$(1/A_A)^{Ne} = \frac{2.119 \times 10^{-5} - 0}{2.119 \times 10^{-7} - 0} (1 - 0.6734) + 0.6734 = 33.33$$

$$N_e = \frac{\log 33.33}{\log (1 / 0.6734)} = 8.868$$

The column height, given by Equation 6.21.7T for tray columns or by Equation 6.21.7P for packed columns, depends on the column diameter, which we will now calculate. From the ideal gas law, the volumetric air flow rate,

$$V_V = \frac{98.39 \ lbmol}{1 \ h} \ \frac{0.7302 \ atm\text{-}ft^3}{1 \ lbmol\text{-}°F} \ \frac{560 \ °R}{1 \ atm} = 4.023 \times 10^4 \ ft^3/h \ (1.14 \times 10^3 \ m^3/h)$$

Assume a superficial velocity of 2 ft/s (0.61 m/s). Calculate a preliminary column cross-sectional area.

$$A = \frac{4.023 \times 10^4}{3600} \ \frac{1}{2} = 5.588 \ ft^2 \ (0.5191 \ m^2)$$

and from Equation 6.23.3 the column diameter,

$$D = \left(\frac{4 (5.588)}{\pi} \right)^{0.5} = 2.667 \ ft \ (0.7981 \ m)$$

Because D > 2.5 ft (0.7620 m), select a tray column.

Next, determine an actual superficial velocity, using Figure 6.18. The flow paramater is out of the range of Figure 6.18. Select 2.0, which means the air flow will be larger than calculated above. The number of equilibrium stages is now 3.727.

$$\frac{1.919 \times 10^4 \ (18)}{98.39 \ (84.94)} \left(\frac{0.07395}{62.00} \right)^{0.5} = 4.181$$

From Equation 6.23.6, the tray spacing is 1.5 ft (0.467 m), and from Equation 6.23.4, k, equals 0.018 m/s (0.05906 ft/s). Because there is no information on foaming for this system, select the lower value of k_V for a foaming liquid. This choice results in a larger column diameter than for a nonfoaming liquid. From Equation 6.23.5,

$$k_V = 0.75 \ (0.018) \ (69/20)^{0.2} = 0.01729 \text{ m/s} \ (0.05673 \text{ ft/s})$$

Thus, from Equation 6.23.3, the maximum air velocity,

$$v_V = 0.01729 \ [\ (62.0 - 0.07395) / 0.07395 \ ]^{0.5} = 0.5003 \text{ m/s} \ (1.641 \text{ ft/s})$$
$$v_S = 0.9(1.641) = 1.477 \text{ft/s} \ (0.4502 \text{ m})$$

From Equation 6.23.1, the revised cross-sectional area for the column,

$$A = \frac{4.023 \times 10^4}{3600} \ \frac{1}{1.477} = 7.566 \text{ ft}^2 \ (0.7029 \text{ m}^2)$$

From Equation 6.23.2, the revised column diameter,

$$D = \left(\frac{4 \ (7.566)}{\pi} \right)^{0.5} = 3.321 \text{ ft} \ (0.9484 \text{ m})$$

Allowing for a safety factor of 15%, obtained from Table 6.30, the column diameter is 3.579 ft (1.091 m). Next, round the column diameter off to the nearest six inches, which is 3.0 ft (1.219 m). Because the diameter is greater than 2.5 ft (0.762 m), select a tray column.

Now, complete the calculation for the column height. The column height is determined by the number of actual trays and the tray spacing plus the height of a section above the top tray and an additional section below the bottom tray. The actual number of trays equals the number of equilibrium stages divided by the column efficiency. The column efficiency is given by Equation 6.21.10T. There is no data for stripping a water solution in Figure 6.17. Bravo [58] states that tray efficiency for steam stripping varies from 25 to 40%. We expect that the efficiency for air stripping is about the same. If we use the average of 25 and 40, according to Equation 6.21.8T, the number of actual stages, $N_A = 3.727 / 0.325 = 11.47$. If we use the same safety factor of 20 % as given in Table 6.30 for fractionators, the number of trays are 14.

From Equation 6.21.7T, the liquid height at the bottom of the column,

$$L_S = \frac{1 \times 10^6 \text{ gal}}{1 \text{ day}} \cdot \frac{1 \text{ day}}{24 \text{ h}} \cdot \frac{1 \text{ h}}{60 \text{ min}} \cdot \frac{5 \text{ min}}{1} \cdot \frac{1 \text{ ft}^3}{7.481 \text{ gal}} \cdot \frac{4}{\pi (4.0)^2 \text{ ft}^2}$$

$= 36.93 \text{ ft } (11.26 \text{ m})$

which is unreasonable.

If we use the second equation for L_S,

$$L_S = 0.06 (14) + 2.0 = 2.840 \text{ m } (9.318 \text{ ft})$$

which is somewhat on the high side when compared to the other rules of thumb.

From Table 6.25, the tray spacing is 1.5 ft (0.4572 m). Thus, from Equation 6.21.7T, the column height,

$Z = 14 (1.5) + 3.0 + 0.25 (4) + 9.318 = 34.31 \text{ ft } (10.58 \text{ m})$

After rounding off, the column height is 34.5 ft (10.5 m). Because of the uncertainty of the column efficiency and other properties, estimates of column diameter and height are usually complemented with testing.

Fractionator Sizing

Occasionally separating multicomponent solutions requires designing a sequence of fractionators. Henley and Seader [31] discuss some aspects of this problem. Once the sequence has been established, then estimate the size of each fractionator. Table 6.27 lists the equations for a short cut method for calculating the height and diameter of fractionators and Table 6.28 outlines the calculation procedure. Like rotary drum filtration, absorbers, and strippers, discussed earlier, the final design may require testing to support the calculations.

As for absorbers and strippers, the height of a fractionator is the sum of the height occupied by trays or packing plus the heights of the top and bottom sections of the fractionator. To determine the height of a fractionator, the first step is to identify and specify the recoveries of the heavy and light key components. The light key component will be recovered to a significant extent in the top product, whereas the heavy key component will be recovered to a significant extent in the bottom product. After specifying recoveries of the key components, the next step is to calculate the recoveries of all other components. The component recoveries are estimated using the Geddes equation [34], Equation 6.27.1 in Table 6.27. Yaws et al. [35] compared the percent recovery of each component, calculated from Equation 6.27.1, with the percent recovery calculated using an exact method and found that the maximum percent deviation was only 0.23% for any one component.

For the equations listed in Table 6.27, it is assumed that the relative volatility is constant, but short cut methods are frequently used when the relative volatility varies. In this case, an average relative volatility is used. King [30] shows that the most appropriate average is the geometric average, defined by Equations 6.27.19. The equations listed in Table 6.27 are restricted to solutions that contain similar compounds, such as alaphatic or aromatic hydrocarbons.

In the short cut method, the number of equilibrium stages needed for a given separation are correlated in terms of the minimum number of stages and the minimum reflux ratio. Fenske [38] derived an expression for the minimum number of actual stages, Equation 6.27.2, by a stage to stage analysis, assuming that the relative volatility is constant. This equation is applicable to multicomponent as well as binary solutions, and is derived in a number of texts, such as by King [30].

Underwood [39] derived Equations 6.27.3 and 6.27.4 for estimating the minimum reflux ratio for a specified separation of two key components. These equations assume constant molar overflow and relative volatility. Underwood showed that at minimum reflux the value of θ in Equations 6.27.3 and 6.27.4 must lie between the relative volatility of the heavy and light key components. If the key components are not adjacent, there will be more than one value of θ. This case is illustrated in an example by Walas [6]. Here, we will assume that the key components are adjacent. As has been pointed out by Walas [6], the minimum reflux ratio calculated by the Underwood equations could turn out to be negative, which means that the equations do not apply for the given separation.

Table 6.27 Summary of Equations for Sizing Fractionators

Subscripts: i = the ith component

F = feed — D = distillate — B = bottom product

LK = light key component — HK = heavy key component

Component Distribution

$$\log (n_{iD}/n_{iB}) = A_C + B_C \log (\alpha_i)_{avg} \qquad (6.27.1A)$$
$$n_{iF} = n_{iD} + n_{iB} \qquad (6.27.1B)$$

Minimum Number of Stages

$$N_M = \frac{\log (x_{LK} / x_{HK})_D (x_{HK} / x_{LK})_B}{\log (\alpha_{LK})_{avg}} \qquad (6.27.2)$$

Minimum Reflux

$$1 - q = \sum_i \frac{(\alpha_i)_{avg} \, x_{iF}}{(\alpha_i)_{avg} - \theta} \qquad (6.27.3)$$

$$R_M + 1 = \sum_i \frac{(\alpha_i)_{avg} \, x_{iD}}{(\alpha_i)_{avg} - \theta} \quad - \quad \text{where } (\alpha_{LK})_{avg} > \theta > (\alpha_{HK})_{avg} \qquad (6.27.4)$$

Optimum Reflux Ratio

$$\frac{R_O}{R_M} = \frac{(1.6 - Y_O)}{6.5} (X_O - 7.5) + 1.6 \qquad (6.27.5)$$

$$Y_O = \frac{(\alpha_{LK})_{avg}}{1.0614 \, (\alpha_{LK})_{avg} - 0.4175} \qquad (6.27.6)$$

$$X_O = \log \ (x_{LK}/x_{HK})_D (x_{HK}/x_{LK})_B (x_{LK}/x_{HK})_F^{0.55 \, (\alpha LK)avg} \qquad (6.27.7)$$

Number of Equilibrium Stages

$$Y_e = 1 - X_e^{Be} \qquad (6.27.8)$$

$$Y_e = (N_e - N_M) / (N_e + 1) \qquad (6.27.9)$$

$$X_e = (R_O - R_M) / (R_O + 1) \qquad (6.27.10)$$

$$B_e = 0.105 \log X_e + 0.44 \qquad (6.27.11)$$

Table 6.27 Continued

Feed Plate Location

$$N_U / N_L = [(x_{HK}/x_{LK})_F (x_{LK,B}/x_{HK,D})^2 (B / D)]^{0.206} \qquad (6.27.12)$$

$$N_e = N_U + N_L \qquad (6.27.13)$$

Column Height

Tray Columns

$$N_A = N_e / E_o \qquad (6.27.14)$$

$$Z = N_A \times \text{Tray Spacing} + L_S + 4.0 \text{ ft} \qquad (6.27.15)$$
Tray Spacing — from Table 6.25

$$L_S = 4 \, V_B \, t_S' \, / \, \pi \, (D')^2 \quad\text{—— D is calculated using Tables 6.23 and 6.24} \qquad (6.27.16)$$

Packed Columns

$$Z = N_e \times (\text{HETS}) + L_S + 4.0 \text{ ft} \qquad (6.27.17)$$

Column Diameter

Follow the procedure outlined in Table 6.23

System Properties

$$\alpha_i = K_i / K_{HK} \qquad (6.27.18)$$

$$(\alpha_i)_{avg} = (\alpha_{iF} \, \alpha_{iD} \, \alpha_{iB})^{1/3} \qquad (6.27.19)$$

$$\alpha_F = \sum_i x_{iF}' \, \alpha_{iF} \quad\text{——} \quad \alpha_{iF} \text{ calculated at } (T_T + T_B)/2 \qquad (6.27.20)$$

$$E_o = f(\mu_F \, \alpha_F) \quad\text{—— Figure 6.17} \qquad (6.27.21)$$

$$K_i = y_i/x_i \qquad (6.27.22)$$

$$K_i = f(T', P') \qquad (6.27.23)$$

$$\mu_F = \sum_i x_{iF}' \, \mu_{iF} \qquad (6.27.24)$$

$$\mu_{iF} = f[(T_T + T_B)/2] \qquad (6.27.25)$$

$$\text{HETS} = D' \quad\text{—— for } D \le 0.5\text{m} \qquad (6.27.26)$$
$$\text{or HETS} = (D')^{0.3} \quad\text{—— for } D > 0.5\text{m}$$

After calculating the minimum reflux ratio and the minimum number of stages, calculate the optimum or actual reflux ratio. According to Henley and Seader [31] for a fractionator containing a large number of stages, $R_O / R_M \approx 1.10$, but for a small number of stages $R_O / R_M \approx 1.50$. In between, use a reflux ratio of $R_O / R_M = 1.30$. Rather than use a rule-of-thumb, we will use the graphical correlation developed by Van Winkle and Todd [40] from computer calculations. Alternatively, calculate the optimum reflux ratio from Equation 6.27.5, which was developed by Olujic [41] by curve fitting Van Winkle and Todd's correlation.

Gilliland [42] correlated the number of equilibrium stages with the minimum number of stages, calculated from the Fenske Equation. Gilliland plotted Y_e, defined by Equation 6.27.9, against X_e, defined by Equation 6.27.10. Gilliland's correlation has been curve fitted by several equations but the simplest of these equations is McCormick's [43] equation, given by Equation 6.27.8. Oliver [44] pointed out that Gilliland's correlation leads to large errors when the number of stages in the stripping section is much larger than the number of stages in the enriching section. Gilliland's correlation requires that the feed be introduced at the optimum stage, calculated from Equation 6.27.12, an empirical equation developed by Kirkbride [45].

The actual number of stages is equal to the number of equilibrium stages divided by the fractionator efficiency(overall column efficiency). Although the tray efficiency will vary, we will use the fractionator efficiency. The fractionator efficiency is obtained from the O'Connel correlation given in Figure 6.17. Vital et al. [46] have reviewed and tabulated fractionator and absorber efficiencies for many systems. These data may help to arrive at a reasonable fractionator efficiency.

Table 6.28 Calculation Procedure for Sizing Fractionators

1. Calculate the feed-bubble-point temperature, and then calculate the K-values for all components at the bubble point. Next calculate the relative volatility of each component relative to the heavy key component.

2. Calculate the constants A_C and B_C in the Geddes equation, Equation 6.27.1, using a specified recovery and relative volatility for the light and heavy key components. There should be one equation for the light key component and another equation for the heavy key component. Then, solve the two equations for A_C and B_C.

3. Using these values of A_C and B_C in Equation 6.27.1, calculate the recovery of the remaining components and hence the composition of the distillate and bottom products.

4. From the composition of the bottom product, calculate the bubble-point temperature.

5. Assume a total condenser. The composition of the vapor from the top tray is equal to the composition of the distillate. Calculate the dew-point temperature of the vapor, which is the temperature at the top tray.

Table 6.28 Continued

6. Calculate the relative volatility, α_i, of the light and heavy key components at the top tray and at the bottom of the column, from Equations 6.27.18 and 6.27.22 (i = LK and i = HK).

7. Calculate the column geometric-average relative volatility, $(\alpha_i)_{avg}$, (feed, distillate, and bottom product) of the light and heavy key components from Equation 6.27.19 (i = LK and i = HK).

8. Calculate the minimum reflux ratio, R_M, from the Underwood equations (Equations 6.27.3 and 6.27.4).

9. Calculate the optimum reflux ratio, R_O, from the Van Winkle and Todd correlation, (Equations 6.27.5 to 6.27.7).

10. Calculate the minimum number of equilibrium stages, N_M, from the Fenske equation, Equation 6.27.2.

11. Calculate the number of equilibrium stages, N_e, from the Gilliland correlation, (Equations 6.27.8 to 6.27.11).

12. Locate the feed point from the Kirkbride equation, Equations (6.27.12 and 6.27.13).

13. Calculate the column diameter, D, using the procedure outlined in Table 6.24.

14. Calculate the mole-fraction average of the relative volatility, α_i, and feed viscosity, μ_i, at the average of the top tray and bottom temperature. Use Equations 6.27.20, 6.27.24, and 6.27.25.

15. Calculate the column overall efficiency, E_O, from Equation 6.27.21.

16. Calculate the length, L_S, at the bottom of the column required for surge capacity from Equation 6.27.16.

17. Calculate the column height, Z, from Equation 6.26.15 for a tray column or from Equation 6.27.17 and 6.27.26 for a packed column.

The height of a tray fractionator is equal to the number of trays times the tray spacing plus additional height above the top tray and below the bottom tray. These additional sections are needed for removal of liquid entrained in the vapor from the top tray and to provide surge capacity for the bottom product. Table 6.25 lists the tray spacing as a function of pressure. Because tray spacing influences the height of a column, it should be kept as small as possible. Tray spacing may be influenced by maintenance considerations. There should be sufficient space between the trays to facilitate inspection and repairs, but occasionally, other considerations affect the spacing. For example, when separating oxygen and nitrogen from liquid air, heat transferred to the fractionator from the surroundings must be minimized, and thus, the fractionator surface area must be a minimum. This consideration results in a tray spacing of as low as 6.0 in (0.152 m) [48].

The height of a packed fractionator is equal to the number of equilibrium stages times the height equivalent to a theoretical stage (HETS). Although this method is not rigorous, Ulrich [50] remarked that it is disquieting to find that the HETS does not vary much in commercial columns after having spend hours learning to calculate combined mass transfer coefficients. For fractionator diameters less than 0.5 m (1.64 ft), Frank [33] recommends the rule of thumb that D = HETS, and for column diameters greater than 0.5 m (1.64 ft), the HETS is given by Equation 6.27.26 [50].

Besides the height occupied by trays or packing, additional height is needed at the top and bottom of the fractionator. Henley and Seader [31] recommend adding 4.0 ft (1.22 m) to the top of the fractionator to minimize entrainment and 10.0 ft (3.05 m) to the bottom for surge capacity. For fractionators or absorbers of about three feet in diameter, Walas [51] recommends that 4.0 ft (1.22 m) be added to the top and 6.0 ft (1.83 m) to the bottom of the column. Ulrich [50] recommends that the volume below the bottom tray be sufficient for 5 to 10 min surge time which results in 1.0 to 4.0 m (3.28 to 13.1 ft) of additional height. Thus, as an approximation add 4.0 ft (1.22 m) to the top of the column and a surge height, L_S, to the bottom of the column. The surge height is calculated from Equation 6.27.16. The diameter of a fractionator or absorber is usually limited to 13.0 ft (3.96 m) and the length to about 200 ft (60.9 m) because of shipping limitations. If lengths larger than 200 ft (60.9 m) are necessary, then two vessels in series could be used. Exceptions to rules-of-thumb sometimes occur. One of the largest fractionators – made in Europe – is 356 ft (109 m) high and 21.0 ft (6.40 m) in diameter [47]. Another large ethylene fractionator built in Deer Park, TX, is 328 ft (100 m) high by 18 ft (5.49 m) in diameter [9]. This column was fabricated in sections and assembled at the site.

For the relationships listed in Table 6.27, assume that the fractionator pressure is constant. If needed, the pressure drop across the column can be estimated by the rules-of-thumb given in Table 6.29.

Safety factors are needed in fractionator design because of uncertainty in system property data, unsuspected trace components in the feed, difference between plant and design conditions – particularly in feed composition and flow rate – and variable operating conditions caused by controllers and by plant upsets [54]. Besides, the reasons for safety factors stated above by Drew [54], the factors should also depend on the uncertainties of the calculation procedure. Different safety factors are required for large and small fractionators as shown in Table 6.30. This occurs because engineering costs for small fractionators are comparable to equipment costs, whereas for larger fractionators equipment costs dominate. Therefore, for large fractionators a more thorough design is justified to save 5 to 10 % of equipment costs, which results in a smaller safety factor.

Table 6.29 Approximate Tray Pressure Drops for Fractionators

Tray Fractionators[a] psi/tray[c]	
Pressure atm	Pressure Drop
< 1.0	0.1
> 1.0	0.05
Packed Fractionators[b] psi/ft[d]	
vacuum	0.1 - 0.2
moderate to high	0.4 - 0.75

a) Source: Reference 6.31
b) Source: Reference 6.33
c) To convert to kPa/tray multiply by 6.848.
d) To convert to kPa/m multiply by 22.47.

Table 6.30: Safety Factors for Fractionator Sizing (Source: Reference 55)

Item	Safety Factors, %	
	Small Column < 4 ft D	Large Column > 4 ft D
Packed Height	---	0 - 15.0
Trays	20.0	10.0
Diameter	15.0	0

Example 6.7 Estimating the Number of Equilibrium Stages

This problem is adapted from a problem given by Fair and Bolles [56] for a de-ethanizer column. A solution of hydrocarbons at its bubble point is pumped into the column at an average pressure of 400 psia (27.6 bar). The composition of the liquid feed is given in Table 6.7.1. Calculate the number of equilibrium stages if the recovery of ethane in the top product is 99%, and the recovery of propylene in the bottom product is also 99%. Also, determine the location of the feed point.

Follow the calculation procedure outlined in Table 6.28 using Equations listed in Table 6.27. The light key (LK) is ethane and the heavy key (HK) is propylene. First, calculate the composition of the top and bottom products. Then, determine the optimum reflux ratio. Next, calculate the number of equilibrium stages. Finally, calculate the location of the feed tray.

To obtain the composition of the top and bottom products, first calculate the relative volatility of each component using the conditions of the feed as a first guess. The relative volatility depends on temperature and pressure. The bubble point of the feed at 400 psia (27.6 bar) and at the feed composition, calculated using ASPEN [57], is 86.5 °F (130 °C). The K-values of the feed are listed in Table 6.7.1. Bubble and dew points could also be calculated using K-values from the DePriester charts [31] and by using the calculation procedures given in Chapter 3. Next, calculate the relative volatility of the feed stream, defined by Equation 6.27.18, for each component relative to the heavy key component.

The relative volatility for each of the feed components in Table 6.7.1, will now be used to calculate the composition at the top tray and the bottom product. First calculate the constants A_C and B_C in Equation 6.27.1. Selecting one mole of feed as the basis of calculation, the moles of ethane and propylene in the distillate and bottom products, using the specified recoveries, are calculated as follows:

$n_{iB} = 0.99\,(0.15) = 0.1485$, moles of propylene in the bottom product

$n_{iD} = 0.01\,(0.15) = 0.0015$, moles of propylene in the top product

$n_{iB} = 0.01\,(0.35) = 0.0035$, moles of ethane in the bottom product

$n_{iD} = 0.99\,(0.35) = 0.3465$, moles of ethane in the top product

Substituting into Equation 6.27.1 for propylene,

$\log(0.0015 / 0.1485) = A_C + B_C \log 1.0$

and for ethane,

$\log(0.3465 / 0.0035) = A_C + B_C \log 2.238$

Table 6.7.1 Preliminary Composition of the Top and Bottom Products for a De-ethanizer

Component	Feed Moles	K_{Fi}	Relative Volatility α_{Fi}	Fraction Recovered in Distillate	Distillate[a] Moles	Bottom[a] Product Moles
CH$_4$	0.05	4.965	7.958	~ 1.0	~ 0.05	~ 0
C$_2$H$_6$ (LK)	0.35	1.396	2.238	0.99	0.3465	3.5×10^{-3}
C$_3$H$_6$ (HK)	0.15	0.6239	1.0	0.01	0.0015	0.1485
C$_3$H$_8$	0.20	0.5488	0.8796	2.332×10^{-3}	4.664×10^{-4}	0.1995
i-Butane	0.10	0.2662	0.4267	6.092×10^{-7}	6.092×10^{-8}	~ 0.10
n-Butane	0.15	0.2213	0.3547	7.397×10^{-8}	1.110×10^{-8}	~ 0.15

a) Basis: one mole of feed

Solving these equations simultaneously for A_C and B_C, we find that $A_C = -1.996$ and $B_C = 11.41$. Using these values of A_C and B_C in Equation 6.27.1 and the component mole balance, $n_{iF} = n_{iD} + n_{iB}$, we can now calculate the moles of methane, propane, n-butane, and i-butane in the distillate and bottom products. The results are given in Table 6.7.1.

After calculating the temperature of the top and bottom products, obtain a new estimate of the column relative volatility for each component. Find the relative volatility of each component in the bottom and top product. Assuming that we have a total condenser, the composition of the vapor rising above the top tray is equal to the composition of the top product. The calculation for the dew-point temperature will give the composition of the liquid on the top tray as well as the temperature. The temperature and liquid composition at the bottom tray is obtained from a bubble point calculation. Next, calculate the relative volatility of each component at the top and bottom tray. Using these values of the relative volatility and the values for the feed, calculate the geometric average volatility, $(\alpha_i)_{avg}$, of each component from Equation 6.26.19. This calculation is summarized in Table 6.7.2

We can now recalculate the composition at the top and bottom trays using the improved values for the relative volatility for each component. The procedure

Table 6.7.2 Summary of the Calculation for the Geometric-Average Relative Volatility

Component	K_{Fi}	α_{Fi}	K_{Ti}	α_{Ti}	K_{Bi}	α_{Bi}	$(\alpha_i)_{avg}$
CH_4	4.965	7.958	3.388	8.225	4.409	3.519	6.130
C_2H_6(LK)	1.396	2.238	0.9150	2.221	2.058	1.642	2.013
C_3H_6(H)	0.6239	1.0	4.119	1.0	1.253	1.0	1.0
C_3H_8	0.5488	0.8796	0.3429	0.8324	1.167	0.9314	0.8802
i-C_4H_{10}	0.2662	0.4267	0.3797	0.1564	0.7515	0.5998	0.4598
n-C_4H_{10}	0.2213	0.3547	0.1364	0.3311	0.6680	0.5331	0.3971

Table 6.7.3 Final Composition of the Top and Bottom Products for a De-ethanizer

Component	Feed Moles	Relative Volatility $(\alpha_i)_{avg}$	Fraction Recovered in Distillate	Distillate Moles	Bottom Product Moles	Distillate Mole Fraction x_{iD}	Bottoms Mole Fraction x_{iD}
CH_4	0.05	6.130	~ 1.0	~ 0.05	~ 0	0.1255	~ 0
C_2H_6 (LK)	0.35	2.013	0.99	0.3465	3.500×10^{-3}	0.8695	5.819×10^{-3}
C_3H_6(HK)	0.15	1.0	0.01	0.0015	0.1485	0.003764	0.2469
C_3H_8	0.20	0.8802	2.350×10^{-3}	4.700×10^{-4}	0.1995	0.001179	0.3317
i-Butane	0.10	0.4598	1.428×10^{-6}	1.429×10^{-7}	~ 0.10	3.583×10^{-7}	0.1663
n-Butane	0.15	0.3970	2.682×10^{-7}	4.023×10^{-8}	~ 0.15	1.010×10^{-7}	0.2494

is the same as the calculation given above using only the feed volatility. Table 6.7.3 summarizes the results. The new compositions could be used to generate a new geometric-mean relative volatility for each component and the calculation can be repeated. Further iteration is not warranted, however, considering the approximate nature of the calculation.

The next step in the procedure is to calculate the optimum or operating reflux ratio. First, calculate the minimum reflux ratio using the Underwood equations, Equations 6.27.3 and 2.27.4. For the calculation use the geometric average volatility of each component listed in Table 6.27.3. Because the feed is at its bubble point, $q = 1$. Thus, Equations 6.27.3 and 6.27.4 becomes

$$\sum_i \frac{(\alpha_i)_{avg}\, y_{iF}}{(\alpha_i)_{avg} - \theta} = \frac{6.130\,(0.05)}{6.130 - \theta} + \frac{2.013\,(0.35)}{2.013 - \theta} + \frac{1.0\,(0.15)}{1.0 - \theta} + \frac{0.8802\,(0.20)}{0.8802 - \theta}$$

$$+ \frac{0.4598\,(0.10)}{0.4598 - \theta} + \frac{0.3970\,(0.15)}{0.3970 - \theta} = 0$$

$$R_M + 1 = \frac{(\alpha_i)_{avg}\, x_{iD}}{(\alpha_i)_{avg} - \theta} = \frac{6.130\,(0.1255)}{6.130 - \theta} + \frac{2.013\,(0.8695)}{2.013 - \theta} + \frac{1.0\,(0.003764)}{1.0 - \theta}$$

$$+ \frac{0.8802\,(0.001179)}{0.8802 - \theta} + \frac{0.4598\,(3.583 \times 10^{-7})}{0.4598 - \theta} + \frac{0.3971\,(1.010 \times 10^{-7})}{0.3971 - \theta}$$

Solving the first of these equations using Polymath, we find that $\theta = 1.297$. Substitute this value of θ into the second equation and solve for R_M to obtain 1.589.

Now, calculate the optimum reflux ratio using Equations 6.27.5 to 6.27.7. From Equation 6.27.6,

$$Y_O = \frac{2.013}{1.0614\,(2.013) - 0.4175} = 1.171$$

and from Equation 6.27.7,

$$X_O = \log \left[\left(\frac{0.3465}{0.0015} \right) \left(\frac{0.1485}{0.0035} \right) \left(\frac{0.35}{0.15} \right) \right]^{0.55\,(2.013)} = 4.399$$

Substituting X and Y into Equation 6.27.5 we obtain

$$\frac{R_O}{1.589} = \frac{(1.6 - 1.171)}{6.5}(4.399 - 7.5) + 1.6$$

Solving, for R_O, we find that $R_O = 2.217$.

To obtain the number of actual stages for the separation, first calculate the minimum number of equilibrium stages from Equation 6.27.2.

$$N_M = \frac{\log(\ 0.3465 / 0.0015)(0.1485 / 0.0035)}{\log 2.013} = 13.14$$

Next, solve Equations 6.27.8 to 6.27.11 to obtain the number of equilibrium stages. From Equation 6.27.10,

$$X_e = \frac{2.217 - 1.589}{2.217 + 1} = 0.1952$$

and from Equation 6.27.11,

$$B_e = 0.105 \log 0.1952 + 0.44 = 0.3655$$

Next, calculate the value of Y from Equation 6.27.8.

$$Y_e = 1 - 0.1952^{0.3655} = 0.4496$$

Finally, from Equation 6.27.9,

$$\frac{N_e - 13.14}{N_e + 1} = 0.4496$$

The number of equilibrium stages, N_e, equals 24.68. Rounding off N_e to the next highest stage, $N_e = 25$.

The feed point location is calculated from the Kirkbride equation, Equation 6.27.12.

$$\frac{N_U}{N_L} = \left[\left(\frac{0.15}{0.35} \right) \left(\frac{5.819 \times 10^{-3}}{3.764 \times 10^{-3}} \right)^2 \left(\frac{0.6015}{0.3984} \right) \right]^{0.206}$$

$N_U + N_L = 25$

Solving these equations simultaneously, $N_U = 11.93$ and $N_L = 13.07$ – rounding off, $N_U = 12$ trays above the feed point and $N_L = 13$ trays below the feed point.

Liquid-Liquid Extractors

Several liquid-liquid extractors have been reviewed by Lo [61]. Extractors are divided into two classes: unagitated, and agitated. Among the unagitated extractors there are the packed and sieve plate designs, which are similar to fractionators and absorbers. Examples of agitated extractors, shown in Figure 6.20, are the rotating disc and Oldshue-Rushton extractor. Another agitated extractor is the Karr reciprocating-plate extractor. For all these extractors, backmixing, which reduces the column efficiency, is a problem. Agitation is needed to increase mass transfer by dispersing one of the phases and increasing turbulence in the continuous phase. In the rotating-disc extractor, the disc is the agitator, in the Oldshue-Rushton column it is flat blade turbine impellers, and in the reciprocating-plate extractor. it is the up-and-down motion of the plate stack. Horizontal stator rings above and below each disc or impeller, shown in Figure 6.20, reduces backmixing.

Extractor Sizing

As for absorbers and strippers, the height of the extractors can be calculated simply by calculating the number of equilibrium stages and multiplying by HETS. Additional height is needed at the top and bottom of the extractor for phase separation. Figure 6.21 shows that the Karr reciprocating-plate extractor is one of the more efficient based on both HETS and throughput. Karr and Lo [62] developed a procedure for scaling reciprocating-plate extractors from small-scale tests. The Karr extractor will be used to illustrate a procedure for sizing extractors.

In Table 6.31, Lo [61] has tabulated the minimum HETS for the methylisobutylketone (MIBK), acetic-acid, water system and the o-xylene, acetic acid water system. The minimum HETS is measured by fixing the geometry of the extractor, holding the throughput constant, and then varying the reciprocating-plate frequency. At low frequencies, the dispersed phase drop size is large and therefore the mass-transfer rate is small, resulting in a large HETS. As the frequency increases, the drop size decreases, and the mass-transfer rate increases decreasing HETS. As shown in Figure 6.22, HETS decreases until flooding occurs. The operating frequency must be less than the minimum frequency to avoid flooding.

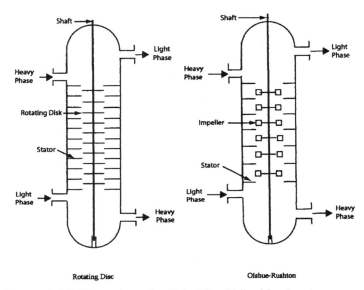

Rotating Disc   Olshue-Rushton

Figure 6.20 Examples of agitated liquid-liquid extractors.

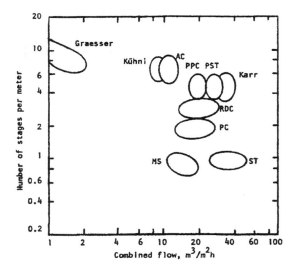

System: Acetone in Water Extracted with Toluene – Toluene Dispersed

Code: AC (Agitated Cell) – PPC (Pulsed Packed Column) – PST (Pulsed Sieve Tray) – RDC (Rotating-Disc Contactor – PC (Packed Column) – MS (Mixer Settler) – ST (Sieve Tray)

Figure 6.21 Comparison of liquid-liquid extractors. (Source Ref. 63 with permission).

Table 6.31 Minimum HETS and Volumetric Efficiency for the Karr, Reciprocating-Plate Extractor (Source: Ref. 61 with permission).

Column diam, in	Amplitude, in	Plate spacing, in	Agitator speed, strokes/min	Extractant	Dispersed phase	Min HETS	Throughput, gal/(h)(ft²)	Volumetric efficiencies V_t/HETS, h^{-1}
I. System: MIBK–Acetic Acid–Water								
1	½	1	360	MIBK	Water	3.1	572	296
			401			2.8	913	523
1	½	1	278	Water	MIBK	4.2	459	175
			152			8.1	1030	204
3	½	1	330	MIBK	Water	4.9	600	196
	½	1	245			6.3	1193	304
	½	2	355			7.5	1837	393
	½	1	320	Water	Water	4.3	548	205
	½	1	230			6.7	1168	280
	½	2	367	Water	Water	5.0	1172	376
			240			7.75	1707	353
12	½	1	430	Water	MIBK	5.8	547	151
(with			285			5.7	1167	328
baffle)	½	1	244	MIBK	MIBK	4.4	599	218
			170			5.6	1193	342
	½	1	250	MIBK	Water	7.2	602	134
			225			7.2	1200	268
			150			14.0	1821	208
	½	1	225	Water	Water	7.0	555	127
			200			9.5	1170	197
			150			11.05	1694	246
	½	1	275	Water	MIBK	9.5	1179	199
	½	1	200	MIBK	MIBK	7.8	595	123
			150			6.2	1202	311
II. System: Xylene–Acetic Acid–Water								
3	1	1	267	Water	Water	9.1	424	75
3	½	1	537	Water	Water	8.2	424	83
3	¼	1	995	Water	Water	7.7	424	88
3	1	2	340	Water	Water	9.1	804	142
36	1	1	168	Water	Water	23.3	425	29*
36	1	1	168	Xylene	Water	20.0	442	36*

The HETS will then be higher then at the minimum point. To estimate the design HETS, select an HETS that is 20% higher than the minimum value. Also, given in Table 6.31 is the maximum volumetric efficiency, defined by Equation 6.31.

$$\eta_V = \frac{V_C + V_D}{HETS} \tag{6.31}$$

Karr and Lo [63] have developed simple scaling rules for the Karr extractor. To scale HETS from one column size to another, requires that the plate spacing, amplitude, and total volumetric flow rate per unit area be kept the same for each extractor. They found for a high interfacial-tension system such as the o-xylene, acetic-acid, water system, that

$$\frac{(HETS)_2}{(HETS)_1} = \left(\frac{D_2}{D_1} \right)^{0.38} \tag{6.32}$$

For a low interfacial-tension system such as MIBK, acetic acid, water, the exponent is 0.36, only slightly different.

To scale the reciprocating frequency, they also developed the following relation.

$$\frac{\omega_2}{\omega_1} = \left(\frac{D_1}{D_2} \right)^{0.16} \tag{6.33}$$

Equations 6.31 to 6.33 have been successfully used to scale many Karr extractors from pilot plant experiments to commercial scale extractors up to 1.53 m (5.02 ft) in diameter [61].

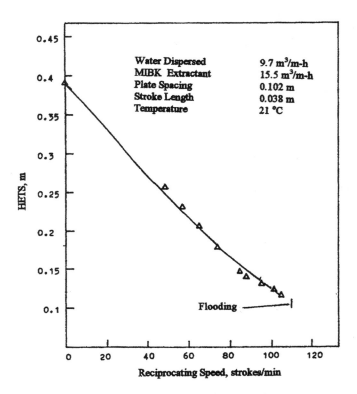

Figure 6.22 Effect of reciprocating-plate frequency on HETS for the Karr column. (Source Ref. 64).

Table 6.32 Summary of Equations for Sizing the Karr, Reciprocating-Plate Extractor

---

y = mass fraction of the key component in the solvent stream
x = mass fraction of the key component in the process stream
Subscripts: F = feed — S = solvent — M = minimum — k = key component
       Refer to Figure 6.16 for meaning of the numerical subcripts.
       k = key component

Minimum Solvent Flow Rate

$$\frac{m_F{}'}{m_{SM}} = \frac{K_k\, x_{1k}{}' - y_{2k}}{x_{1k}{}' - x_{2k}} \qquad (6.32.1)$$

Mass Balance

$$y_{1k} = y_{2k}' + (m_F' / m_S)(x_{1k}' - x_{2k}) \qquad (6.31.2)$$

$$x_{2k} = (1 - \varepsilon')\, x_{1k}' \qquad (6.32.3)$$

Operating Solvent Flow Rate

$$m_F'/m_S = C'\,(m_F'/m_{SM}) \qquad (6.32.4)$$

Number of Equilibrium Stages

$$(1/A_E)^{Ne} = \frac{x_{1k}' - y_{2k}'/K_k}{x_{2k} - y_{2k}'/K_k}\,(1 - A_E) + A_E \qquad (6.32.5)$$

Extractor Height

$$Z_E = N_e\,(HETS) + D \qquad (6.32.6)$$

Extractor Diameter

$$A = [\,(m_F' / \rho_F') + (m_S / \rho_S')\,] / J_T \qquad (6.32.7)$$

$$A = \pi D^2 / 4 \qquad (6.32.8)$$

$$\frac{(HETS)}{(HETS)_1} = \left(\frac{D}{D_1}\right)^{0.38} \qquad (6.32.9)$$

System Properties

$$K_k = f(T') \qquad (6.32.10)$$

$$J_T = f(\omega',\ \text{extractor geometry}) \quad\text{---}\quad \text{Table 6.31} \qquad (6.32.11)$$

$$HETS / D^{1/3},\ f(\text{interfacial tension}') \quad\text{---}\quad \text{Figure 6.23} \qquad (6.32.12)$$

$$A_E = (m_F' / m_S) / K_k \qquad (6.32.13)$$

Variables

m_{SM} - K_k - y_{2k} - x_{2k} - m_{SM} - m_S - Z - N_e - HETS - D - J_T - A - A_E

Table 6.32 lists the equations for sizing a Karr extractor, which is only a rough approximation for a preliminary process design. Table 6.33 outlines the calculating procedure. Tests using the actual solution, solvent, and equipment are necessary to arrive at an accurate extractor size. Again, it is assumed that the solutions are dilute so that the operating and equilibrium curves are linear. Thus, the Kremser equation, Equation 6.32.5, can be used to calculate the number of equilibrium stages. The subscript V refers to the light phase and the subscript L to the heavy phase. In Table 6.32, Equations 6.32.1 to 6.31.5 are for mass transfer from the heavy phase to the light phase. Before using the Kremser equation, the operating solvent flow rate is required, which can be calculated from Equations 6.32.1 and 6.31.4. After specifying the recovery of the key component, the exit composition of the solvent stream is calculated from Equation 6.32.2. After calculating the column diameter from Equations 6.32.7 and 6.32.8, use Equation 6.32.6 to calculate the height of the extractor.

Although the size of the end sections of an extractor, where phase separation occurs, could be calculated by a method similar to the one described in the section on decanter sizing, a more approximate method will be used. Karr and Lo [62] give the dimensions of the extractor used in their studies. The diameter of the end section is 50 % greater than the column diameter, and its height is a little less than the column diameter. For a pulsed-column extractor, Valle-Riestra [53] used a continuous-phase flux of 0.5 gal/min-ft^2 (3.40 m/min) and a height/diameter ratio of 1.0 to size the end sections of an extractor. The cross-sectional area of the extractor, and hence, the diameter is calculated by dividing by the total volumetric flow rate (the sum of the volumetric flow rates for both phases) by the total volumetric flow rate per unit of extractor cross-sectional area, obtained from Table 6.31. Then, add the diameter to the product of N_e and HETS, as shown in Equation 6.32.6, to obtain the total column height.

The HETS for an extractor can be estimated by using the scaling rules developed by Karr and Lo [62] and experimental values of HETS summarized in Table 6.31. First, determine if the extraction system is a low interfacial-tension system or a high interfacial-tension system. Next, select a value of HETS from Table 6.31 from the following systems:

low interfacial-tension – MIBK, acetic acid, water system
high interfacial-tension – o-xylene, acetic acid, water system

Then, scale this value of HETS for the extractor diameter using Equation 6.32.9. A simpler procedure for obtaining HETS, however, is to use the correlation given by Henley and Seader [31], shown in Figure 6.23. the correlation is acceptable for both a low and high interfacial-tension system. The problem, however, is that interfacial-tension data may not be available.

To complete sizing the Karr column requires sizing the electric motor. The size of the electric motor to disperse one of the phases is small. Walas [51] states that a 1.5 hp (1.12 kW) motor is sufficient to agitate a Karr extractor 30 in (0.762

m) in diameter and 20 ft (6.70 m) high. These data can be use as a guide to estimate the motor power.

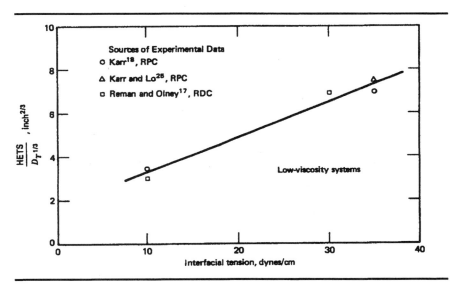

Figure 6.23 Effect of Interfacial Tension on HETS for the RDC and RPC Extractors (Source: Reference 6.31 with permission).

Table 6.33 Calculation Procedure for Sizing the Karr, Reciprocating-Plate Extractor

Refer to Figure 6.16 for meaning of the numerical subscripts.

1. Calculate mass fraction of the key component in the leaving heavy phase, $x_{2\,k}$, from Equation 6.32.3.

2. Calculate the maximum slope of the operating line from Equation 6.32.1.

3. Calculate the ratio of the feed mass flow rate to the solvent mass flow rate, m_F/m_S, from Equation 6.32.4, and the operating solvent flow rate, m_S.

4. Calculate the mass fraction of the key component in the entering light phase, y_{1k}, in the light phase from Equation 6.32.2.

5. Calculate the extraction factor, A_E, from Equation 6.32.13.

6. Calculate the number of equilibrium stages, N_e, from Equation 6.32.5.

Table 6.3.3 Continued

7. Calculate the extractor cross-sectional area, A, and the diameter, D, from Equations 6.32.7, 6.32.8 and 6.32.11.

8. Find $(HETS)_1$ at D_1 from Equation 6.32.12.

9. Calculate $(HETS)$ at D from Equation 6.32.9.

10. Calculate the extractor height, Z, using Equation 6.32.6.

Example 6.8: Sizing a Karr Reciprocating-Plate Extractor

To illustrate the procedure for sizing a Karr extractor, we will use a process design described by Drew [69]. The design requires separating a solution of methylene chloride and methanol. The first step in the process is to contract.

Data
Feed Compostion:

Methylene Choride	2185 lb/h (991 kg/h), 0.9851 mass fraction
Methanol	33 lb/h (15.0 kg/h), 0.01488 mass fraction
Total flow rate	2218 lb/h (1010 kg/h)

Methanol Recovery $\varepsilon = 95\,\%$ by weight
Methanol Distribution Coefficient (water/methylene chloride) = 2.0, estimated by
    Drew (6.69)
Density in lb/ft^3 (kg/m^3)

Methylene Chloride	82.41 (1320)
Methanol	48.7 (780)
Water	62.43 (999)

$C = 0.5$ (in Equation 6.32.4)

To size the extractor, follow the procedure given in Table 6.33 using the equations listed in Table 6.32. Because the methylene chloride solution is heavier than water, it is introduced at the top and the water at the bottom of the extractor. Refer to Figure 6.16 for the meaning of the numerical subscripts.
    From Equation 6.32.4,
$x_{2K} = (1 - 0.95)\,0.01488 = 7.440 \times 10^{-4}$

    From Equation 6.32.1,

$$\frac{m_F}{m_{SM}} = \frac{2.0\,(0.01488) - 0}{0.01488 - 7.44\mathrm{x}10^{-4}} = 2.105$$

where m_{SM} is the minimum solvent flow rate.
From Equation 6.32.4, the operating feed to solvent ratio,

$$\frac{m_F}{m_S} = 0.5\,\frac{m_F}{m_{SM}} = 0.5\,(2.105) = 1.053$$

$m_S = m_F\,/\,1.053 = 2218\,/\,1.053 = 2106$ lb/h (955 kg/h)

where m_S is the operating solvent flow rate.

Substitute $x_{2K} = 7.440\mathrm{x}10^{-4}$ and $m_F\,/\,m_S = 1.053$ into Equation 6.32.2 for the methanol balance. The methanol mass fraction in the exit water stream,

$y_{1K} = 0 + 1.053\,(0.01488 - 7.440\mathrm{x}10^{-4}) = 0.01489$

From Equation 6.32.13,

$A_E = 1.053\,/\,2 = 2.\,0.5265$
Now, calculate the number of equilibrium stages from Equation 6.32.5.

$$(1/\,0.52655)^{Ne} = \frac{0.1489 - 0}{7.44\mathrm{x}10^{-4} - 0}\,(1 - 0.5265) + 0.5265$$

$N_e = 7.103$

Rounding off N_e, we obtain 4 equilibrium stages.
To calculate the extraction height from Equation 6.32.6, first calculate HETS. HETS is correlated with interfacial tension in Figure 6.23. The interfacial tension does not appear to be available for this system. Twifik [70] correlated HETS with dimensionless groups, but his correlation also requires the interfacial tension. We will use the data given in Table 6.31 for MIBK. Table 6.31 gives data for several extractor diameters. Select the 12 in (0.305 m) diameter extractor, which is expected to be close to the calculated diameter. For the 12 in (0.3048 m) extractor there are several values of HETS at varying agitator speeds and throughputs. Select the extractor that gives the maximum volumetric efficiency. The minimum HETS is 5.6 in (0.142 m), and the total volumetric throughput is 1193

gal/h-ft² (48.3 m/h). To calculate the column cross-sectional area from Equation 6.32.7, requires the volumetric flow rates of both the light and heavy phases.

$$\frac{m_F}{\rho_F} = \frac{2185}{82.41} + \frac{33}{48.7} = 27.19 \text{ ft}^3/\text{h} \ (203.4 \text{ gal/h}, 0.770 \text{ m}^3/\text{h})$$

$$m_S / \rho_S = (7.481) (2106 / 62.43) = 252.4 \text{ gal/h} \ (0.995 \text{ m}^3/\text{h})$$

From Equations 6.32.7 and 6.32.8,

$$A = \frac{203.4 + 252.4}{1193} = 0.3821 \text{ ft}^2 \ (0.0355 \text{ m}^2)$$

$$D = (4 A / \pi)^{1/2} = [4 (0.3821) / 3.142]^{1/2} = 0.6975 \text{ ft} \ (8.370 \text{ in}, 0.213 \text{ m})$$

Next correct HETS for column diameter from Equation 6.32.9. Because D is less than 30 in (0.762 m), select a standard pipe size. From piping tables [66], select a Schedule 10S pipe, which has an inside diameter of 10.42 in (0.265 m).

$$HETS = 5.6 \left(\frac{10.42}{12} \right)^{0.38} = 5.307 \text{ in} \ (0.135 \text{ m})$$

Because 5.307 in. is a minimum value, increase it by 20% to avoid flooding. Therefore, the design HETS is 6.368 in (0.162 m). From Equation 6.32.6, the extraction height,

$$Z_E = 4 (6.368) = 25.47 \text{ in} \ (0.6469 \text{ m})$$

Rounding the height to the nearest 3 in (0.0762 m), $Z_E = 27$ in. (0.6858 m). Because this is a short extractor and because of the assumptions made, increase the extraction height to 6 ft (1.97 m). The extra cost would not be substantial.

Now, add top and bottom sections to separate the phases. The diameter of both settlers is 50% greater than the extractor diameter, and the height of each settler is equal to the settler diameter. Therefore, the height of both settlers,

$$Z_S = 2 (1.5) (10.42) = 31.26 \text{ in} \ (0.794 \text{ m})$$

To join the settlers to the extractor requires reducers, which are about a foot long. The total height of the column,

$$Z = Z_E + Z_S + \text{reducers} = 27.0 + 31.26 + 24.0 = 82.26 \text{ in} \ (6.86 \text{ ft}, 2.09 \text{ m})$$

Round off Z to 7 ft (2.13 m).

Because of the assumptions and approximations made in this problem, the final design of the column must be confirmed by testing in a pilot plant. Cusack and Karr [71] discuss the need for pilot plant testing.

NOMENCLATURE

A	area or projected area
A_A	absorption factor
A_E	extraction factor
A_F	filter area or cross-sectional area for flow
A_I	interfacial area in a decanter
A_T	total rotary drum area
B	bottoms flow rate
C	concentration, mass per unit volume
C_D	drag coefficient
D	packing size or drop diameter
D	diameter or distillate flow rate
D_M	average diameter
E_o	column efficiency
F	fraction of drum area required for filtering
F_B	buoyant force
F_D	drag force
F_G	gravitational force

g	acceleration of gravity
H	height
H_D	dispersion zone thickness
HETS	height equivalent to a theoretical stage
J_T	total volumetric flow rate per unit area
k	flooding factor
k_V	entrainment factor
K	liquid-vapor or liquid-liquid equilibrium ratio
L	length
L_D	the length of the dispersion layer
L_S	length of the lower section of a column, or decanter length required for the dispersed phase to settle
m	mass or molar flow rate
m_D	mass of dry cake
m_L	mass flow rate of liquid, molar flow rate, or mass of a liquid drop
m_S	mass of dry filter cake
m	vapor mass flow rate or molar flow rate
M_L	molecular weight of a liquid
M_V	molecular weight of a vapor
$n_{i\,B}$	number of moles of component i in the bottoms
$n_{i\,D}$	number of moles of component i in the distillate
N_A	number of actual stages
N_e	number of equilibrium stages

N_L	number of trays below the feed tray
N_M	minimum number of trays
N_U	number of trays above the feed tray
P	pressure or perimeter
P_o	internal or operating pressure, or the pressure at the surface of a filter cake
P_S	pressure drop across a filter cake
P_V	pressure produced by a vacuum pump
q	a measure of the thermal condition of the feed
R	radius or reflux ratio
Re	Reynolds Number
R_h	hydraulic radius
R_M	minimum reflux ratio
R_O	optimum reflux ratio
s	specific surface (the surface area per unit volume of particle)
S	stress
t_C	corrosion allowance, thickness
t_D	time for a drop to reach a liquid-liquid interface
t_F	filtering time
t_S	shell thickness or surge time
t_H	head thickness
t_R	residence time

T   temperature

v   velocity

v_d terminal velocity of a dispersed phase drop

v_S superficial velocity

v_V vapor velocity

V   volume

V_B volumetric flow rate of bottom-product

V_D volumetric flow rate of the dispersed phase

V_L volumetric flow rate of the liquid, or light phase

V_V volumetric flow rate of vapor

x   mole fraction in the liquid phase or distance

x_{LK} mole fraction of the light key component in the liquid

x_{HK} mole fraction of the heavy key component liquid

y   mole faction in the gas or vapor phase

Z   column height

Z_T tray spacing

Greek

α relative volatility, or specific resistance

$(\alpha_i)_{avg}$ geometric mean

ε weld efficiency, fraction absorbed or stripped, void fraction (porosity)

ε_H head weld efficiency

ε_S shell weld efficiency

η_V volumetric efficiency

μ viscosity

μ_C viscosity of the continuous phase

θ dispersed phase parameter

ρ density

ρ_S solid density

σ surface tension

ω reciprocating frequency

Subscripts

B   bottom of a fractionator

C   continuous phase

D   dispersed phase or distillate

F   fractionator

H   heavy phase

HK   heavy key component

i   interface or the ith component

k   key component

L   liquid or light phase

LK   light key component

m   minimum

s   solvent

T   temperature or top tray

V   vapor

REFERENCES

1. Rase, H.F. Barrow, M.H., Project Engineering of Process Plants, John Wiley & Sons, New York, NY,1964.
2. Gumm, W.G. Turner, J. E., The Nondestructive Testing Spectrum, Chem. Eng., 83, 17, 64, 1976.
3. Wallace, A.E., Webb, W. P., Cut Costs with Realistic Corrosion Allowances, Chem. Eng., 88, 17, 123, 1981.
4. Gerunda, A., How to Size Liquid-Vapor Separators, Chem. Eng., 88, 9, 81, 1981.
5. Aerstin, F., Street, G., Applied Chemical Process Design, Plenum Press, New York, NY, 1978.
6. Walas, S. M., Chemical Process Equipment, Butterworth Publishers, Stoneham, MA, 1988.
7. Markovitz, R. E., Choosing the Most Economical Vessel Head, Chem. Eng., 78, 16, 102, 1971.
8. Mulet, A., Corripio, A.B., Evans, L. B., Estimate Cost of Pressure Vessels via Correlations, Chem. Eng., 88, 20, 145, 1981.
9. Anonymous, News Features, Chem. Eng., 84, 26, 84, 1977.
10. Sivals, R., Pressure Vessel Design Manual, Sivals Inc., Odessa, TX, No Date.
11. Younger, A.H., How to Size Future Process Vessels, Chem. Eng., 62, 5, 201, 1955.
12. Holmes, T.L., Chen. G. K., Design and Selection of Spray/Mist Elimination Equipment, Chem. Eng., 91, 21, 82, 1984.
13. York, O.H., Poppele, E.W., Wire Mesh Mist Eliminators, Chem. Eng. Prog., 59, 6, 45, 1959.
14. Watkins, R.N., Sizing Separators and Accumulators, Hydrocarbon Proc., 46, 11, 252, 1967.
15. Sibulkin, M., A Note on the Bathtub Vortex and the Earth's Rotation, Amer. Sci., 71, 4, 352, 1983.
16. Patterson, F.M., Vortexing Can Be Prevented, Oil Gas J., 67, 31, 118, 1969.
17. Jacobs, L.J., Penney, W. R., Phase Separation, Handbook of Separation Processes, R. W. Rousseau, ed., John Wiley & Sons, New York, NY, 1987
18. Drown, D.C., Thomson, W. J., Fluid Mechanic Considerations in Liquid-Liquid Settlers, Ind. Eng. Chem. Process Des. Dev., 16, 2, 197, 1977.

19. Selker, A.H., Sleicher, C. A., Factors Effecting which Phase will Disperse when Immiscible Liquids are Stirred Together, Can. J. Chem. Eng., $\underline{43}$, 6, 298, 1965.

20. Barton, R.L., Sizing Liquid-Liquid Separators, Chem. Eng., $\underline{81}$, 13, 111, 1974.

21. Bailes, P.J., Godfrey, J.C., Slater, M.J., Designing Liquid/Liquid Extraction Equipment, The Chem. Eng., No. 370, 331, 1981.

22. Hooper, W. B., Jacobs, L. J., Decantation, Handbook of Separation Techniques for Chemical Engineers, P.A. Schwitzer, ed., McGraw-Hill, New York, NY, 1979.

23. McCabe, L.W. Smith, J.C., Harriott, P., Unit Operations of Chemical Engineering, 6th ed., McGraw-Hill, New York, NY, 2001.

24. Flood, J.E. Porter, H.F., Rennie, F.W., Filtration Practice Today, Chem. Eng., $\underline{73}$, 13,163, 1966.

25. Chalmers, J.M., Elledge, L.R., Porter, H. F., Filters, Chem. Eng., $\underline{62}$, 6, 191, 1955.

26. Bird, R.B., Stewart ,W.E., Lightfoot, E.N.,Transport Phenomena, John Wiley & Sons, New York, NY, 1960.

27. Tiller, F.M., Filtration Theory Today, Chem. Eng., $\underline{73}$, 13, 151, 1966.

28. Svarovsky, L., Advanced in Solid Liquid Separations I, Chem. Eng., $\underline{86}$, 14, 62, 1979.

29. Treybal, R.R., Mass-Transfer Operations, McGraw-Hill, 3rd ed., New York, NY, 1980.

30. King, C.J., Separation Processes, McGraw-Hill, New York, NY, 1981.

31. Henley, E.J., Seader, J.D., Equilibrium-Stage Separation Operations in Chemical Engineering, John Wiley & Sons, New York, NY, 1976.

32. Harrison, M.E., France, J.J., Troubleshooting Distillation Columns, Part 3: Trayed Columns, Chem. Eng., $\underline{96}$, 5, 126, 1989.

33. Frank, O., Shortcuts for Distillation Design, Chem. Eng., $\underline{84}$, 6, 111, 1977.

34. Geddes, R.L., A General Index of Fractional Distillation Power for Hydrocarbon Mixtures, AIChE J., $\underline{4}$, 4, 389, 1958.

35. Yaws, C.L., Patel, P.M., Pitts, F.H., Fang, C.S., Estimate Multicomponent Recovery, Hydrocarbon Proc., $\underline{58}$, 2, 99, 1979.

36. McNulty, K.J., Effective Design for Absorption and Stripping, Chem. Eng., $\underline{101}$, 11, 92, 1994.

37. Brochure, Buyer's Guide to Chemical Process Equipment from Pfaudler, Bulletin 1138, The Pfaudler Co., Rochester, NY, No Date.

38. Fenske, M.R., Fractionation of Straight-Run Pennsylvania Gasoline, Ind. Eng. Chem., $\underline{24}$, 5, 482, 1932.

39. Underwood,E.R., Fractional Distillation of Multicomponent Distillation - Calculation of Minimum Reflux Ratio, J. Inst. Petrol., $\underline{32}$, 274, 614, 1946.

40. Van Winkle, M.C., Todd, W., Optimum Fractionation Design by Simple Graphics Methods, Chem. Eng., $\underline{78}$, 21, 136, 1971.

41. Olujic, Z., Optimum Reflux Ratio, Chem. Eng., $\underline{88}$, 21, 184, 1981.

42. Gilliland, E.R., Multicompnent Rectification: Minimum Reflux Ratio, Ind. Eng. Chem., 32, 9, 1101, 1940.
43. McCormick, J.E., A Correlation for Distillation Stages and Reflux, Chem. Eng., 59, 13, 75, 1988.
44. Oliver, E.D., Diffusional Separation Processes: Theory, Design & Evaluation, John Wiley & Sons, New York, NY, 1966.
45. Kirbride, G.G., Process Design Procedure for Multicomponent Fractionators, Petrol. Refiner, 23, 9, SP321, 1944.
46. Vital, T.J., Grossel, S.S., Olsen, P.I., Part 1, Estimating Separation Efficiency, Hydrocarbon Proc., 63, 10, 147, 1984.
47. Anonymous, Update, Chem. Eng. Progr., 86, 10, 14, 1990.
48. Smith, B.D., Design of Equilibrium Stage Processes, McGraw-Hill Book Co., New York, NY, 1963.
49. Branan, C.R., Rules of Thumb for Chemical Engnineers, Gulf Publishing, Houston, TX, 1994.
50. Ulrich, G.D., A Guide to Chemical Engineering Process Design and Economics, John Wiley & Sons, New York, NY, 1984.
51. Walas, S.M., Rules of Thumb, Chem. Eng., 94, 4, 75, 1987.
52. Fair, J.R., Distillation, Handbook of Separation Process Technology, R.W. Rousseau, ed., John Wiley & Sons, New York, NY, 1987.
53. Valle-Riestra, J.F., Project Evaluation in the Chemical Process Industries, McGraw-Hill, New York, NY, 1983.
54. Drew, J.W., Distillation Column Startup, Chem. Eng., 90, 23, 221, 1983.
55. Drew, J.W., Two-Tier Safety Factors, Letter to the Editor, Chem. Eng., 91, 4, 5, 1984.
56. Fair, J.R, Bolles, W.L., Modern Design of Distillation Columns, Chem. Eng., 79, 9, 156, 1968.
57. Aspen Plus Steady State Simulation, Aspen Technology Inc., Cambridge, MA, 2000.
58. Bravo, J.L., Design Steam Strippers, Chem. Eng. Progr., 90, 12, 56, 1994.
59. Kremser, A., Theoretical Analysis of Absorption Process, Natl. Petrol. News, 22, 21, 42, 1930.
60. Vatatavuk, W.M., Neveril, R.B., Part XIII: Cost of Gas Absorbers, Chem. Eng., 89, 20, 135, 1982.
61. Lo, T.C., Commercial Liquid-Liquid Extraction Equipment, Handbook of Separation Techniques for Chemical Engineers, 2nd ed., P. A. Schneitzer, ed. , McGraw-Hill, New York, NY, 1988.
62. Karr, A., Lo, T.C., Scaleup of Large Diameter Reciprocating-Plate Extraction Columns, Chem. Eng. Progr., 72, 11, 68, 1976.
63. Stichlmair, J., Leistungs-und Kostenvergleich verschiedener Apparatebauarten fur die Flussig/Flussig-Extraktion, Chem. Ing. Tech., 52, 3, 253, 1980.

64. Choi, H.D., The Effect of Operating Variables on the Extraction Efficiency of a Reciprocating-Plate Extractor, Masters Thesis, Stevens Institute of Technology, Hoboken, NJ, 1979.
65. Henley, E.J., Seader, J.D., Separation Process Principles, John Wiley & Sons, New York, NY, 1998.
66. Perry's Chemical Engineers' Handbook, Perry, R. C., Green, D.W., eds., McGraw-Hill, New York, 1997.
67. Bennett, C.O., Myers, J. E., Momentum, Heat, and Mass Transfer, 2nd ed., McGraw-Hill, New York, NY, 1974.
68. Shukla, H.M., Hicks, R.E., Process Design Manual for Stripping Organics, NTIS PB84-232628, US Dept. of Commerce, Aug. 1984.
69. Drew, J.W., Design for Sovent Recovery, Chem. Eng. Progr., 71, 2, 92, 1975.
70. Tawfik, W.V., Optimization of Fuel Grade Ethanol Recovery System Using Solvent Extraction, PhD Thesis, Georgia Institute of Technology, Atlanta, GA, 1986.
71. Cussack, R., Karr, A., A Fresh Look at Liquid-Liquid Extraction, Chem. Eng., 98, 4, 112, 1991.
72. Scheiman, A.D., Size Vapor-Liquid Separators Quickly by Nomograph, Hydrocarbon Process. & Pet. Refiner, 42, 10, 165, 1963.
73. Sigales, B. More on the Design of Reflux Drums, Chem. Eng., 82, 20, 87, 1975.
74. Humphrey, J..L., Keller, G.E., Separation Process Technology, McGraw-Hill, New York, NY, 1997.
75. Frank, O., Personal Communication, Consulting Engineer, Convent Station, NJ, Jan. 2002.
76. Student Contest Problem, American Institute of Chemical Engineers, New York, NY, 1987.
77. Brochure, Goodloe Column Packings, Bulletin 891 B-2, Otto H. York Co., Parsippany, NY, 1996.
78. Brochure, Separation Columns for Distillation and Absorption, Sulzer Brothers Lmt., Winterthur, Switzerland, 1988.

7

Reactor Design

There are numerous reactor types, but in this chapter the objective is to consider only a few common types. These are: batch, continuous stirred tank, homogenous plug flow and fixed bed catalytic reactors. To size other reactor types and for a more thorough treatment of reactor design than presented here, the reader can consult books written on reactor design, such as Fogler [16], Smith [23], and Forment and Bischoff [31].

REACTOR SELECTION

Because of the variety of reactors available, some engineers believe that reactor classification is not possible. No matter how incomplete a classification may be, however, the designer needs some guidance, even though there may be some reactor types that do not fit into any classification. Accordingly, we will classify reactors using the following criteria:

1. form of energy supplied
2. phases in contact
3. catalytic or noncatalytic
4. batch or continuous
5. packed or suspended bed

In Chapter 1, reactions were classed according to the form of energy supplied to the reaction: thermochemical, biochemical, electrochemical, photochemical, plasma, and sonochemical. Table 7.1 gives an example of each reaction type. Since thermochemical reactions are the most common, we will consider them in detail in this chapter.

Mixtures of alkyl halides and chlorinated aromatic side chains are produced industrially in photochemical reactors. For example, reacting methane with chlorine, using mercury arc lamps, produces a mixture of the four isomers of chloromethane [1].

Samdani and Gilges [2] list a number of commercial processes for electrochemically synthesizing organic compounds. An example is the conversion of glucose to gluconic acid. Gluconic acid, sold as a 50% aqueous solution, is used in metal pickling and as a protein coagulant in the production of tofu (soy bean curd), as well as in many other applications [3].

A sonochemical reaction is an indirect way of conducting a thermochemical reaction. Ultrasound causes cavitation in liquids, elevating the temperature in microscopic cavities in the liquid, which promotes chemical reaction. There appears to be no commercial application of ultrasonic energy to conduct chemical reactions. Pandit and Moholkar [4] list several organic reactions conducted in the laboratory. A possible future application is the destruction of chlorinated hydrocarbons in wastewater or ground water [5].

A plant operated by Huls in Marl, Germany, uses an electric-arc plasma reactor to produced acetylene [6]. A plasma is an electrically conductive but electrically neutral gas. In this process, a hydrocarbon and hydrogen mixture flows into a reaction chamber where the hydrocarbon is cracked into acetylene, ethylene, hydrogen, and soot.

Table 7.1 Energy Sources for Chemical Reaction

Energy Source	Product Example
Thermochemical	Ammonia
Biochemical	Ethanol
Electrochemical	Gluconic Acid
Photochemical	Chloromethanes
Plasma	Acetylene
Sonochemical	Fumaric Acid
	(Laboratory Scale)

The next consideration is classifying reactors according to the phases in contact. These are:

1. gas-liquid
2. liquid-liquid
3. gas-solid
4. liquid-solid
5. gas-liquid-solid

After specifying the energy form, the catalyst and the phases in contact, the next task is to decide whether to conduct the reaction in a batch or continuous mode. In the batch mode, the reactants are charged to a stirred-tank reactor (STR) and allowed to react for a specified time. After completing the reaction, the reactor is emptied to obtain the products. This operating mode is unsteady state. Other unsteady-state reactors are: (1) continuous addition of one or more of the reactants with no product withdrawal, and (2) all the reactants added at the beginning with continuous withdrawal of product. At steady-state, reactants flow into and products flow out continuously without a change in concentration and temperature in the reactor.

Table 7.2 Summary of Reactor Types

Operating Mode →	Batch	Continuous				
Reactor Type →	Tank	Tank	Tank Battery		Tubular	
Flow type →	Agitated	Agitated	Cocurrent	Counter-current	Cocurrent	Counter-current
Phases[a]						
Gaseous	R	C	C	N	C	N
Liquid	C	C	C	N	C	N
Gas-Liquid[b]	C	C	R	C	R	C
Liquid-Liquid	C	C	C	C	R	C
Gas-Solid	C	C	R	C	R	C
Liquid-Solid	C	C	R	C	R	C
Gas-Liquid-Solid	C	C	R	C	C	C

a) C indicates common reactor operation, R indicates rare, and N indicates never.
b) Gas bubbling through a liquid.

Source: Adapted from Ref. 7.

To guide the reactor selection process, Walas [7] has classified reactions according to the operating mode (batch or continuous), reactor type (tank, tank battery, tubular), flow type (back mixed, multistage back mixed), and the phases in contact. This reactor classification in Table 7.2 indicates if a particular reactor arrangement is commonly used, rarely used, or not feasible.

Economics determines whether to use a continuous flow or a batch reactor. Generally, if the residence time is large and the production rate small, select a batch reactor. This relationship is shown in Figure 7.1, which can be used to obtain a preliminary selection of a reactor. When the application is located in overlapping areas or near a boundary, make a careful analysis to determine the most economic choice.

There are two ideal models for developing reactor-sizing relationships: the plug flow and the perfectly stirred-tank models. In the plug-flow model, the reactants flowing through the reactor are continuously converted into products. During reaction there is no radial variation of concentration, backmixing or forward mixing. In a perfect STR, the reactants are thoroughly mixed so that the concentration of all species and temperature are uniform throughout the reactor and equal to that leaving the reactor.

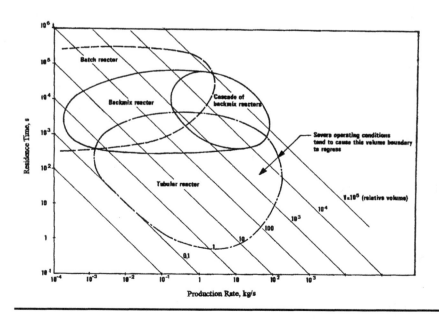

Figure 7.1 Application areas for several reactor types. From Ref. 8.

STIRRED-TANK REACTOR SELECTION

The operating mode of a stirred-tank reactor may be either continuous or batch. A STR consists of a vessel to contain the reactants, a heat exchanger, a mixer, and baffles to prevent vortex formation and to increase turbulence, enhancing mixing.

To evaluate and select a STR, consider the following factors:

1. mixing
2. heat transfer
3. jacket pressure drop
4. cleaning

Sufficient power must be supplied to the liquid to approach the ideal model of a thoroughly-mixed reacting system. Inadequate mixing results in a longer average residence time and thus a larger reactor volume than for the ideal model. Designing a mixing system requires selecting and sizing the impeller, baffles, and electric motor. For a preliminary design, all that is necessary is to estimate the mixer power.

An important consideration when sizing a STR is heating or cooling the reactor contents. There are several heat exchangers, which are classified as either an internal or external heat exchanger. The internal heat exchangers are immersed directly into the reacting liquid and consist of spiral coils, harp coils, and hollow or plate baffles. We will only consider spiral coils when designing an STR.

The external heat exchanger may either be a jacket or a shell-and-tube heat exchanger. For the latter, the reactor contents circulate through an external flow loop containing the heat exchanger. The jacket types, as illustrated in Figure 7.2, consist of the simple jacket – with or without a spiral baffle or nozzles for promoting turbulence – the partial pipe coil, and the dimple jacket. The simple jacket consists of an outer cylinder enclosing part of the reactor. A heat-transfer fluid flows in the annular area surrounding the reactor, as shown in Figure 7.2. If the heat-transfer rate is limited by the jacket heat-transfer coefficient, then increase the turbulence in the jacket by using a spiral baffle or nozzles. The spiral baffle is wound around and welded to the reactor. The baffle channels the fluid from the jacket entrance to the jacket exit. Channeling the fluid increases its velocity and turbulence, resulting in a higher heat transfer coefficient. The partial pipe coil is formed by cutting a pipe along its longitudinal axis. Then, the coil is wrapped around the reactor in a helix and welded onto the reactor shell. The dimple jacket consists of hemispherical dimples pressed into a thin plate, which is then wrapped around and welded onto the reactor. The jacket area covers about 80% of the reactor surface, consisting of a bottom elliptical head and a cylindrical shell.

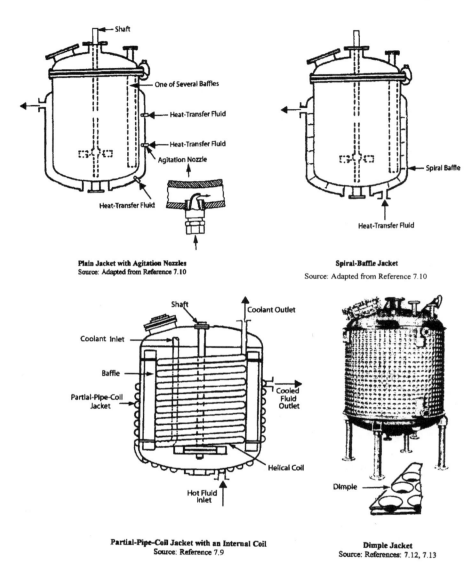

Plain Jacket with Agitation Nozzles
Source: Adapted from Reference 7.10

Spiral-Baffle Jacket
Source: Adapted from Reference 7.10

Partial-Pipe-Coil Jacket with an Internal Coil
Source: Reference 7.9

Dimple Jacket
Source: References: 7.12, 7.13

Figure 7.2 Examples of stirred-tank reactors. With permission.

The factors that influence the selection of a heat exchanger are:

1. heat-transfer coefficients
2. jacket pressure
3. reactor pressure
4. jacket pressure drop
5. cleanliness
6. cost

Figure 7.3 compares calculated overall heat-transfer coefficients for several reactor heat exchangers, using water for both the jacket and reactor fluid. The figure shows that the highest heat-transfer coefficient is obtained with internal coils and the lowest with the simple jacket (called the conventional jacket in Figure 7.3) without a spiral baffle or agitation. It is assumed that the flow rate for the internal coil is the coil flow rate and not the jacket flow rate, as plotted in Figure 7.3. Heat-transfer coefficients for the half-pipe coil, agitated, and baffled jackets are comparable.

The jacket pressure and reactor pressure also influences jacket selection. If the jacket pressure is large the reactor wall thickness becomes large, reducing

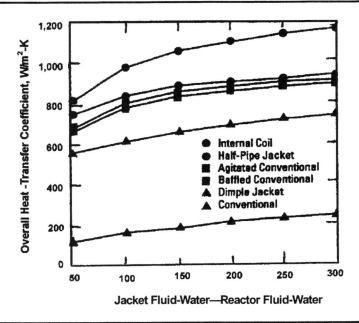

Figure 7.3 Comparison of STR heat exchangers. From Ref. 20 with permission.

heat transfer [12]. Markovitz [12] has given the following rules for selecting the jacket type:

for $\leq$ 500 gal (1.89 m^3)	use the simple jacket
for > 500 gal (1.89 m^3)	use the dimple or half-pipe coil
if the reactor pressure is greater	
than twice the jacket pressure	use the simple jacket
for a jacket pressure $\leq$ 300 psi (20.7 bar)	use the dimple
for a jacket pressure > 300 psi (20.7 bar)	use the half-pipe coil jacket
but < 1000 psi (68.9 bar)	use the half-pipe coil jacket
for steam the pressure is $\leq$ 750 psi (51.7 bar)	use the half-pipe coil jacket

Besides heat transfer and structural considerations, pressure drop across the jacket is also important because it affects both pump and power costs. For the dimple and partial-pipe-coil jackets, the pressure drop will be higher than in the simple jacket because of the increased turbulence. The pressure drop in the dimpled jacket is approximately 10 to 12 times higher than in the simple jacket [12]. For this reason, the liquid velocity in the dimpled jacket is limited to about two feet per second. There is no limitation on the number of inlet and outlet connections for the partial-pipe coil. Thus, to reduce the fluid velocity and hence the pressure drop, the process engineer will split the heat-transfer fluid into zones, as shown in Figure 7.2. The partial pipe-coil jacket is more versatile – it can be used with both high and low temperature heat-transfer fluids. If the heat-transfer coefficient inside the reactor is small compared to the jacket heat-transfer coefficient, then consider using the simple jacket. Because it is difficult to clean dimple jackets, they should not be used with dirty fluids. Also, do not use the dimple jacket for applications requiring high temperature organic heat-transfer fluids, which may degrade to form solids. The solids will deposit on the dimples, fouling the jacket.

The most frequently used internal heat exchanger is the spiral coil. Manufacturers fabricated internal coils by bending straight lengths of pipe. The number of coil banks that can be placed in a reactor depends on the minimum coil radius, which is about 8 to 12 in (0.203 to 0.305 m). Below the minimum coil radius, the pipe will crush during coiling. A common pipe diameter is 2 in. (50.8 cm). The outer coils are less efficient in transferring heat than the inner coils, which are close to the impeller, because the heat transfer coefficient decreases from the inner coil to the outer coil. Hicks and Gates [14] described the design of polymerization reactors using three banks of coils.

CONTINUOUS STIRRED-TANK REACTOR SIZING

Sizing continuous stirred-tank reactor (CSTR) requires selecting a standard reactor, given in Table 3, from a manufacturer. Table 7.4 lists the relations for calculating the reaction volume, heat transfer area, and the mixer power for CSTRs.

Table 7.5 gives the calculation procedure. Any reaction kinetics, indicated by Equation 7.4.4, can be used in the procedure. For each reactor in the series, we assume

1. perfect mixing
2. constant volume
3. constant temperature
4. constant density
5. constant heat capacities
6. equal mixer power for each reactor

Table 7.3 Standard Stirred Tank Reactors (Source Ref. 13).

Rated Capacity[a] gal	Actual Capacity[a] gal	Jacket Area[b] ft^2	Outside Diameter[c] in	Straight Shell[c] in
500	559	75	54	51
750	807	97	60	60
1000	1075	118	66	66
1200	1253	135	66	78
1500	1554	155	72	81
2000	2083	191	78	93
2500	2756	230	84	105
3000	3272	256	90	108
3500	3827	283	96	111
4000	4354	304	102	111
5000	5388	353	108	123
6000	6601	395	120	120
8000	8765	466	132	132
10,000	10,775	540	144	135

a) To convert gal to m^3, multiply by 3.785x10^{-3}.
b) To convert ft^2 to m^3, multiply by 9.29x10^{-2}.
c) To convert in to m, multiply by 2.54x10^{-2}.

Table 7.4 Summary of Equations for Sizing CSTRs

First Subscript: entering stream or CSTR number –– n
leaving stream –– n + 1
Second Subscript: reactant A

Table 7.4 Continued

Mole Balance

$$m_{n, A}' = m_{n+1, A} + (x_{n+1, A}' - x_{n, A}') \, m_{n, A} \qquad (7.4.1)$$

Energy Equation

$$(\Delta h_n) \, m_n' + (\Delta H^o_R) \, (x_{n+1, A} - x_{n,A}') \, m_{n, A} = Q_n + (\Delta h_{n+1}) \, m_{n+1} \qquad (7.4.2)$$

Rate Equations

$$- r_{n, A} \, V_r = (x_{n+1, A} - x_{n, A}') \, m_{n, A}' \qquad (7.4.3)$$

$$r_{n, A} = f(c_{n+1, A}) \qquad (7.4.4)$$

$$c_{n+1, A} = m_{n+1, A} / V_V' \qquad (7.4.5)$$

$$V_R = f(V_r) \ -\!- \text{ from Table 7.3} \qquad (7.4.6)$$

$$Q_J = U_J \, A_J \, (T_J - T_R') \qquad (7.4.7)$$

$$T_J = (T_{J1}' + T_{J2}')/2 \qquad (7.4.8)$$

$$A_J = f(V_r) \ -\!- \text{ Table 7.3} \qquad (7.4.9)$$

If $Q_n \le Q_J$ -- then $A_R = A_J$ (7.4.10)
If $Q_n \ge Q_J$ -- then calculate Q_C

$$Q_C = U_C \, A_C \, (T_C - T_R') \qquad (7.4.11)$$

$$T_C = (T_{C1}' + T_{C2}') / 2 \qquad (7.4.12)$$

$$A_C = 4.6 \, V_r^{2/3} \ -\!- \ A_C \, (m^2), \, V_r (m^3) \qquad (7.4.13)$$

If $Q_n \le Q_C$ -- then $A_R = A_C$ (7.4.14)
If $Q_n \ge Q_C$ and $Q_n \le Q_J + Q_C$ -- then $A_R = A_J + A_C$
If $Q_n \ge Q_J + Q_C$ -- then $A_R = A_E'$

$$P_n = p \, V_r \qquad (7.4.15)$$

$$p = f(\text{application}') \ -\!- \text{ Table 7.7} \qquad (7.4.16)$$

System Properties

$$(\Delta h_n)\, m_n' = \Sigma_i\, c_{pi}'\, m_{n,\,i}'\, (T_R' - T_o') \qquad (7.4.17)$$

$$(\Delta h_{n+1})\, m_{n+1} = \Sigma_i\, c_{pi}'\, m_{n+1,\,i}\, (T_R' - T_o') \qquad (7.4.18)$$

$$\Delta H^o{}_R = H^o{}_C' - (H^o{}_B' - H^o{}_A') \qquad (7.4.19)$$

$$k = A'\, \exp\,(-\,E'\,/\,R'\,T_R') \qquad (7.4.20)$$

$$U_J = f(\text{reaction solution}', \text{jacket fluid}') \;\; -\!- \;\; \text{Table 7.6} \qquad (7.4.21)$$

$$U_C = f(\text{reaction solution}', \text{coil fluid}') \;\; -\!- \;\; \text{Table 7.6} \qquad (7.4.22)$$

Unknowns

$m_{n+1,\,A}$ - $x_{n+1,A}$ - Δh_n - Δh_{n+1} - $\Delta H^o{}_R$ - $r_{n,\,A}$ - V_r - $c_{n+1,A}$ - V_R - Q_J - U_J - A_J - T_J -
Q_n - U_C - A_C - T_C - A_R - Q_C - P_n - p - k

Table 7.5 Calculation Procedure for Sizing CSTRs

---

1. Obtain the reaction volume, V_r , from Equations 7.4.1, 7.4.3 to 7.4. 5 and 7.4.20.

2. Select a standard reaction volume, V_R (rated capacity), from Equation 7.4.6.

3. Calculate the actual conversion, $x_{n,\,A}$, using the rated capacity, Equations 7.4.1 and 7.4.3 to 7.4.5.

4. Next calculate the heat-transfer rate in each reactor, Q_n, from Equations 7.4.2 and 7.4.17 to 7.4.19.

5. Determine if the jacket area, A_J, is sufficient.

6. Calculate the jacket heat-transfer rate, Q_J, from Equations 7.4.7 to 7.4.9 and 7.4.21.

7. Determine if Q_J is sufficient from Equation 7.4.10.

8. If not, determine if the coil area, A_C, is sufficient.

9. Calculate the coil heat-transfer rate, Q_C, from Equations 7.4.11 to 7.4.13 and 7.4.22.

Table 7.5 Continued

10. Determine if Q_C is sufficient from Equation 7.4.14.

11. If not, determine if the jacket + coil area is sufficient.

12. Calculate the jacket plus coil heat-transfer rate, $Q_J + Q_C$.

13. Determine if $Q_J + Q_C$ is sufficient from Equation 7.4.14.

14. If not, then an external heat exchanger is necessary. The area may be estimated from the approximate method outlined in Chapter 4.

15. Finally, calculate the mixer power, P_n, by from Equations 7.4.15 and 7.4.16.

Table 7.6 Approximate STR Overall Heat-Transfer Coefficients Source Ref. 7.33[a]).

Coil/Agitated Liquid		
Coil Side	Agitated Liquid	U^b Btu/h-°F-ft^2
Steam	Aqueous Solution	90 – 160
Steam	Organic Solution	60 – 130
Steam	Heavy Oil	30 – 60
Hot Water	Aqueous Solution	90 – 130
Hot Water	Organic Solution	60 – 100
Cooling Water	Aqueous Solution	80 – 120
Cooling Water	Organic Solution	50 – 90
Brine	Aqueous Solution	60 – 100
Brine	Organic Solution	50 – 90
Organic Oil	Heavy Organic	60 – 110
Jacket/Agitated Liquid		
Steam	Aqueous Solution	70 – 130
Steam	Organic Solution	60 – 110
Cooling water	Aqueous Solution	60 – 110
Cooling Water	Organic Solution	50 – 80
Organic Oil[c]	Heavy Organic	30 – 50

a) For additional data see Reference 11.
b) To convert to W/m^2-K multiply by 5.678.

Table 7.7 Approximate Mixer Power for Stirred-Tank Reactors

Application	Power[c] hp/1000 gal
Blending[a]	0.2 - 0.5
Homogeneous Reaction[a]	0.5 - 1.5
Reaction with Heat Transfer[a]	1.5 - 5.0
Liquid-Liquid Mixtures[a]	5.0
Liquid-Gas Mixtures[a]	5.0-10.0
Slurries[a]	10.0
Fermentation[b]	3.0-10.0
Emulsion Polymerization[b]	6.0-7.0
Suspension Polymerization[b]	3.0-10.0
Solution Polymerization[b]	15.0-40.0

a) Source: Reference 7
b) Source: Reference 15
c) To convert to W/m^3 multiply by 197.0.

Constant density implies that the volumetric flow rate from reactor to reactor is constant. The relationships listed in Table 7.4 apply to any number of CSTRs in series. The subscript, n, refers to the reactor number and also to the number of the entering stream. The subscript, n + 1, refers to the number of the leaving stream. Equations 7.4.1 to 7.4.3, are the mole balance for reactant A, the energy equation, and the rate equation.

STRs are usually never completely filled unless top withdrawal of the liquid is required. At the top of the reactor, we will allow some empty volume, called head space. Blaasel [15] recommends allowing 15% head space for reactors less than 1.9 m^3 (500 gal) and 10% head space for reactors greater than 1.9 m^3 (500 gal). After calculating the reaction volume, then add the headspace according to these rules to obtain the reactor volume. After calculating the reactor volume, select a standard reactor from a manufacturer. A standard reactor is less expensive than a reactor made-to-order. Table 7.3 lists standard-size reactors, which will vary somewhat from manufacturer to manufacturer. In Table 7.3, the rated capacity is the reaction volume, and the actual volume includes the headspace. Because the manufacturer has allowed for headspace in this case, we need not allow headspace according to the above rules.

To transfer heat, size either a STR with a jacket or one with internal coils. Try the jacketed reactor first because it is the least costly. The available heat-transfer area consists of the cylindrical surface of the reactor and the dished bottom. Only 80% of the total surface area of an STR is available for heat transfer. The upper head contains nozzles, a port for the mixer, lugs for support, and usually a sight glass, as shown in Figure 7.2.

We will use a spiral coil to illustrate the calculation procedure. First, consider a jacketed STR. If the jacket heat-transfer area is insufficient, then consider an internal heat exchanger and finally a shell-and-tube external heat exchanger. For the latter case, the reacting solution is pumped out of the reactor continuously, cooled in a heat exchanger, and then returned to the reactor. If the jacketed reactor does not provide sufficient heat-transfer area, then try using internal helical coils. If more than one coil is used, then the heat transfer coefficient must be reduced by 30% for each additional coil [14]. Thus, if the reaction requires three coils, then the coil near the reactor wall will only have 40% of the heat-transfer coefficient of the coil closest to the impeller. Frank [33] believes that this reduction in the heat-transfer coefficient may be too pessimistic. Each coil requires spacing between the reactor wall and other coils. To minimize interfering with liquid recirculation, the coils should not extend completely to the surface of the liquid or the bottom of the tank. Hicks and Gates [14] recommend locating the top of the coil at least one sixth of the diameter of the reactor below the liquid surface. They also recommend locating the bottom coil at one-sixth the coil diameter above the bottom of the STR.

The jacket temperature, T_J, in Equation 7.4.8, equals the average of the jacket inlet and outlet temperatures. For a coil also use the average of the inlet and outlet temperatures. First, determine if there is sufficient heat-transfer area by assuming a simple jacket. The area of the jacket is given in Table 7.3. The area will be about the same for simple, pipe coil, and dimple jackets. If the jacket area is insufficient, then determine if coils will provide the additional surface area. The reactor volume should be compensated for the volume occupied by the coils.

Example 7.1 Sizing a CSTR for Synthesizing Propylene Glycol

This problem is an adaptation of a problem taken from Fogler [16]. Propylene glycol is produced by hydrating propylene oxide using a solution of 0.1 % sulfuric acid in water as a catalyst. The reaction is

$$CH_2 \text{---} CH \text{---} CH_3 + H_2O \rightarrow CH_2 \text{---} CH \text{---} CH_3$$
$$\lfloor \underline{\quad O \quad} \rfloor \qquad\qquad\qquad \lfloor OH \quad \lfloor OH$$

An equi-volumetric solution of methanol and propylene oxide flows into a CSTR. At the same time, a 0.1% sulfuric acid solution also flows into the CSTR at a rate of 2.5 times the combined flow rate of propylene oxide and methanol. The coolant is chilled water. Size the reactor, determine the heat exchanger type and area, and calculate the mixer power.

Data

Methanol volumetric flow rate	800 ft^3/h (22.7 m^3/h)
Propylene oxide volumetric flow rate	800 ft^3/h (22.7 m^3/h)
Acid solution volumetric flow rate	4000 ft^3/h (113 m^3/h)
Feed inlet temperature	75 °F (23.9 °C)
Reaction temperature	100 °F (37.8 °C)
Chilled water inlet temperature	5 °C (41 °F)
Chilled water exit temperature	15 °C (59 °F)
Required propylene oxide conversion	0.37

Thermodynamic properties are summarized in Table 7.1.1, and reaction properties are given below. Fogler [16] estimated the heat capacity for propylene glycol using a rule-of-thumb. The rule states that the majority of low-molecular-weight, oxygen-containing organic liquids have a heat capacity of 0.6 cal/g °F ± 15 % (35 Btu/lbmol-°F)

Reaction Properties

Pre-exponential factor, A	16.96 x 10^{12} h^{-1}
Activation energy, E	32,400 Btu/lbmol (75,330 kJ/kgmol)

Follow the calculation procedure outlined in Table 7.5. Using the equations listed in Table 7.4, first calculate the reaction volume. Then select a standard reaction volume (rated capacity) from Table 7.3. The actual capacity (reactor volume)

Table 7.1.1 Thermodynamic Properties for Proplyene Glycol Synthesis

Component	Molecular Weight	Density^a g/cm^3	Heat Capacity^b Btu/lbmol-°F	Standard Enthalpy of Reaction^{c,d} Btu/lbmol
Propylene Oxide	58.08	0.859	35	−66,600
Water	18.02	0.9941	18	−123,000
Propylene Glycol	76.11	1.036	46	−226,000
Methanol	32.04	0.7914	19.5	—

a) To convert g/cm^3 to kg/m^3 multiply by 1000.
b) To convert Btu/lbmol-°F to kJ/kgmol-°K multiply by 4.187.
c) At 25 °C (77 °F)
d) To convert Btu/lbmol to kJ/kgmol multiply by 2.325.

of the reactor is greater than the rated capacity to allow for some headspace. If the rated capacity from Table 7.3 is greater than the calculated reaction volume, calculate the actual conversion, $x_{n, A}$. The conversion will increase because of the increased reaction volume and therefore residence time.

The first step is to calculate limits for the reaction volume. One CSTR will give the maximum volume and a plug-flow reactor will give the minimum volume. The total reaction volume for multiple CSTRs will lie somewhere between these two limits. After calculating the reaction volume, calculate the required heat transfer and the heat-transfer area. Then, either select a jacket, a coil, jacket plus a coil, or an external heat exchanger.

First, calculate the reaction volume. Assuming that the density does not change significantly during reaction, the total volumetric flow rate at the reactor exit,

$$V_v = 800 + 800 + 4000 = 5600 \text{ ft}^3 /h \ (159 \text{ m}^3/h)$$

In this case the subscript A in Table 7.4 refers to propylene oxide. The molar flow rate of propylene oxide is

$$m_{1,A} = \frac{800}{58.08} \ 0.859 \ (62.43) = 738.7 \text{ lbmol/h} \ (335 \text{ kgmol/h})$$

After substituting the reaction parameters into Arhenius's equation, Equation 7.4.20, we obtain

$$T = 100 + 459.7 = 559.7 \ {}^{\circ}R \ (311 \text{ K})$$

$$k = 16.92 \text{ x } 10^{12} \exp \frac{- 32400}{1.987} \ \frac{1}{559.7} = 3.766 \text{ h}^{-1}$$

For one CSTR, n = 1 in Equations 7.4.1 and 7.4.3 to 7.4.5.

$$m_{1A} = m_{2A} + (x_{2A} - x_{1A}) \ m_{1A}$$

$$- r_{2A} \ V_r = (x_{2A} - x_{1A}) \ m_{1A}$$

$$r_{2A} = - k \ c_{2A}$$

$$c_{2A} = m_{2A} / V_V$$

Substitute $x_{1A} = 0$, $x_{2A} = 0.37$, $V_V = 5600$ ft^3/h, $m_{1A} = 738.7$ lbmol/h and k = 3.766/h^{-1} into these equations. Then, solve the equations using POLYMATH [22]. The reaction volume is 873.3 ft^3 (24.7 m^3), which is the maximum reaction volume.

The minimum reaction volume will be for a plug flow reactor, which, for a first order reaction is

$$V_r = \frac{V_V}{k} \ln \frac{1}{1 - x_{2A}} = \frac{5600}{3.766} \ln \frac{1}{1 - 0.37} = 687.0 \text{ ft}^3 \ (19.5 \text{ m}^3)$$

For two CSTRs, generate a set of equations for n = 1 and n = 2. For the first CSTR, when n = 1, the equations are the same as written above. For the second CSTR, n = 2 in Equations 7.4.1 and 7.4.3 to 7.4.5.

$$m_{2A} = m_{3A} + (x_{3A} - x_{2A}) \, m_{2A}$$

$$- r_{3A} V_r = (x_{3A} - x_{2A}) \, m_{2A}$$

$$r_{3A} = - k \, c_{3A}$$

$$c_{3A} = m_{3A} / V_V$$

With $x_{3A} = 0.37$, and using the same values of x_{1A}, V_V, m_{1A}, and k as for one CSTR, the eight equations are solved simultaneously using POLYMATH [22]. For two CSTRs, the total reaction volume is 772.9 ft^3 (21.9 m^3), which is in between the reaction volumes of a single CSTR (873.3 ft^3) (24.7 m^3) and the plug flow reactor (687.0 ft^3) (19.5 m^3). The difference in reaction volume between one and two CSTRs is only 11.5 %, which is not substantial. Select a single CSTR.

The next step is to select a standard CSTR. For the calculated reaction volume of 873.3 ft^3 (6533 gal, 24.7 m^3), select a standard reaction volume from Table 7.3 of 8000 gal (30.3 m^3). Also, from Table 7.3 the reactor volume is 8765 gal (33.2 m^3) to allow for headspace. Now, we have a number of options available. One option is not to fill the standard reactor up to the rated volume but only up to the calculated reaction volume of 6533 gal (24.7 m^3) and maintain the volumetric flow rate at 5600 ft^3/h (21.2 m^3/h). This means that the conversion will be 0.37 as specified. CSTRs have a minimum operating reaction volume to avoid imperfect mixing. Mixing depends on the properties of the reaction mixture, the impeller design and speed, and the internal design of the CSTR. The minimum operating reaction volume for good mixing should be determined by consulting with the manufacturer of the CSTR. A second option is to fill the reactor up to 8000 gal (30.3 m^3) and increase the volumetric flow rate to keep the residence time and therefore the conversion constant. A third option is to again fill the reactor up to maximum capacity, and maintain the volumetric flow rate at 5600 ft^3/h (21.2 m^3/h). In this case, the residence time increases because of the increased reaction volume, increasing the conversion. Choosing the third option, substitute 1069 ft^3 (8000 gal, 30.3 m^3) for the reaction volume and the same values for x_{1A}, V_V, m_{1A}, and k into the equations above for n = 1, and solve for x_{2A}. Using POLYMATH [22], the conversion is now 0.4182. This design will give some flexibility. If the demand for product increases, the feed rate can be increased, but the conversion will decrease. The original required conversion is 0.37.

The mole balance can now be completed for one CSTR. The inlet molar flow rate for propylene oxide is calculated above. The inlet molar flow rate of methanol,

$$m_{1M} = \frac{800}{32.04}\, 0.7914\,(62.43) = 1{,}234 \text{ lbmol/h (560 kgmol/h)}$$

and the inlet molar flow rate of water,

$$m_{1W} = \frac{4000}{18.02}\, 0.9941\,(62.43) = 13{,}780 \text{ lbmol/h (6240 kg/mol/h)}$$

The inlet flow rates (stream 1) are entered into Table 7.1.2. For x_{2A} = 0.4182, the outlet flow rates are also entered into Table 7.1.2.

Next, select a heat exchanger and calculate the heat transfer area. First, calculate the required heat transfer, Q_n, from an energy balance. Obtain the enthalpy of reaction from Equation 7.4.19 and the standard enthalpies of reaction listed in Table 7.1.1.

$$\Delta H^\circ_R = -\,222{,}600 - (-\,123{,}000 - 66{,}600)$$

$$\Delta H^\circ_R = -33{,}000 \text{ Btu/lbmol } (-76{,}760 \text{ kJ/kgmol})$$

The enthalpy flowing into and out of the reactor for each component is calculated relative to 25 °C (77 °F), using heat capacities from Table 7.1.1. The results are contained in Table 7.1.3.

Solve for the required heat transferred, Q_n, using the energy equation, Equation 7.4.2 in Table 7.4. Substituting the enthalpy of reaction, the enthalpy into the reactor, and the enthalpy out of the reactor, obtained from Table 7.1.3, we find that

$$Q_n = -\,2{,}085{,}000 - 33{,}000\,(308.9) + 9{,}465{,}000)$$

$$Q_n = -2.814 \times 10^6 \text{ Btu/h } (-2.97 \times 10^6 \text{ kJ/h})$$

Thus, heat must be transferred out of the reactor to maintain the reaction temperature at 100 °F (37.8 °C).

Next, calculate the heat transfer for a jacket, Q_J, for the 8000 gal (30.3 m^3) standard reactor from Equations 7.4.7 to 7.4.9. The average jacket temperature,

$$T_J = (5 + 15)/2 = 10 \text{ °C (50 °F)}$$

Selecting an approximate overall heat-transfer coefficient is a problem because of insufficient data. Although there are correlations available for calculating the individual heat-transfer coefficients and hence the overall heat-transfer coefficients, at the preliminary stage of the process design, we try to avoid detailed calculations. The best we can do is to select a coefficient that best matches the conditions in the CSTR. Because the jacket liquid is water, and the reactor liquid is a dilute aqueous solution, we find that from Table 7.6, U_J varies from 60 to 110 Btu/h-ft^2-°F (341 to 625 W/m^2-°F) The average value is 85 Btu/h ft^2 °F (483 W/m^2-K). From Equation 7.4.9, we find that the standard 8000 gal (30.3 m^3) reactor has a jacket area of 466 ft^2 (43.3 m^2). From Equation 7.4.7, the heat that can be transferred to the jacket,

$$Q_J = 85 \ (466) \ (100 - 50) = 1.981 \text{x} 10^6 \text{ Btu/h } (2.09 \text{x} 10^6 \text{ kJ/h})$$

which is insufficient according to Equation 7.4.10 because we are required to remove 2.814x10^6 Btu/h (2.97x10^6 kJ/h), but the jacket is only capable of removing 1.981x10^6 Btu/h (2.08x10^6 kJ/h).

Next, determine if the heat-transfer rate for a coil, Q_C, will be sufficient. From Table 7.6, the closest match we can find for an overall heat-transfer coefficient is for an aqueous solution in a coil and water in the reactor. From Table 7.6, U_C varies from 80 to 120 Btu/h-ft^2-°F (454 to 681 W/m^2-K), the average being 100 Btu/h-ft^2-°F (568 W/m^2-K). The heat-transfer area for a coil is given by Equation 7.4.13.

$$A_C = 4.6 \ [3.785 \text{x} 10^{-3} \ (8000)]^{2/3} = 44.69 \text{ m}^2 \ (480.9 \text{ ft}^2)$$

From Equation 7.4.11, the heat transfer rate for a coil,

$$Q_C = 100 \ (480.9) \ (100 - 50) = 2.405 \text{x} 10^6 \text{ Btu/h } (2.537 \text{x} 10^6 \text{ kJ/h})$$

Clearly, a coil alone is also insufficient. Now, if we add the jacket and coil heat transfer rates,

$$Q_C + Q_J = 1.981 \text{X} 10^6 + 2.405 \text{X} 10^6 = 4.386 \text{x} 10^6 \text{ Btu/h } (4.63 \text{x} 10^6 \text{ kJ/h})$$

which is sufficient. Therefore, the solution is to use both a coil and a jacket to remove the enthalpy of reaction.

The final step is to calculate the mixer power requirement. From Equation 7.4.16, the application that matches this design is reaction with heat transfer. From Table 7.7, the required power varies from 1.5 to 5 hp/1000 gal. The average power is 3.25 hp/1000 gal (640 W/m^3). Then, according to Equation 7.4.15 the mixer power,

P = (3.25 hp/1000 gal) (8000 gal) = 26 hp (19.4 kW)

From Table 5.10, a standard-size electric motor is 30 hp (22.4 kW), which results in a safety factor of 15.4%.

Table 7.1.2 Mole Balance for a CSTR Producing Propylene Glycol

Propylene oxide conversion = 0.4182

Stream No	Temperature °F (°C)	Flow Rates[a] (lbmol/h)			
		C_3H_6O	$C_3H_6(OH)_2$	CH_3OH	H_2O
1	75 (23.9)	738.7	0	1, 234	13,780
2	100 (68.0)	429.8	308.9	1, 234	13,470

[a] To convert to kgmol/h divide by 2.205.

Table 7.1.3 Energy Balance for a CSTR Producing Propylene Glycol

Component	Enthalpy In[a], Btu/h	Enthalpy Out[a], Btu/h
C_3H_6O	738.7 (35)(68 − 75) = − 181,000	429.8 (35) (100 − 68) = 481,400
$C_3H_6(OH)_2$	0	308.9 (46) (100 − 68) = 454,700
$CH_3(OH)$	1,234(19.5)(68 − 75)= − 168,400	1,234 (19.5) (100 − 68) = 770,000
H_2O	13,780(18.0)(68 −75) =−1,736, 000	13,470 (18.0) (100 − 68) = 7,759,000
Total	−2,085,000	9,465,000

a) To convert to kJ/h multiply by 1.055.

SIZING BATCH REACTORS

The equipment for batch reactors is identical to that of CSTRs. Table 7.8 lists the equations for sizing batch reactors, and Table 7.9 outlines the calculation procedure. First, calculate the reaction volume, V_r, by using Equations 7.8.3 to 7.8.6. This calculation requires an estimate of the batch time, defined by Equation 7.8.5, which is the sum of the times for charging, heating, reacting, discharging, cooling, emptying, and cleaning. These times are given in Table 7.8 for a polymerization reaction. No other time data seems to be available. Next, find the reactor volume, V_R, using Equation 7.8.7. The reactor volume is greater than the reaction volume because of an allowance for headspace.

After calculating the reactor volume, the next step is to calculate the heat-transfer area. The reactant concentration, and therefore the heat-transfer rate decreases as the reaction proceeds. We have to calculate the heat-transfer area when the heat-transfer rate is a maximum, which is at initial conditions. First, calculate the initial rate of reaction, r_{Ao}, from Equation 7.8.4, and then calculate the heat transferred using Equations 7.8.1, 7.8.2 and 7.8.18 to 7.8.21. Next, determine the heat-exchanger type using Equations 7.8.11 and 7.8.15.

Table 7.8 Summary of Equations for Sizing Batch Reactors

Energy Equation

$$Q_R = r_{Ao} V_R \Delta h_R \qquad (7.8.1)$$

$$\Delta h_R = \Delta h_1 + \Delta H^\circ_R + \Delta h_2 \qquad (7.8.2)$$

Rate Equations

$$t_R = f(k', x_A') \qquad (7.8.3)$$

$$r_{Ao} = f(k', c_{Ao}') \qquad (7.8.4)$$

$$t_B = t_F' + t_H' + t_R + t_C' + t_E' \qquad (7.8.5)$$

$$V_r = m_1' t_B / \rho' \qquad (7.8.6)$$

$$V_R = f(V_r) \ -- \ \text{Table 7.3} \qquad (7.8.7)$$

$$Q_J = U_J A_J (T_J - T_R') \qquad (7.8.8)$$

$$T_J = (T_{J1}' + T_{J2}') / 2 \qquad (7.8.9)$$

$$A_J = f(V_R) \;\; -\!\!- \;\; \text{Table 7.3} \tag{7.8.10}$$

If $Q_R \le Q_J$ $\;-\!\!-\;$ then $A_R = A_J$ $\tag{7.8.11}$
If $Q_R \ge Q_J$ $\;-\!\!-\;$ then calculate Q_C

$$Q_C = U_C \; A_C \; (T_C - T_R') \tag{7.8.12}$$

$$T_C = (T_{C1}' + T_{C2}')/2 \tag{7.8.13}$$

$$A_C = 4.6 \; V_R{}^{2/3} \;\; -\!\!- \;\; A_C \, (m^2), \; V_R \, (m^3) \tag{7.8.14}$$

If $Q_{Ro} \ge Q_J$ and $Q_{Ro} \le Q_C$ $\;-\!\!-\;$ then $A_R = A_C$ $\tag{7.8.15}$
If $Q_{Ro} \ge Q_J$ and $Q_{Ro} = (Q_J + Q_C)$, then $A_R = A_J + A_C$
If $Q_{Ro} \ge (Q_J + Q_C)$ $\;-\!\!-\;$ then $A_R = A_E'$

$$P = p \; V_R \tag{7.8.16}$$

$$p = f(\text{application'}) \;\; -\!\!- \;\; \text{Table 7.7} \tag{7.8.17}$$

System Properties

$$\Delta h_1 = \sum_i c_{pi}' (T_R' - T_o') \tag{7.8.18}$$

$$\Delta h_2 = \sum_i c_{pi}' (T_R' - T_o') \tag{7.8.19}$$

$$\Delta H^\circ_R = H^\circ_C{}' - (H^\circ_B{}' + H^\circ_A{}') \tag{7.8.20}$$

$$k = A' \exp(-E'/R'T_R') \tag{7.8.21}$$

$$U_J = f(\text{reaction solution'}, \text{jacket fluid'}) \;\; -\!\!- \;\; \text{Table 7.6} \tag{7.8.22}$$

$$U_C = f(\text{reaction solution'}, \text{coil fluid'}) \;\; -\!\!- \;\; \text{Table 7.6} \tag{7.8.23}$$

Unknowns

$Q_{Ro} - r_{Ao} - A_R - V_R - \Delta h_R - t_B - t_R - V_r - V_R - Q_J - U_J - A_J - T_J - Q_C - U_C - A_C - T_C - P - p - \Delta h_1 - \Delta h_2 - \Delta h^\circ_R - k$

Table 7.9 Calculation Procedure for Sizing Batch Reactors

1. Calculate the reaction volume, Vr , from Equations 7.8.3, 7.8.5, 7.8.6, and 7.8.21.

2. Select a standard reactor size (rated capacity) from Equation 7.8.7

3. Calculate the initial heat-transfer rate, Q_{Ro}, from Equations 7.8.1, 7.8.2, 7.8.4, 7.8.18 to 7.8.20.

4. Determine if the jacket area, A_J, is sufficient.

5. Calculate the jacket heat-transfer rate, Q_J, from Equations 7.8.8 to 7.8.10 and Equation 7.8.22.

6. Determine if Q_J is sufficient from Equation 7.8.11.

7. If not, determine if the coil area, A_C, is sufficient.

8. Calculate coil heat-transfer rate, Q_C, from Equations 7.8.12 to 7.8.14 and Equation 7.8.23.

9. Determine if Q_C is sufficient from Equation 7.8.15.

10. If not, determine if the jacket + a coil areas are sufficient.

11. Calculate the jacket + coil heat-transfer rate, $Q_J + Q_C$.

12. Determine if $Q_J + Q_C$ is sufficient from Equations 7.8.15.

13. If not, then an external heat exchanger is necessary. The area may be estimated by using the approximate method outlined in Chapter 4.

14. Calculate the mixer power required from Equations 7.8.16 and 7.8.17.

Table 7.10 Cycle Times for a Batch Polymerization Reactor. (Source Ref. 16).

Activity	Time, h
Charge feed to the reactor	1.5 - 3.0
Heat to reaction temperature	1.0 - 2.0
Carry out reaction.	Varies
Empty and clean reactor	0.5 - 1.0

Example 7.2 Sizing a Batch Reactor for Producing Drying Oil

This problem is adapted from a problem given by Smith [23]. To illustrate the
method for sizing a batch reactor outlined in Table 7.8, consider the production of
drying oil from acetylated Castor oil. Drying oils are added to paints to aid the
formation of a protective coating when drying. Acetylated Castor oil (AO) de-
composes according to the first order reaction,

$$(AO) \, (l) \rightarrow CH_3COOH \, (g) + drying \, oil \, (l)$$

When heating castor oil, drying oil and acetic acid forms. During the reaction the
acid evaporates from the solution. Calculate the reactor volume, the type and area
of the heat exchanger, and the mixer power.

Data
Reaction temperature	300 $^\circ$C (572 $^\circ$F)
Acetic acid equivalent in AO	0.156 g of acetic acid/g of AO
Molecular weight of acetic acid	60
Heat capacity of reacting mixture	0.60 Btu/lb-$^\circ$F (2.5 kJ/kg-K)
Heat of reaction	15,000 cal/gmol (27,000 Btu/lbmol)
Conversion	95%
Average feed rate	1000 lb/h (453.6 kg/h)
AO Density	0.9 g/cm^3 (56.2 lb/ft^3, 900 kg/m^3)

Reaction Properties
The reaction is a first order with respect to a pseudo concentration of acetic acid in
acetylated castor oil, i.e., moles of acetic acid per unit volume of castor oil.
$r_A = k \, c_A$ r_A = moles of acid/unit volume-unit time
Activation energy   44,500 cal/gmol (80,100 Btu/lbmol)
Pre-exponential factor 1.937×10^{15} min^{-1}

Follow the calculation procedure given in Table 7.9. First, calculate the reactor volume. Then, calculate the heat-transfer area and the mixer horsepower.

Because the reaction is first order, Equation 7.8.3 becomes $r_A = k \, c_A$. If the
change of volume during the reaction is small, the reaction time, Equation 7.8.4,
for a first order reaction is

$$t_R = (1 \, / \, k) \ln [1 \, / \, (1 - x_A)]$$

From Equation 7.8.21, with $A = 1.937 \times 10^{15}$ min^{-1} and $E = 80,100$ Btu/lbmol,
and at 300 $^\circ$C (1032 $^\circ$R),

$$k = 1.937 \times 10^{15} \exp [- 80,100 / 1.987 \, (1032)] = 0.02102 \, min^{-1}$$

Then, for 95% conversion the reaction time,

$t_R = (1 / 0.02102) \ln [1 / (1 - 0.95)] = 142.5$ min (2.375 h)

From Equation 7.8.5, calculate the batch time, t_B. Because we do not have values for t_F, t_H, t_C, and t_{BC} for this reaction, we will have to make use of the times given in Table 7.10 for polymerization reactions. Except for the charging time and cooling times, select the worst case from Table 7.10. We will assume that it takes the same time to cool the reactor as it does to heat the reactor. We have some control over the time it takes to charge the reactor. By adjusting a control valve, assume that we can charge the reactor in 1.5 hours. Thus, the batch time, from Equation 7.8.5,

$t_B = 1.5 + 2.0 + 2.375 + 2.0 + 1.0 = 8.875$ h

Now, calculate the reaction volume from Equation 7.8.6.

$V_r = 1000 (8.875) / 56.2 = 157.9$ ft^3 (1181 gal, 4.47 m^3)

Next, select a standard (4.54 m^3) reactor size, from Equation 7.8.7. From Table 7.3, we find that there is a 1200 gal standard reactor. To allow for some flexibility select a 1500 gal (5.68 m^3) reactor. Even if the production rate requires 1181 gal, the reactor will be filled to 1500 gal, which increases the production rate.

Now, we have to decide on how to remove the enthalpy of reaction – using a jacket, a coil, a coil and a jacket or an external heat exchanger. First, check if a jacket will suffice. Because the reaction is an unsteady-state process, the heat transfer will vary with time. Initially, the reaction rate will be a maximum because the concentration of acetylated castor oil (AO) is at its maximum value. As the reaction proceeds, the concentration of acetylated oil will decrease, as will the heat-transfer rate. In this problem Δh_R is given. Thus, we do not need Equations 7.8.2 and 7.8.18 to 7.8.20. From Equation 7.8.4, calculate the initial rate of reaction.

$$r_{Ao} = k\, c_{Ao} = 0.02102 \; \frac{1}{\text{min}} \; 60 \, \frac{\text{min}}{\text{h}} \; 0.156 \, \frac{\text{lb acid}}{\text{lb AO}} \; 56.20 \, \frac{\text{lb AO}}{\text{ft}^3} = 11.06 \, \frac{\text{lb acid}}{\text{ft}^3\text{-h}}$$

Now, from Equation 7.8.1 calculate Q_R. The reaction volume now equals 1500 gal.

$$Q_R = \frac{450.0 \; \text{Btu}}{1 \; \text{lb acid}} \; \frac{11.06 \; \text{lb acid}}{1 \; \text{ft}^3\text{-h}} \; \frac{1500 \; \text{gal}}{1} \; \frac{1 \; \text{ft}^3}{7.481 \; \text{gal}}$$

$Q_R = 997,900$ Btu/h (1.05×10^6 kJ/h)

Next, select a heat-transfer fluid. If we select steam at 700 °F (371 °C) the jacket pressure will be 3094 psia (213 bar), which is much too high. At this pressure, the jacket and reactor-wall will be very thick, resulting in a costly reactor. Also, it is not good practice to use high-pressure steam for heating. If we select Dowtherm A vapor, the jacket pressure will be 106.8 psia (7.37 bar) at 700 °F (343 °C) [24]. The maximum temperature allowed for Dowtherm A is 750 °F (399 °C) [24].

Next, estimate the overall heat-transfer coefficient for the jacket, U_J. Assuming $D_o/D_I \approx 1$ and $D_o/D_{LM} \approx 1$, where D_{LM} is the log-mean diameter. Then U_J is approximately given by

$$\frac{1}{U_J} = \frac{1}{h_i} + \frac{1}{h_w} + \frac{1}{h_{fi}} + \frac{1}{h_{fo}} + \frac{1}{h_o} + \frac{x_w}{k}$$

The heat transfer-coefficients and fouling factors are listed in Table 7.2.1. Because of the acetic acid, select SS316 as the material of construction. The thermal conductivity, k, of SS316 and the wall thickness of the reactor, x_w, are given in Table 7.2.1.

$$\frac{1}{U_J} = \frac{1}{250} + \frac{1}{1700} + \frac{1}{360} + 0.001 + \frac{0.5}{113}$$

$U_J = 78.18$ Btu/h-ft^2-°F (444 W/m^2-K)

Using Equations 7.8.8 to 7.8.12, the heat transfer rate for the 1500 gal reactor containing 155 ft^2 (14.4 m^2) of jacket area is

$Q_J = 78.18$ (155.0) (700.0 − 572.0) = 1.551×10^6 Btu/h (1.64×10^6 kJ/h)

which is acceptable because the heat absorbed by the reaction is 9.979×10^5 Btu/h (1.05×10^6 kJ/h).

Finally, calculate the mixer power. Using Equation 7.8.17, we find that for a reaction with heat transfer the power required varies from 1.5 to 5 hp/1000 gal. The average is 3.25 hp/1000 gal (640 W/m^3) . Thus, from Equation 7.8.16,

P = (3.25 / 1000) (1500) = 4.875 hp (364 kW)

From Table 5.10, the nearest standard size electric motor is 5 hp (373 kW). The safety factor for this selection will only be 2.5 %. Therefore, select the next larger-size motor, which is 7.5 hp (559 kW). The safety factor for this selection is 53.8 %.

Table 7.2.1 Heat-Transfer Coefficients for a Batch Reactor

Heat-Transfer Coefficient[a] Btu/h -ft^2-°F		Source	Remarks
h_i	250	Reference 10	assuming the worst case
h_W	1,700	Reference 10	
h_{fi}	0		the inside surface is cleaned after each batch
$1/h_{fo}$	0.001	Reference 24	
h_o	360	Reference 24	condensing Dowtherm A
k^b	113		for SS316

a) To convert to W/m^2-K multiply by 5.678.
b) Units of the thermal conductivity are Btu-in/ft^2-h-°F.

PACKED-BED CATALYTIC REACTORS

Catalysts change the reaction mechanism and therefore the rate of the reaction. If the reaction rate increases, the reaction volume will decrease, reducing the cost of the reactor. Many chemical syntheses are impractical without using a catalyst.

Catalytic Pellet Selection

If a pure catalyst is structurally weak and cannot be formed into a pellet or is too expensive to use as a pellet, then the catalyst is deposited as a thin film on an inert support. Because the reaction rate is proportional to the catalyst surface area, the pellet must be porous to achieve a large surface area.

Besides chemical properties of the catalyst, the mechanical properties of the support material must also be considered when selecting a catalyst. Support materials are mostly alumina, silica, activated carbon or diatomaceous earth, but alumina is more widely used than the other materials [18]. Pellets are usually molded or extruded into spheres, cylinders, or rings. Extrusion is a lower cost operation than molding [18]. The most common pellet diameters are 1/32, 1/16, and 1/8 in (0.794, 1.59, and 3.18 mm). Pellets should have a high compressive strength to resist crushing and abrasion and a low pressure drop to minimize compressor and power costs. Because pellets are packed in a bed, the bulk crushing strength of the pellets limits the bed height. Trambouze et al. [8] define bulk crushing strength as the stress that produces 0.5 % fines as determined by compressing the pellets in a press. Pellet strengths vary from 1.0 to 1.3 MPa (145 to 189 psi) for several pellets tabulated by Trambouze et al. [8].

Selecting a pellet size, shape, and porosity (void fraction in the pellet) is a trade-off between achieving high reactivity, high crushing strength, and low pressure drop. Promoting high reactivity requires a porous pellet with a large internal surface area, which requires small pores. Small pores, however, lower the diffusion rate, reducing the pellet activity. The rate of diffusion increases with increasing pore size, but the increased pore size reduces surface area and therefore reactivity. Consequently, there is an optimum pore size that maximizes pellet reactivity. Reactor reactivity increases if the pellet diameter is reduced, allowing more pellets to be packed into a reactor, but then the pressure drop is increased. Low pressure drop is achieved using large pellets, but then this reduces the catalyst surface area for a unit volume of reactor. Also, crushing strength decreases with increasing porosity particularly when the porosity is above 50% [17].

Packed-Bed Reactor Selection

Catalyst pellets are contained in a reactor, as shown in Figure 7.4, in a single bed, multiple beds in a single shell, several packed tubes in a single shell, or a single bed with imbedded tubes. Deviation from the simple single bed may be required because of the need to add or remove heat, to redistribute the flow to avoid channeling, or to limit the bed height to avoid crushing the catalyst. In all the reactors shown in Figure 7.4, the reacting gases flow downward through the bed instead of upward to avoid fluidization and minimize entrainment of catalyst in the exit gases.

The simplest packed-bed reactor is the adiabatic, single-bed reactor shown in Figure 7.4a. According to Trambouze et al. [8], it is the most frequently used reactor type. If the reactants must be cooled to limit catalyst fouling or deactivation, then select one of the other reactor types. In the reactor shown in Figure 7.4b, part of the feed stream is diverted and mixed with hot gases from the upper bed before entering the lower bed. The methanol-synthesis reactor, discussed in Chapter 3, uses this method of cooling. Adding an excess of one of the reactants or an inert gas could also reduce the temperature rise of the reactants. These gases are heat sinks, absorbing the enthalpy of reaction. In the reactor shown in Figure 7.4c, the catalyst is packed in tubes, and a heat-transfer fluid flows in the

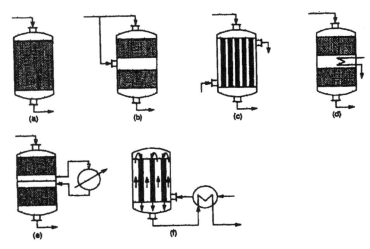

Figure 7.4 Examples of packed-bed reactor arrangements. From Ref. 7 with permission.

shell to add or remove heat. If the heat-transfer fluid is water, then steam can be generated for use in the process. Alternatively, the reacting gases could be cooled with an external or internal intercooler as shown in Figures 7.4d, 7.4e, and 7.4f.

The reactors shown in Figures 7.4a, 7.4b, 7.4d, and 7.4e are really a series of adiabatic reactors. In another arrangement, feed gas cools the reacting gases as illustrated in Figure 7.4f. Here, feed gas is preheated in an external interchanger by cooling the exit gas, and then the feed gas is further heated in the reactor by flowing upward, countercurrent to the downward flow of the hot reactants.

Approximate Reactor Sizing

After selecting a reactor type and catalyst configuration, the next step is to calculate the reactor volume. Before undertaking a detailed calculation, we need to estimate the reactor volume. A quick estimate is sometimes needed to check an exact calculation or to prepare a budget for a proposal. For packed bed or homogenous reactors, the space velocity is a way of rapidly sizing reactors. Space velocity is defined as the ratio of the volumetric feed flow rate to the reaction volume or the ratio of mass feed flow rate to the catalyst mass. The volu-

volume or the ratio of mass feed flow rate to the catalyst mass. The volumetric feed-gas flow rate is calculated at a standard temperature and pressure. Thus, the space velocity is defined by:

GHSV = hourly volumetric feed-gas flow rate/reaction volume
LHSV = hourly volumetric liquid-feed flow rate/reaction volume
WHSV = hourly mass feed flow rate/catalyst mass

The units of space velocity are the reciprocal of time. Usually, the hourly volumetric feed-gas flow rate is calculated at 60 °F (15.6 °C) and 1.0 atm (1.01 bar). The volumetric liquid-feed flow rate is calculated at 60 °F (15.6 °C). Space velocity depends on the design of the reactor, reactor inlet conditions, catalyst type and diameter, and fractional conversion. Walas [7] has tabulated space velocities for 102 reactions. For example, for the homogeneous conversion of benzene to toluene in the gas phase, the hourly-volumetric space velocity is 815 h^{-1}. This means that 815 reactor volumes of benzene at standard conditions will be converted in one hour. Although space velocity has limited usefulness, it allows estimating the reaction volume rapidly at specified conditions. Other conditions require additional space velocities. A kinetic model is more useful than space velocities, allowing the calculation of the reaction volume at different operating conditions, but a model requires more time to develop, and frequently time is not available.

Table 7.11 lists equations for sizing a reactor using space velocity, and Table 7.12 outlines a calculation procedure. First, calculate the reaction volume, using a space velocity. Then, calculate the reactor cross-sectional area, using a superficial gas velocity. Ulrich [9] states that the superficial velocity varies from 0.005 to 1 m/s (0.00164 to 3.29 ft/s). Forment and Bischoff [31] used 1 m/s (3.29 ft/s). Fulton and Fair [27] used 1 ft/s (0.3048 m/s) for a methanation reactor for the synthesis of phthalic anhydride from o-xylene. We will use about 0.3048 to 1.0 m/s (1.0 to 3.28 ft/s). From the cross-sectional area, calculate the reactor diameter, which should be rounded off in six-inch increments. Then, calculate the reactor length by summing up bed length and allowing about three additional feet for inert ceramic balls at the top and bottom of the bed. Next, round off reactor length to the nearest three-inch increment. The balls promote a uniform velocity across the catalyst bed and prevent a dished-shaped depression from forming at the top because of the jet action of the incoming flow. The bed itself, however, is the prime flow distributor. Alternatively, or in addition to the balls, add a baffle plate at the reactor entrance to deflect the jet of incoming gases. The height of the bed is limited to at least 1/2 D to promote uniform flow distribution and not more than 25 ft (7.62 m) to avoid crushing the catalyst.

Table 7.11 Summary of Equations for Sizing Packed-Bed Reactors Using Space Velocities

Rate Equations

$$V_B = F' / \rho_B' \, S_{CW}' \;\; -- \;\; \text{or} \; V_B = V_{VS}' / S_{CV}' \;\; -- \;\; \text{or} \; V_B = V_V' / S_{CL}' \qquad (7.11.1)$$

Transport

$$\Delta p = (\Delta p)_B' \, L_R \;\; -- \;\; (\Delta p)_B \approx 0.11 \;\; \text{psi/ft of bed (0.0252 bar/m)} \qquad (7.11.2)$$

$$V_V' = v_S' \, A_B \;\; -- \;\; v_S \approx 1.0 \; \text{m/s} \; (\, 3.28 \; \text{ft/s}) \qquad (7.11.3)$$

Geometric Relations

$$A_B = \pi D^2 / 4 \;\; -- \;\; \text{maximum } D \approx 13.5 \; \text{ft (4.11 m)} \qquad (7.11.4)$$

$$V_B = W_B / \rho_B' \qquad (7.11.5)$$

$$L_B = V_B / A_B \;\; -- \;\; L_B \text{ minimum} = 1/2 \, D$$
$$-- \;\; L_B \text{ maximum} \approx 25 \; \text{ft (7.62 m)} \qquad (7.11.6)$$

$$L_R = L_B + L_I' \;\; -- \;\; L_I \approx 3 \; \text{ft (0.914 m)} \qquad (7.11.7)$$

Unknowns

$$V_B - \Delta p - A_B - D - W_B - L_B - L_R$$

Table 7.12 Calculation Procedure for Sizing a Packed-Bed Reactor Using Space Velocity

1. Calculate the bed volume, V_B, from Equation 7.11.1.

2. Calculate the bed area, A_B, from Equation 7.11.3.

3. Calculate the reactor diameter, D, from Equation 7.11.4. Round off D in 6 in (0.152 m) increments, starting at 30 in (0.762 m). If D is less than 30 in (0.762 m), use standard pipe.

4. After rounding off D, calculate the actual bed area from Equation 7.11.4.

5. Calculate the bed length, L_B, from Equation 7.11.6.

6. Calculate the reactor length, L_R, from Equation 7.11.7. Round off L_R in 3 in (0.25 ft; 0.0762 m) increments (for example, 5.0, 5.25, 5.5, 5.75 etc.).

7. Calculate the reactor pressure drop, Δp, from Equation 7.11.2.

8. Calculate the actual reaction volume from Equation 7.11.6, using the corrected bed diameter and length.

9. Calculate the catalyst weight, W_B, from Equation 7.11.5.

Finally, estimate the pressure drop across the bed to complete the design of the reactor system. To promote uniform flow distribution across the bed, Trambouze et al. [8] recommend a pressure drop per unit length of bed of at least 2500 Pa/m (0.11 psi/ft). To the pressure drop across the bed, add an additional pressure drop equivalent to about 3 ft (0.914 m) of bed height [21] to account for pressure losses caused by the vessel nozzles, distributor (balls or other devices), and bed supports, if needed.

Example 7.3 Packed-Bed, Catalytic, Reactor Sizing Using Space Velocity

In 1973, because of a natural gas shortage, the US evaluated two methods of transporting natural gas from overseas producers. One method was to liquefy the natural gas (LNG). LNG is produced by well established processes and then shipped in cryogenic tankers at −161 °C (−258 °F). The other method was to convert the natural gas to methanol, as discussed by Winter and Kohle [26], by a process similar to the one described in Chapter 3. Then, the methanol would be shipped to the US and converted back to methane in two catalytic reactors in series. The first reactor converts methanol to a mixture of gases, which contains methane. The composition of the gases leaving this reactor, which is given in Table 7.3.1, becomes the input to the second reactor. In the second reactor, some of the carbon monoxide and dioxide in the mixture is converted to additional methane. Table 7.3.1 gives the gas analysis out of the second reactor.

After the second reactor, the methane is separated from the mixture before entering the natural-gas pipeline. Estimate the reactor size using the space velocity given below.

Data (Source: Ref. 27).
Catalyst   nickel deposited on kieselguhr
Catalyst size   1/8 in tablets (3.18 mm)
Bed void fraction   0.38
Bulk density   90 lb/ft³ (1440 kg/m³)
Space velocity   3000 h⁻¹ (at 60 °F, 1 atm) (289 K, 1.01 bar)
Molecular weight in   20.4

Table 7.3.1 Reactor Composition

Component	Molecular Weight	Reactor Composition Mole Fraction	
		Input	Output
H_2O	18.02	0.2861	0.30877
CH_4	16.04	0.4558	0.48139
H_2	2.0	0.0771	0.03730
CO	28.01	0.1140	0.00015
CO_2	44.0	0.1696	0.17253
Temperature, K		527.6	588.7
Pressure, bar		27.92	

Source Ref. 27.

Flow rate in   20,350 lbmol/h (9230 kgmol/h)
Superficial velocity   1 ft/s (0.3048 m/s)

Follow the procedure outlined in Table 7.12 using the equations listed in Table 7.11. First calculate the molar density at 60 °F and 1 atm because the space velocity is given at standard conditions. From the ideal gas law, the molar density at standard conditions,

$$\rho_S = \frac{p}{R\,T} = \frac{1.01}{0.08314}\frac{1}{289} = 0.04204 \text{ kgmol/m}^3 \ (2.63 \times 10^{-3} \text{ lbmol/ft}^3)$$

The volumetric flow rate at standard conditions,

$$V_{VS} = \frac{9230}{0.04204} = 2.196 \times 10^5 \text{ m}^3/\text{h} \ (7.75 \times 10^6 \text{ ft}^3/\text{h})$$

According to Equation 7.11.1,

$$V_B = \frac{V_{VS}}{S_{CV}} = \frac{219600}{3000} = 73.20 \text{ m}^3 \ (2580 \text{ ft}^3) \text{ of catalyst}$$

Next, calculate the cross-sectional area of the bed by first calculating the molar gas density. Although the temperature, pressure, and molar flow rate will vary through the reactor, use the reactor inlet conditions to calculate the molar density.

$$\rho = \frac{27.92}{0.08314} \frac{1}{527.6} = 0.6365 \text{ kgmol/m}^3 \text{ (0.0397 lb/ft}^3)$$

The volumetric flow rate in the bed,

$$V_V = \frac{9230}{0.6365} = 14{,}500 \text{ m}^3/\text{h} \ (5.12 \text{x} 10^5 \text{ ft}^3/\text{h})$$

From Equation 7.11.3 and using the superficial velocity of 1.0 ft/s (0.3048 m/s) given by Fulton and Fair [27], the cross sectional area of the bed,

$$A_B = \frac{14500 \text{ m}^3}{1} \ \frac{1 \text{ h}}{\text{h}} \ \frac{1 \text{ s}}{3600 \text{ s}} \ \frac{}{0.3048 \text{ m}} = 13.21 \text{ m}^2 (142 \text{ ft}^2)$$

Next, calculate the bed diameter to determine if it exceeds the shipping limit of 13.5ft (4.11 m) specified in Equation 7.11.4.

$$D = [(4 / 3.142) (13.21)]^{1/2} = 4.101 \text{ m} (13.5 \text{ ft})$$

When adding the vessel-wall thickness the reactor diameter will be greater. At a design pressure of 500 psig (34.5 barg), Fulton and Fair [27] calculate a wall thickness of 4 in (10.2 cm). To keep below the shipping diameter of 13.5 ft (4.11 m), use an inside diameter of diameter of 12.5 ft (3.81 m).

The actual bed cross-sectional area is

$$A_B = \frac{3.142}{4} \ \frac{(3.810)^2}{1} = 11.40 \text{ m}^2 (37.4 \text{ ft}^2)$$

From Equation 7.11.6, the bed length,

$$L_B = \frac{73.20}{11.40} = 6.421 \text{ m} (21.1 \text{ ft})$$

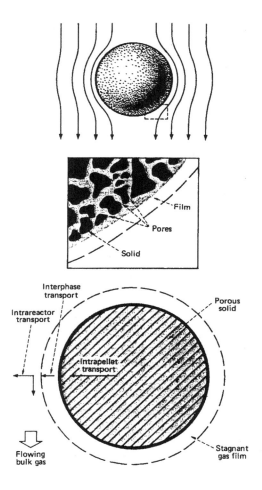

Figure 7.5 Mechanisms of mass transfer for catalytic pellets. (Source Ref. 18 with permission).

Round off the bed height to 22 ft (6.71 m), which, according to Equation 7.11.6, is below 25 ft (7.62 m), the maximum bed height allowed to avoid crushing the pellets.

From Equation 7.11.7, the reactor length

$L_R = (22 + 3) = 25$ ft (7.62 m)

According to the rule given in Step 6 in Table 7.12, there is no need to round off the reactor length.

From Equation 7.11.2, an estimate of the pressure drop is 0.11 psi/ft (0.0249 bar/m) of bed. Allowing for a pressure drop of 3 ft (0.914 m) of bed height for internals, the pressure drop across the reactor,

$$\Delta p = 0.11 \ (22 + 3) = 2.75 \text{ psi } (0.190 \text{ bar})$$

From Equation 7.11.6, the actual bed volume,

$$V_B = 22.0 \ (3.142 \ / \ 4) \ (12.5)^2 = 2700 \text{ ft}^3 \ (765 \text{ m}^3)$$

Finally, calculate the catalyst mass from Equation 7.11.5.

$$W_B = 90 \ (2700) = 2.430 \text{x} 10^5 \text{ lb } (1.10 \text{x} 10^5 \text{ kg})$$

Plug-Flow Reactor Model

First, select a reactor arrangement and catalyst configuration. The next step is to select a reactor model for calculating the reaction volume. An exact model of reactor performance must include mass transfer of reactants from the fluid to the catalyst sites within the pellet, chemical reaction, and then mass transfer of products back into the fluid. Table 7.13 lists the steps, and Figure 7.5 illustrates the processes involved. Here, only simple models are of interest to estimate the reaction volume for a preliminary design. The reaction volume is that volume occupied by the catalyst pellets and the space between them. We must provide additional volume for internals to promote uniform flow and for entrance and exit sections. The total volume is called the reactor volume. After calculating the reactor volume, the next step is to determine the reactor length and diameter.

A simple model is the one-dimensional, plug-flow, pseudo-homogeneous model. In this model, we will consider the fluid and solid phases as a single phase. For this model to apply we must fulfill the following conditions:

1. adiabatic operation
2. flat velocity profile
3. no axial dispersion
4. no radial dispersion
5. pseudo-homogeneous assumption

Table 7.13 Steps in a Catalytic Reaction (Source: adapted from Ref. 16).

1. Mass transfer of reactants from the fluid to the pore entrances of the catalyst pellet
2. Diffusion of reactants through the porous catalyst to the internal catalytic surface
3. Adsorption of reactants on the catalyst surface
4. Reaction on the catalyst surface
5. Desorption of products from the surface
6. Diffusion of products from the interior of the pellet to the pore entrance
7. Mass transfer of products from the pore entrance to the fluid

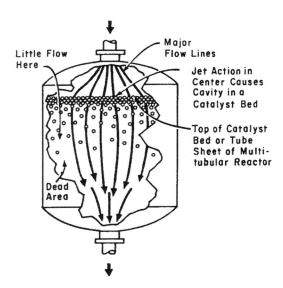

Figure 7.6 Flow pattern in a packed bed reactor. (Source Ref. 19 with permission).

The first condition of adiabatic operation is achieved by providing sufficient insulation. To fulfill the second condition of maintaining a flat velocity profile at each bed cross section requires preventing flow maldistribution. Flow maldistribution is either bypassing or channeling of the flow, creating stagnant areas within the reactor as shown in Figure 7.6. The result is a reduction in conversion. Poor pellet distribution and a dished catalyst bed, caused by the entering jet of gas, can cause flow maldistribution. Distributing the flow evenly at the inlet and outlet of the reactor prevents flow maldistribution. To avoid bypassing, Trambouze et al. [8] recommend a pressure drop per unit length of packing of 2500 Pa/m (0.11 psi/ft). They also recommend a bed diameter to mean pellet diameter of 10 to reduce wall effects. To provide even flow distribution requires inlet and outlet flow distributors and layers of inert balls of varying diameters at the top and bottom of the bed as illustrated in Figure 7.7. The upper layers of large balls also prevents dishing of the catalyst bed. Tarhan [25] estimated that back mixing is essentially eliminated when the ratio of the bed length to the mean pellet diameter, L/d, is equal to or greater than fifty. Most industrial reactors satisfy this condition [25]. The mean pellet diameter is defined as the diameter of a sphere that has the same volume as the pellet.

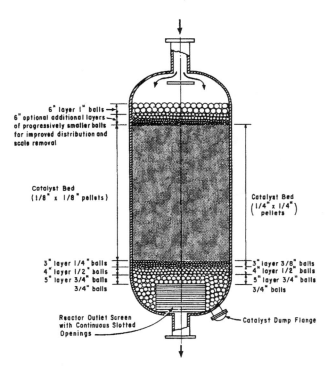

Figure 7.7 Packed bed reactor design. From Ref. 19 with permission.

The third and fourth condition are fulfilled by Tarhan [25]. "Axial dispersion is fundamentally local backmixing of reactants and products in the axial, or longitudinal direction in the small interstices of the packed bed, which is due to molecular diffusion, convection, and turbulence. Axial dispersion has been shown to be negligible in fixed-bed gas reactors. The fourth condition (no radial dispersion) can be met if the flow pattern through the bed already meets the second condition. If the flow velocity in the axial direction is constant through the entire cross section and if the reactor is well insulated (first condition), there can be no radial dispersion to speak of in gas reactors. Thus, the one-dimensional adiabatic reactor model may be actualized without great difficulties."

The pseudo-homogeneous assumption means that both the solid and fluid phases are are considered a single phase. Therefore, we avoid considering mass and heat transfer from and to the catalytic pellets. This model assumes that the component concentrations and the temperature in the pellets are the same as those in the fluid phase. This assumption is approximated when the catalyst pellet is small and mass and heat transfer between the pellets and the fluid phase are rapid. The reaction rate for this model, called the global reaction rate, includes heat and mass transfer. If heat and mass transfer are made insignificant, then the reaction rate is called the intrinsic reaction rate.

Equations for sizing packed-bed reactors are listed in Table 7.14, and a calculating procedure is outlined in Table 7.15. The procedure for calculating the reactor dimensions is similar to that given for the space-velocity method. In this procedure, however, the calculation of the reaction volume is more accurate than the method using space velocity. First, the reaction volume for adiabatic operation is calculated by solving the mole and energy balances along with the kinetic equation. Also, instead of using a rule-of-thumb, we use the Ergun equation, Equation 7.14.5 in Table 7.14, derive by Bird et al. [32], to calculate the superficial velocity in the bed. To calculate the velocity, fix the pressure drop across the bed, $(\Delta p)_B$ to insure good flow distribution as given in Equation 7.14.5. This equation requires calculating the average viscosity and density of a gas mixture, as given by Equations 7.14.14 and 7.14.15. Pure component viscosities are estimated using the corresponding state approach outlined by Bird et al. [32].

Table 7.14 Summary of Equations for Sizing a Packed-Bed Reactor – One-Dimensional, Plug-Flow, Pseudo-Homogeneous, Model

Mole Balance

$$r_A \, dW_C = m_{Ao}' \, dx_A \tag{7.14.1}$$

Energy Equation

$$\Delta h_R \, m_{Ao}' \, dx_A + m_T' \, c_P \, dT = 0 \tag{7.14.2}$$

Rate Equation

$$r_A = k \, f(p_i) \tag{7.14.3}$$

$$y_i = p_i / P' \tag{7.14.4}$$

Transport Equations

$$(\Delta p)_B' = 150 \, [\mu \, v_S / (D_p')^2] \, (1 - \varepsilon')^2 / (\varepsilon')^3 + 1.75 \, [\, \rho \, (v_S')^2 / D_p' \,] \, (1 - \varepsilon') / (\varepsilon')^3$$

$$\text{--- } 0.11 \text{ psi/ft} < (\Delta p)_B < 0.2 \text{ psi/ft} \, [\, 2470 < (\Delta p)_B > 4490 \text{ Pa/m} \,] \tag{7.14.5}$$

$$\Delta p = [(\Delta p)_B] \, (L_B + 3 \text{ ft}) \tag{7.14.6}$$

$$V_V = v_S \, A_B \tag{7.14.7}$$

Geometric Relations

$$A_B = \pi \, D^2 / 4 \text{ --- maximum } D \approx 13.5 \text{ ft (4.11 m)} \tag{7.14.8}$$

$$V_B = W_B / \rho_B' \tag{7.14.9}$$

$$L_B = V_B / A_B \quad \text{--- } L_B \text{ minimum} = \tfrac{1}{2} \, D$$
$$\text{--- } L_B \text{ maximum} \approx 25 \text{ ft (7.62 m)} \tag{7.14.10}$$

$$L_R = L_B + L_I' \quad \text{--- } L_I \approx 3 \text{ ft (0.914 m)} \tag{7.14.11}$$

System Properties

$$k = A' \exp (E'/ R' \, T') \tag{7.14.12}$$

$$c_P = \Sigma \, y_i' \, c_{Pi}' \tag{7.14.13}$$

$$\mu = \Sigma \, y_i' \, \mu_i' \tag{7.14.14}$$

$$\rho = \Sigma \, y_i' \, \rho_i' \tag{7.14.15}$$

$$P' \, V_V = m_T' \, R' \, T' \tag{7.14.16}$$

$$\Delta h_R = \Delta h_1' + \Delta H^o_R{}' + \Delta h_2' \tag{7.14.17}$$

Unknowns

$$r_A - W_C - x_A - \Delta h_R - \Delta p - c_P - \mu - \rho - k - A_B - V_B - W_B - D - L_B - y_I - v_S - V_V$$

Table 7.15 Packed-Bed Reactor-Sizing Calculation Procedure – One-Dimensional, Plug-Flow, Pseudo-Homogeneous, Model

1. Calculate the average heat capacity, c_p, at reactor inlet conditions from Equation 7.14.13.

2. Calculate the mass of catalyst required, W_B, for the specified conversion, $x_{A,}$, from Equations 7.14.1 to 7.14.4, 7.14.12, and 7.14.17.

3. Calculate the average viscosity, μ, at inlet conditions from Equation 7.14.14.

4. Calculate the average density, ρ, at inlet conditions from Equation 7.14.15.

5. Calculate the superficial gas velocity, v_S, from Equation 7.14.5.

6. Calculate the inlet volumetric flow rate, V_V, from Equation 7.14.16.

7. Calculate the bed area, A_B, from Equation 7.14.7.

8. Calculate the bed diameter, D, from Equation 7.14.8. Round off D in 6 in (0.152 m) increments, starting at 30 in (0.762 m). If D is less than 30 in (0.762 m), use standard pipe.

9. After rounding D, calculate the actual bed area using the actual D from Equation 7.14.8.

10. Calculate the actual superficial velocity from Equation 7.14.7.

11. Calculate the actual bed pressure drop for a unit length, $(\Delta p)_B$, from Equation 7.14.5.

12. Calculate the bed length, L_B, from Equation 7.14.10. Calculate minimum and maximum L_B, and, if necessary, adjust L_B.

13. Calculate the reactor length, L_R, from Equation 7.14.11. Round off L_R in 3 in (0.25 ft, 0.0762 m) increments (for example, 5.0, 5.25, 5.5, 5.75 etc.).

14. Calculate the total bed pressure drop, Δp, from Equation 7.14.6.

15. Calculate the actual bed volume from Equation 7.14.10.

16. Calculate the catalyst mass using the actual bed volume from Equation 7.14.9.

**Example 7.4   Packed-Bed, Catalytic, Reactor Sizing Using the Plug Flow
Model**

Styrene is produced by dehydrogenation of ethylbenzene in an adiabatic, fixed-
bed reactor. Although Sheel and Crowe [29] list ten reactions and several products, the major reaction is the conversion of ethylbenzene to styrene, according
to the following equation.

$$\phi \, C_2 H_5 \rightarrow \phi \, C_2 H_3 + H_2$$

At first, we only need an estimate of reactor size so that we will only consider this reaction. Because the reaction is endothermic and the number of
moles increases during reaction, conversion increases by conducting the reaction
at a high temperature, a low pressure, and with the addition of an inert diluent.
Steam is selected as the diluent because it also suppresses carbon formation,
preheats the feed to the reaction temperature, and acts as a heat source, preventing a sharp drop in temperature during the course of reaction. Without steam,
ethylbenzene will pyrolize, forming carbon which coats the catalyst.

Although thermodynamics favors a high reaction temperature, the rate of
formation of by-products increases rapidly with increasing temperature. Thus,
the actual reaction temperature is a trade-off between high conversion to styrene
and minimizing by-product formation. The catalyst selected (unspecified by
Sheel and Crowe [29]), gives an acceptable conversion at a low temperature
where side reactions are minimized.

Estimate the reactor length, diameter, the mass of catalyst, and the pressure
drop across the reactor. When determining the amount of catalyst, assume that
the reactor pressure is constant. The first step is to calculate the mass of catalyst
required to convert the ethyl benzene to styrene. Then, calculate the volume occupied by the catalyst pellets using the bulk density. Finally, determine the reactor dimensions.

Although temperature, pressure, and composition change across the reactor, system properties will be calculated at inlet conditions. Changes in temperature and system properties through the bed will be moderated because of the
large excess of steam.

Data		Reference
Ethyl benzene flow rate	9000 lb/h (4082 kg/h)	29
Ethyl benzene molecular weight	106.16	
Steam flow rate	8000 lb/h (8165 kg/h)	29
Water molecular weight	18.016	
Mixed feed temperature	600 °C (1110 °F)	29
Inlet pressure	2.33 atm (34.25 psi, 2.362 bar)	
Final conversion	0.45	28

Bed void fraction	0.445	29
Bulk density	1300 kg/m^3 (81.16 lb/ft^3)	31
Equivalent catalyst diameter	0.005 m (0.0164 ft)	31

$\Delta h_R = 1.20737 \times 10^5 + 4.56\ T$ — T in K	31
$r_A = k\ (p_E - p_S\ p_H\ /\ K_P)$ — T in K	31
$k = 12600\ \exp\ (-11000\ /\ T)$ — T in K	28
$K_P = 0.027\ \exp\ [0.21\ (T - 773)]$ — T in K	28

To solve this problem, follow the procedure outlined in Table 7.15 using the Equations listed in Table 7.14.

Substitute the molar flow rate of ethyl benzene, m_{Ao} = 4082 kg/h (9000 lb/h), into Equation 7.14.1, and rearrange the equation to obtain

$$dW_C\ /\ dx_A = m_{Ao}\ /\ r_A = (4082/106.16)\ /\ r_A = 38.45\ /\ r_A$$

The average heat capacity, c_p = 244.5 kJ/kgmol (105 Btu/lbmol), is calculated at the reactor inlet conditions, using heat capacities taken from Reid et al. [30] and Equation 7.14.13. The changes in temperature and composition through the reactor will not significantly change the heat capacity.

The enthalpy of reaction (Equation 7.14.17), Δh_R, given by Froment and Bischoff [31], is calculated below. We will also assume that the temperature will not change significantly throughout the reactor. The large excess of steam will moderate the decrease in temperature. Letting T = 873.2 K (1570 °R), the reactor inlet temperature,

$$\Delta h_R = 120737 + 4.56\ T = 120737 + 4.56\ (873.2)$$

$$\Delta h_R = 1.247 \times 10^5\ \text{kJ/kgmol}\ (5.36 \times 10^4\ \text{Btu/lbmol})$$

Substituting ΔH_R, m_{Ao}, and c_P into the energy equation, Equation 7.14.2, we obtain

$$dT/dx_A = \frac{-\Delta h_R\ m_{Ao}}{c_P\ m_T} = \frac{-124700}{1}\frac{\text{kJ}}{\text{kgmol}}\ \frac{1}{244.5}\frac{\text{kmol-K}}{\text{kJ}}\ \frac{38.45\ \text{kgmol}}{}$$

$$dT/dx_A = -19610\ /\ m_T$$

According to the chemical equation there will be an increase in the molar flow rate as the reaction proceeds. The total molar flow rate, m_T, is equal to molar flow rate into the reactor plus the increase in moles caused by the reaction. Therefore,

$m_T = 491.6 + 38.45\ x_A$

Now, evaluate the mole fraction, y_i, for each component from Equation 7.14.4. Take one kgmol of ethyl benzene as the basis for the calculation. The kgmol of steam per kgmol of incoming styrene is 11.78. Therefore, for the fractional conversion of ethyl benzene, x_A,

Ethyl Benzene	$1 - x_A$
Steam	11.78
Styrene	x_A
Hydrogen	x_A
Total	$12.78 + x_A$

$y_E = (1 - x_A) / (12.78 + x_A)$

$y_S = x_A / (12.78 + x_A)$

$y_W = 11.78 / (12.78 + x_A)$

$y_H = x_A / (12.78 + x_A)$

After Substituting these equations in terms of the conversion into Equation 7.14.3, we find that

$$r_A = k\ f(p_i) = k\left[p_E - \frac{p_S\ p_H}{K_P}\right] = k\left[\frac{(1 - x_A)\ P}{12.78 + x_A} - \frac{x_A{}^2\ P^2}{K_P\ (12.78 + x_A)^2}\right]$$

where the subscripts, E = ethyl benzene, S = styrene, and H = hydrogen.

From the above equations,

$dT/dx_A = -19610 / m_T$

and

$m_T = 491.6 + 38.45\ x_A$

Also, from the problem statement,

$k = 12600\ exp\ (-11000 / T)$

and

$$K_P = 0.027 \exp [\, 0.21 \, (T - 773)]$$

Solving these last five equations simultaneously using Polymath, at a conversion of $x_A = 0.45$, the catalyst mass is 4164 kg (9180 lb) and the final temperature is 856.0 K (1540 °R). The decrease in temperature is only 17.2 K (31 °R), which verifies the original assumption that the temperature decrease would be small.

Next, calculate the reactor dimensions. First, calculate the superficial velocity using the Ergun Equation (Equation 7.14.5). This equation requires calculating the average viscosity and density. The mole fraction average viscosity at the inlet conditions is 2.408×10^{-5} Pa-s (0.0241 cp). Also, the mole fraction average of the gas density at inlet conditions is 0.7996 kg/m³ (0.499 lb/ft³). The recommended pressure drop range across the bed to insure good flow distribution is given by Equation 7.14.5. The smaller the reactor diameter, the greater the superficial velocity, and the greater the pressure drop. If we select an average value of $(\Delta p)_B$ of 0.155 psi/ft (3550 Pa/m), the calculated superficial velocity from Equation 7.14.5 is 1.274 m/s (4.180 ft/s).

Now, calculate the reactor diameter. First, calculate the volumetric flow rate using Equation 7.14.16.

$$V_V = \frac{491.4 \text{ kmol}}{1 \text{ h}} \; \frac{1 \text{ h}}{3600 \text{ s}} \; \frac{0.08314 \text{ bar-m}^3}{1 \quad \text{kmol-K}} \; \frac{873.2 \text{ K}}{2.362 \text{ bar}} = 4.195 \text{ m}^3/\text{s} \; (148.1 \text{ ft}^3/\text{s})$$

Next, calculate the bed area using Equation 7.14.7.

$$A_B = 4.195 \, / \, 1.274 = 3.293 \text{ m}^2 \; (116.3 \text{ ft}^3/\text{s})$$

Finally, calculate the bed diameter using Equation 7.14.8.

$$D = [4 \, (3.293) \, / \, 3.142]^{1/2} = 2.047 \text{ m} \; (6.716 \text{ ft})$$

According to Step 8 in Table 7.14, round off the diameter to 7.0 ft (2.134 m).

Because the bed diameter has increased, the superficial velocity will decrease, and therefore the bed pressure drop will decrease, according to Equation 7.14.7. The actual bed area,

$$A_B = 3.142 \, (2.134)^2 \, / \, 4 = 3.577 \text{ m}^2 \; (38.59 \text{ ft}^2)$$

and from Equation 7.14.7 the actual superficial velocity,

$v_S = 4.195 / 3.577 = 1.173$ m/s (3.85 ft/s)

The actual pressure drop from Equation 7.14.5, when D = 7.0 ft, is $(\Delta p)_B =$ 3018 Pa/m (0.133 psi/ft).

From Equation 7.14.9, the bed volume,

$V_B = 4164 / 1300 = 3.203$ m^3 (113.1 ft^3)

and from Equation 7.14.10 the bed length,

$L_B = V_B / A_B = 113.1 / 38.59 = 2.931$ ft (0.893 m)

which, according to Equation 7.14.10, is below the recommended-minimum bed height of ½ (7.0) = 3.5 ft (1.07 m). Increase L_B to 3.5 ft (1.07 m), which will allow for a safety factor. The packed volume of a commercial styrene reactor, reported by Scheel and Crowe [29], has a bed height of 5.28 ft (1.61 m) and a diameter of 6.40 ft (1.95 m). The packed volume for the reactor is 168.9 ft^3 (4.78 m^3), whereas the calculated reaction volume is 113.1 ft^3 (3.20 m^3). The difference is 55.8 ft^3 (1.58 m^3), which is not completely unreasonable, considering the assumptions that were made. Also, it is not known if the height of the bed for the commercial reactor includes ceramic balls for promoting uniform flow distribution. If it does, then the height of the catalyst bed will be less, bringing the calculated height in closer agreement with the commercial reactor. The present calculation puts us into the right ballpark, and the final decision on the reactor dimensions will be based on pilot-scale tests.

Allowing space for internals, the reactor length,

$L_R = L_B + L_I = 3.5 + 3.0 = 6.5$ ft (1.98m)

According to Step 13 in Table 7.15, L_R requires no rounding.

The total pressure drop across the reactor is calculated from Equation 7.14.6 is

$\Delta p = (0.133) (3.5 + 3) = 0.8645$ psi (0.0596 bar)

The actual bed volume,

$V_B = L_B A_B = 1.07 (3.577) = 3.827$ m^3 (135 ft^3)

From Equation 7.14.9 the catalyst mass,

$W_B = 1300 (3.827) = 4975$ kg (1.10×10^4 lb)

NOMENCLATURE

English

A	area or pre-exponential factor
c	concentration
c_{Ao}	initial concentation of A
c_P	heat capacity
D	diameter
D_P	particle diameter
E	activation energy
F	mass flowrate into a packed bed
h_{fi}	inside fouling heat-transfer coefficient
h_{fo}	outside fouling heat-transfer coefficient
h_i	inside film heat-transfer coefficient
h_o	outside film heat-transfer coefficient
h_w	reactor wall heat-transfer coefficient
k	reaction rate constant or thermal conductivity
K_P	chemical equilibrium constant
L_B	length of packed bed
L_I	length required for reactor internals
L_R	reactor length
m	molar flow rate
m_T	total molar flow rate

p	power per unit volume, pressure or partial pressure
P	power
Q	heat-transfer rate
Q_{Ro}	initial heat-transfer rate
r	rate of reaction
r_{Ao}	initial rate of reaction of A
R	gas constant
S_{CW}	WHSV (weight hourly space velocity) — lb feed/h per lb of catalyst
S_{CV}	GHSV (gas hourly space velocity) — ft^3 gas/h per ft^3 of catalyst
S_{CL}	LHSV (liquid hourly space velocity) — ft^3 liquid/h per ft^3 of catalyst
t	time
t_B	batch time
t_C	time required to clean a batch reactor
t_E	time required to empty a batch reactor
t_F	time required to load a batch reactor
t_H	time required to heat a batch reactor to the reaction temperature
t_R	reaction time
T	temperature
U	overall heat-transfer coefficient
v_S	superficial velocity
V_B	bed volume
V_r	reaction volume

V_R	reactor volume
V_V	volumetric flow rate
V_{VS}	volumetric flow rate at standard conditions
W_B	mass of packed bed
x	conversion
x_W	wall thickness
y	mole fraction

Greek

Δh_1	change in enthalpy from reactor inlet to standard conditions
Δh_2	change in enthalpy from standard conditions to reactor outlet conditions
Δh_R	change in enthalpy from reactor inlet to reactor outlet conditions
H^o_R	standard enthalpy of reaction
ε	void fraction
μ	viscosity
ρ	mass or molar density

Subscripts

A	reactant A
B	reactant B or packed bed
C	coil or reactant C
E	external
i	i^{th} component
I	internal

J   jacket

n   CSTR number and the number of the entering stream

P   particle

R   reference

W   wall

REFERENCES

1. Riegel's Handbook of Industrial Chemicals, Kent J.A., Ed., 8th ed., Van Nostrand-Reinhold, New York, NY, 1983.
2. Samdani, G., Gilges, K., Electrosynthesis: Positively Charged, Chem. Eng., 98, 5, 37, 1991.
3. Lowenheim, F.A., Moran, M.K., eds., Faith, Keyes and Clark's Industrial Chemicals, 4th ed., Wiley-Interscience, New York, NY, 1975.
4. Pandit, A.B., Moholkar, V.S., Harness Cavitation to Improve Processing, Chem. Eng. Progr., 92, 7, 57, 1996.
5. Cheung, H.M., Bhatnagar, A., Jansen, G., Sonochemical Destruction of Chlorinated Hydrocarbons in Dilute Aqueous Solution, Env. Sci. & Tech, 25, 8, 1510, 1991.
6. Zanetti, R.J., Plasma: Warming up to New CPI Applications, Chem. Eng., 90, 26, 14, 1983.
7. Walas, S.M., Chemical Process Equipment, Butterworths, Boston, MA, 1988.
8. Trambouze, P., Van Landeghem, H,, Wauquier, J.P., Chemical Reactors-Design/Engineering/Operation, Gulf Publishing, Houston, TX, 1988.
9. Ulrich, G.D., A Guide to Chemical Engineering Process Design and Economics, John Wiley & Sons, New York, NY, 1984.
10. Bollinger, D.H., Assessing Heat Transfer in Process-Vessel Jackets, Chem. Eng., 89, 19, 95, 1982.
11. Frank, O., Personal Communication, Consulting Engineer, Convent Station, NJ, Jan. 2002.
12. Markowitz, R.E., Picking the Best Vessel Jacket, Chem. Eng., 78, 26, 156, 1971.
13. Brochure, Half-Pipe Coil Jacket Reactors, Brighton Corp., Cincinnati, OH, June 1971.
14. Hicks, R.W., Gates,L.E., Fluid Agitation in Polymer Reactors, Chem. Eng. Progr.,71, 8, 74, 1975.

15. Blaasel, V.D., Preliminary Chemical Engineering Plant Design, 2nd ed., Van Nostrand-Reinhold, New York, NY, 1990.
16. Fogler, H.S., Elements of Chemical Reaction Engineering, Prentice Hall, Englewood Cliffs, NJ, 1992.
17. Bartholomew, C.H., Hecker, W.C., Catalytic Reactor Design, Chem. Eng., 101, 6, 70, 1994.
18. Fulton, J.W., Selecting the Catalyst Configuration, Chem. Eng., 93, 9, 97, 1986.
19. Rase, H.F., Chemical Reactor Design for Process Plants, vol. 2, Case Studies and Design Data, John Wiley & Sons, New York, NY, 1977.
20. Garvin, J., Understanding the Thermal Design of Jacketed Vessels, Chem. Eng. Progr., 95, 6, 61, 1999.
21. AlChE Student Contest Problem, American Institute of Chemical Engineers, New York, NY, 1982.
22. Shachham, M., Shachham O., POLYMATH, (Version 4.0), CACHE Corp., Austin TX, 1996.
23. Smith, J.W., Chemical Engineering Kinetics, 3rd ed., McGraw-Hill, New York, NY, 1981.
24. Anonymous, Dowtherm Heat Transfer Fluids, The Dowtherm Co., Midland, MI, 1967.
25. Tarhan, T.M., Catalytic Reactor Design, McGraw Hill, New York, NY, 1983.
26. Winter, C., Kohle, A., Energy Imports: LNG Vs MeOH, Chem. Eng., 80, 26, 233, 1973.
27. Fulton, J.W., Fair, J.W., Manufacture of Methanol and Substitute Natural Gas, Monsanto Co., St. Louis, MO, Sept.1974.
28. Chen, N.H., Process Reactor Design, Allyn and Bacon, Boston, MA, 1983.
29. Sheel, J.G.P., Crowe, C.M., Simulation and Optimization of an Existing Ethylbenzene Dehydration Reactor, Can. J. Chem. Eng., 47, 4, 183, 1969.
30. Reid, R.C., Prausnitz, J.M., Sherwood, T.K., The Properties of Gases and Liquids, 3rd ed., McGraw-Hill, New York, NY, 1977.
31. Forment, G.F., Bischoff, K.B., Chemical Reactor Analysis and Design, 2nd ed. John Wiley & Sons, New York, NY, 1990.
32. Bird, R.B., Stewart, W.B., Lightfoot, E.N., Transport Phenomena, John Wiley & Sons, New York, NY, 1960.
33. Markovitz, R.E., Process and Project Data Pertinent to Vessel Design, Chem. Eng., 84, 21, 123, 1977.

8

Design of Flow Systems

Flow-system design is one of the most frequently occurring design problems encountered by process engineers. Fluids flow through, reactors, separators, heat exchangers, and other process units. Not only is the flow system one of the salient features of a chemical plant, but it is also frequently encountered in research and development. Kern [30], starting in December 1974, has discussed several aspects of flow system design in a twelve-part series published by Chemical Engineering.

Just as the electrical engineer selects resistors, capacitors, and transistors when designing an electric circuit, the chemical engineer selects valves, pumps, and flow meters to produce a flow system. The procedure followed in the design of a flow system is to determine:

1. pipe-fitting type
2. valve type and size
3. materials of construction
4. pipe size
   length
   wall thickness
   diameter
5. flow-meter type and size
6. pump type and size
7. piping supports

Flow-system design is one of the last steps in the design of a chemical plant. After designing and locating all equipment, then the process engineer can complete the flow system design. In the following sections, we will consider the above elements of a flow system in some detail.

PIPE FITTINGS

The major functions of pipe fittings are to:

1. change the direction of flow
2. reduce or enlarge pipe size
3. split or combine fluid streams
4. facilitate disconnecting piping from equipment
5. access the flow system for temperature, pressure, flow rate, and liquid level measurements, and for sampling process streams

Several common threaded pipe fittings are shown in Figure 8.1, but welded fittings and piping are frequently used. In plants, threaded piping is mostly used for water, steam, and natural gas [31]. For threaded fittings, a thread sealant must be used to prevent leakage. For welded fittings, flanges are used to connect pipe to equipment. In this case, gaskets are needed for sealing. Because welded connections are less likely to leak, process piping is always welded [31].

Ninety-degree, forty-five degree, and the street elbow, shown in Figure 8.1, change the direction of flow. The reducing coupling or reducer and bushing change the pipe size, and a coupling joins two lengths of piping of the same size. The pipe tee and the pipe cross combine or split fluid streams. They are also used to gain access to the flow system for sampling the fluid and to measure process variables. When removing equipment from the flow system for repair or replacement, pipe unions are required for threaded piping and flanges for welded piping. Even if removal of equipment is not necessary, a little reflection will show that for threaded piping a union is a necessity when making a connection between two fixed points. If we do not use a union, one end of a pipe will unscrew while screwing the other end into a fitting.

VALVE TYPE

Before selecting a valve, the function of each valve type must be considered first. Several valve types, listed in Table 8.1, are used for on-off service, prevention of back flow, and throttling. Figure 8.2 to 8.4 shows only a few examples of valve types. For a discussion of many other valve types see Reference 8.2.

The simplest valve function is on-off service. Examples of this valve type are gate and ball valves, shown in Figure 8.2. A ball valve is used for tight shut-off. One application is a drain valve on a tank, where it is required to have the

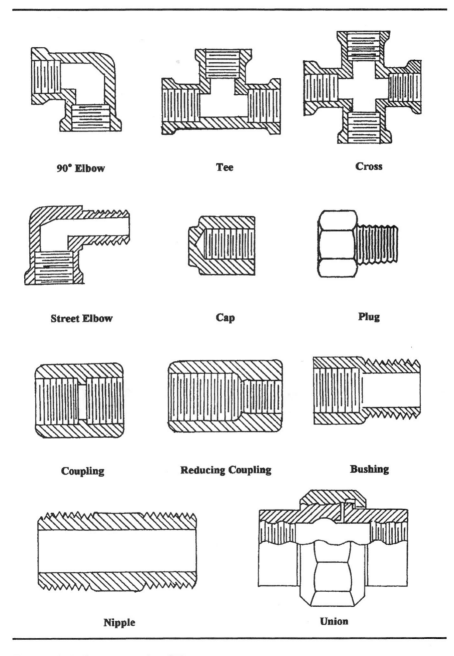

Figure 8.1 Common pipe fittings.

Table 8.1 Classification of Valves

On-Off	Prevention of Back Flow	Throttling	
		Automatic or Manual	Regulators (Self-Operated)
Gate Slide Ball Solenoid Toggle	Ball Check Swing Check Piston Check	Globe Needle Butterfly Diaphragm Pinch	Pressure Flow Rate Temperature Level

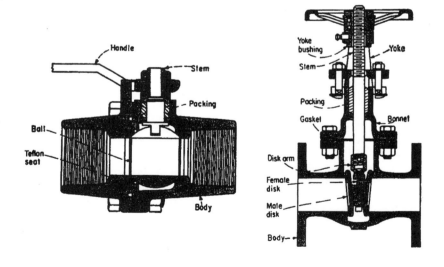

Ball
Source: Reference 8.2

Gate
Source: Reference 8.2

Figure 8.2 Examples of on-off valves. From Ref. 32 with permission.

valve shut off tightly while in service to prevent valuable or dangerous liquids from leaking. Occasionally, a tank requires cleaning or repairs. Then the valve is completely opened to empty the tank quickly. Another application is to isolate equipment from the flow system for replacement or repair.

　　Check valves prevent back flow. The ball check valve, shown in Figure 8.3, is particularly simple in that it has no moving parts requiring close clearances bet-

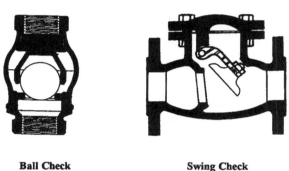

Ball Check
Source: Reference 3.32

Swing Check
Source: Reference 3.32

Figure 8.3 Examples of check valves. From Ref 32 with permission.

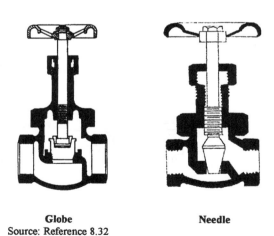

Globe
Source: Reference 8.32

Needle

Figure 8.4 Examples of throttling valves. From Ref 32 with permission.

ween parts, and thus it is very reliable. The life of this valve is prolonged because the ball continually rotates and thus wears evenly. An example of the use of a check valve is when pumping a liquid into a pressurized vessel. If the power delivered to a pump fails, the liquid in the vessel will flow back through the feed line and damage the pump. To prevent this, install a check valve in the feed line.

To control the flow rate, which is called throttling, requires either a manually operated needle or globe valve (Figure 8.4) or an automatic control valve (Figure 8.5). The control valve in Figure 8.5 contains flanged connections. Needle valves are usually used in experimental work to make manual adjustments of flow rate. Globe valves are commonly used for adjusting the flow rate in utility supply lines.

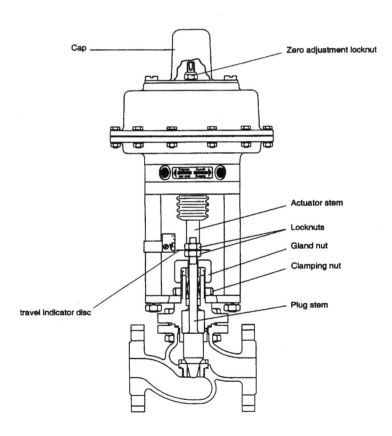

Cap
Zero adjustment locknut
Actuator stem
Locknuts
Gland nut
Clamping nut
Plug stem
travel indicator disc

Figure 8.5 Example of an automatic control valve. From Ref. 15.

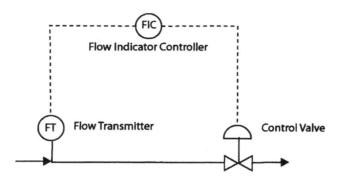

Figure 8.6 An automatic control loop.

Automatic valves are part of a control loop, which is shown in Figure 8.6. The loop contains a primary element, which measures the controlled variable, such as temperature, pressure, flow rate, and liquid level. The operation of a control loop is the same regardless of what variable is controlled. In the case of flow-rate control, the controller obtains the flow rate from transmitter a flow meter and compares the measured flow rate with a value that has been preset in the controller. If the flow rate is greater than the preset value, the controller increases air pressure on top or bottom of a diaphragm in the valve. Then, the valve partially closes to reduce the flow rate. On the other hand, if the flow rate is below the preset value, the controller will act to reduce the air pressure on the diaphragm, and hence the valve opens wider. Electric motors can also operate automatic control valves.

The self-actuated or self-operated control valve is called a regulator. Regulators require no external power source to operate, such as air, but operate entirely from the energy obtained from the flowing fluid. The entire control loop is built into the valve. Because of their low cost, consider regulators first for control applications. Regulators are available for pressure, flow rate, temperature, and liquid-level control. Figure 8.7 shows a pressure regulator for controlling steam pressure. Compressing the upper spring of the regulator by turning the hand wheel in a clockwise direction sets the outlet pressure. This is opposite to the required direction to open manually-operated valves. When the spring at the top of the valve is compressed, a thin diaphragm located directly below the spring moves the diaphragm downward, opening a small pilot valve. Steam enters a passage above the pilot valve and then flows through the dashed passage, shown in Figure 8.7, to a piston located in the lower chamber. Steam pressure pushes the piston up, opening the main valve to let steam into the downstream side of the valve. A small amount of steam flows in the passage located on the downstream side of the

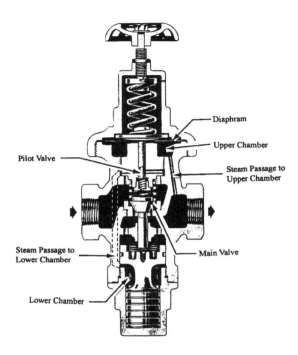

Figure 8.7 An example of a steam regulator. From Ref. 3.

valve that leads to an upper chamber directly below the diaphragm. The steam pressure pushes the diaphragm upward to relieve some of the compression of the spring. Then, the pilot valve partially closes, letting less steam into the piston chamber, and the main valve partially closes decreasing the outlet pressure. A balance will finally be achieved, and the main valve will reach an equilibrium position, allowing a steady flow of steam into the system at a desired outlet pressure.

Figure 8.8 shows a typical installation for a steam-pressure regulator. Steam normally is "wet", i.e., it contains droplets of water that could interfere with the operation of the regulator. A steam separator installed before the regulator, removes condensate from the steam. Also, a strainer placed before the separator prevents dirt from depositing in the separator and regulator. Pipe unions are located at convenient positions so that both the steam separator and regulator can be easily removed for repairs or replacement. If uninterrupted operation of the process is required, a throttling valve is installed in the bypass line with two on-off valves before and after the regulator and steam separator. Thus, the steam can be

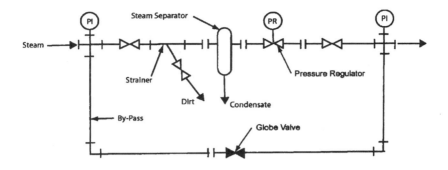

Figure 8.8 Installation of a steam-pressure regulator. From Ref. 3.

regulated manually by using the throttling valve while replacing the steam separator or regulator.

Steam traps are also self-regulating valves that are required after all steam heat exchangers and in long pipelines where steam can condense. Steam traps maintain steam pressure in the heat exchanger and discharge water and noncondensable gases, such as air. If the water and gases accumulate in the heat exchanger, heat transfer will be reduced. Reference 22 discusses several types of steam traps. We will only discuss two types, which are shown in Figure 8.9. The first type is a balanced-pressure thermostatic trap, which contains a bellows filled with a liquid that evaporates when heated and condenses when cooled. As cool condensate and air flows toward the trap, the vapor in the bellows condenses, the bellows contracts, and the valve opens. Then, steam pushes the mixture of air and water out of the trap. When steam reaches the trap, the liquid in the bellows evaporates, the bellows expands, and the valve closes. Condensate and air again accumulate in the trap and the cycle repeats.

A second type is the thermodynamic trap. When there is condensate and air in the trap, the disc shown in Figure 8.9 is in the raised position and steam will push the mixture out of the trap. After all the condensate and air leave the trap, the steam flows under the disc at a high velocity because of the constricted passage. The kinetic energy of the steam increases, and according to Bernoulli's equation the pressure must decrease. The pressure on top of the disc is now greater than below the disc and the disc drops on the seat, closing the trap. When condensate and air again accumulates in the trap, the cycle repeats.

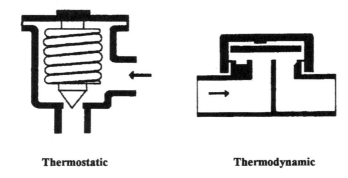

Thermostatic   Thermodynamic

Figure 8.9 Examples of steam traps. From Ref. 22.

PRELIMINARY DESIGN OF A FLOW SYSTEM

After the function of fittings and valves and the principles of the automatic control loop are understood, then you can make a preliminary design of the flow system. An example is the flow system for a continuous stirred-tank reactor shown in Figure 8.10. After designing and locating the reactor and feed tank, we can then design the flow system. This design entails evaluating and selecting fittings, valves, pumps, and instrumentation. The final design requires sizing these components. In Figure 8.10, two reactants are continuously pumped into the reactor, but we will only consider one feed system.

Starting at the feed tank, first install a flanged joint at the outlet so that we can easily disconnect the piping from the tank. Then, connect a tee to the flanged connection. One branch of the tee leads to a shut-off valve for emptying the tank, and the other branch leads to the pump. Flanged connections and shut-off valves are placed before and after the pump so that it can easily be removed from the system for repairs or replacement. Pumps can fail. Consequently, it is good practice to install pressure gages before and after the pump. Pressure gages, which are designated as PI for pressure indicator, are placed before and after a pump to help the operator to troubleshoot. Also, it is common practice to have a spare pump in case the operating pump fails. To control the flow rate of reactants to the reactor, set the required flow rate on a flow-indicator-controller (FIC). A flow meter measures the flow rate, and the controller corrects for any deviation from the required flow rate by automatically opening or closing the control valve. Because a control valve is a mechanical device, it could fail. Therefore, you want to keep the

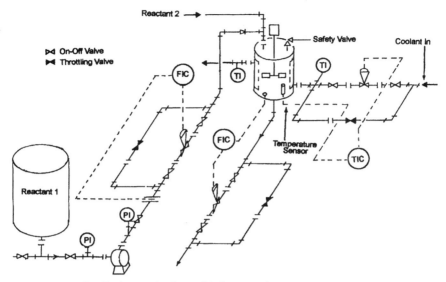

Reactant 2 →

On-Off Valve
Throttling Valve

Safety Valve   Coolant In

FIC

TI

TI

FIC

Reactant 1

Temperature
Sensor

TIC

PI

PI

Figure 8.10 Preliminary design of a flow system.

system in operation while the control valve is repaired or replaced. To accomplish this, install a bypass line around the control valve with flanged connections and shut-off valves before and after the control valve. With this arrangement, the flow rate can be controlled manually with a manual throttling valve in the bypass line. For threaded piping, we must have a union in the bypass line.

In most chemical reactors, temperature is a critical variable that must be controlled. Cooling water circulates in the reactor jacket, removing the enthalpy of reaction. To control the reaction temperature, the cooling-water flow rate to the jacket is controlled. Set the desired temperature on the temperature-indicator-controller (TIC), which is measured by a temperature sensor installed in the reactor. The control valve automatically corrects any deviations from the desired temperature by adjusting the cooling-water flow rate into the jacket.

To prevent flooding or emptying of the reactor, requires a liquid-level controller (LC). In this case, the pressure exerted by the liquid in the reactor measures the liquid level. The operation and installation of the liquid-level control valve is the same as the flow and temperature control valves.

If the reaction is exothermic, there is a possibility that the reaction may run away, creating excessive pressures in the reactor. Because the reaction rate varies exponentially with temperature, the effect can be very rapid, and a safety valve prevents an excessive pressure increase. As soon as the pressure in the reactor reaches a preset value, the safety valve opens, dumping the reactor contents into a holding tank. Also, control valves can be designed to fail wide open if the air supply fails so that the cooling-water flow rate is a maximum to prevent the reactor from overheating.

MATERIALS OF CONSTRUCTION

Selecting materials of construction is an important aspect of designing flow systems. The process engineer, more than any other engineer, must handle corrosive as well as dangerous fluids. We will not discuss corrosion here. The interested reader can refer to Fontana and Greene [4] for further details.

The designer, in order to increase the reliability of his design, should critically examine all parts of his flow system to determine what parts contact the fluid. This is particularly true of pumps and valves where critical parts may be overlooked, for example, seals. The designer should also be aware that some organic solvents attack polymeric materials, such as rubber and plastics. Thus, in addition to selecting metals to avoid corrosion, the designer checks the compatibility of polymeric materials with solvents. Erosion of piping and fittings by the process fluid must also be considered. Solids suspended in fluids may cause excessive wear of piping, pumps, and valves. Even for a pure liquid, as the velocity approaches 10 ft/s (3.05 m/s) [31], erosion will occur. Corrosion data for a given fluid may be obtained from Craig and Anderson [5] or by consulting equipment manufacturers. The Chemical Engineering Handbook [1] also contains some data on corrosion.

MACROSCOPIC MECHANICAL ENERGY BALANCE

The most important relationship in designing flow systems is the macroscopic mechanical-energy balance, or Bernoulli's equation. Not only is it required for calculating the pump work, but it is also used to derive formulas for sizing valves and flow meters. Bird, et al. [6] derived this equation by integrating the microscopic mechanical-energy balance over the volume of the system. The balance is given by

$$\frac{\Delta(v^2/\alpha)}{2\,g_C} + \frac{g}{g_C}\Delta z + \int_1^2 \frac{dp}{\rho} + W + E = 0 \tag{8.1}$$

The units of each term are $ft\text{-}lb_F/lb_M$, where pound force is lb_F, and pound mass is lb_M. The conversion factor, g_C, equals $32.2\ lb_M\text{-}ft/s^2\text{-}lb_F$. In the first term, the kinetic energy term, the factor α corrects for the velocity profile across the pipe. For laminar flow in a pipe the velocity profile is parabolic and $\alpha = 1/2$. If the velocity profile is flat, $\alpha = 1$. For very rough pipes and turbulent flow α may reach a value of 0.77 [7]. Unless the kinetic energy term in the mechanical energy balance becomes large compared to the other terms, it suffices to let $\alpha = 1$ for turbulent flow, which occurs in many engineering applications.

The second term in the mechanical energy balance is the change in potential energy. The third term is "pressure work," and its evaluation depends on whether the fluid is compressible or incompressible. The last two terms are the work done by the system, W, and the friction loss, E. For an incompressible fluid, the density may be removed from the integral sign. Then, Equation 8.1 becomes

$$\frac{\Delta(v^2/\alpha)}{2\,g_C} + \frac{g}{g_C}\,\Delta z + \frac{\Delta p}{\rho} + W + E = 0 \qquad (8.2)$$

VALVE SIZING

Valve size is not necessarily the same size as the pipe to which it will be connected. It is frequently less. The valve orifice size and the shape of the valve plug, shown in Figure 8.11, determine the valve size. To size a valve, the flow rate

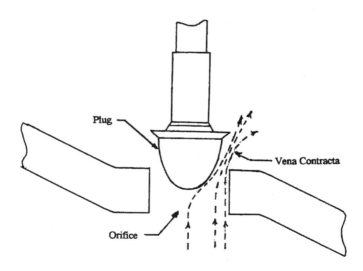

Figure 8.11 Throttling valve plug and orifice.

and the required pressure drop across the valve must be determined. The formula required for valve sizing depends on the properties of the fluid and the flow regime. These factors are:

1. liquid or gas flow
2. laminar or turbulent flow
3. flashing
4. cavitation
5. incompressible or compressible flow
6. choked flow
7. non-ideal gas effects
8. effects of piping arrangement
9. limit on outlet velocity to prevent shock waves and noise

We will only consider turbulent flow of an incompressible fluid, which also includes the flow of gases – if the pressure drop is small – as well as the flow of liquids. Formulas for other cases are discussed in References 8, 9, and 20. Reference 20 summarizes valve-sizing formulas in an attempt to standardize them.

Figure 8.12 shows the various pressure drops through a throttling valve. When the fluid enters the valve, there is a small drop in pressure cause by frictional losses. As the fluid passes through the small opening of the valve, the fluid

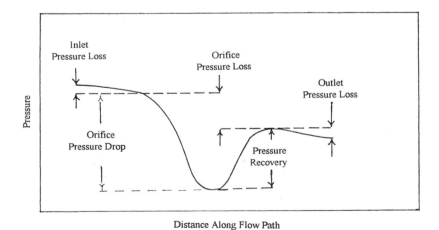

Figure 8.12 Pressure profile across a throttling valve.

velocity rapidly increases. Simultaneously, the pressure drops rapidly, as illustrated in Figure 8.12. A further increase in velocity and decrease in pressure may occur because of the formation of a vena contracta which is the contraction of the jet flowing from the orifice, as illustrated in Figure 8.11. For liquids, if the pressure reaches the vapor pressure of the liquid, vaporization will occur. After the vena contracta, the pressure increases and the fluid velocity decreases, because of an increase in the cross-sectional area of the valve. The pressure at the outlet of the valve will not reach its value at the inlet because there is a pressure loss caused by friction. If the pressure rise is rapid, any vapor bubbles formed in the valve will collapse instantaneously releasing large amounts of energy in a small area, which may be sufficient to dent the metal. This phenomena is called cavitation, i.e., cavitation is the formation of vapor bubbles followed by their sudden collapse. Dissolved gases will also cause cavitation, such as air dissolved in water.

The problem that we must consider next is to relate the pressure drop across the valve to flow rate and valve size. After applying Bernoulli's equation, Equation 8.2, across the valve we obtain, for an incompressible fluid,

$$E = \frac{(p_1 - p_2)}{\rho} \tag{8.3}$$

because the change in kinetic energy and potential energy is small, and the work done is zero.

The friction loss term, E, is given by the empirical expression

$$E = K \frac{v_O^2}{2 g_c} \tag{8.4}$$

where, K, an experimentally determined factor, is the friction-loss factor for a valve.

Combining Equations 8.3 and 8.4 to eliminate E and solving for, the fluid velocity in the valve orifice, v_O, we find that

$$v_O = \left(\frac{2 g_c (p_1 - p_2)}{K \rho} \right)^{1/2} \tag{8.5}$$

Multiply Equation 8.5 by A_O to obtain the volumetric flow rate through the valve, and let $\rho = \rho_w \eta$, where η is the specific gravity of the fluid. If a valve coefficient, C_V, is defined by

$$C_V = 7.48 \, (12) \, (60) \, A_O \left(\frac{2 g_c}{K \rho_w} \right)^{1/2} \tag{8.6}$$

and because $Q = A_O \, v_O$, the volumetric flow rate,

$$Q = C_V \left(\frac{(p_1 - p_2)}{\eta} \right)^{1/2} \tag{8.7}$$

which is a formula used by valve manufacturers to size valves for incompressible fluids. Because the valve coefficient, C_V, contains the orifice area, C_V is not a constant for any particular valve but varies with the position of the valve stem and hence the valve plug. The units used for C_V by valve manufacturers are gal/min for flow rate and lb_F/in^2 for pressure. Thus, C_V has units of gal-in/min-$lb_F^{1/2}$

Sizing valves requires calculating C_V for the design flow rate and then selecting an appropriate valve from a manufacturer. The valve coefficient contained in manufacturers' catalog is the maximum coefficient. If the valve were sized at the normal operating flow rate, the system would then be out of control if an upset should occur. To avoid this, Chalfin [9] recommends sizing a control valve for a flow rate that is 30% greater than the normal operating flow rate.

The designer, to insure good process control, specifies the pressure drop across the valve. At low-pressure drops, the valve characteristic curve is distorted resulting in poor control. Boger [10] and Moore [11] discuss this effect. The valve characteristic curve is a plot of the valve opening against flow rate. There are several rules of thumb in the engineering literature for assigning the pressure drop across a control valve. Sandler and Lukiewicz [29] recommend a pressure drop of 30 to 50% of the frictional pressure drop – also called the dynamic pressure drop – in the system and a minimum of 5 to 10 psi (0.345 to 0.67 bar). Forman [12] states the assigned pressure drop is not an arbitrary value like 5 psi (0.345 bar). He recommends a pressure drop of 33% of the frictional pressure drop for a linear valve and 50% for an equal-percentage valve [12]. For a valve that has a linear characteristic curve, the flow rate varies linearly with valve opening. For a valve that has an equal-percentage characteristic curve, the flow rate varies non-linearly with valve opening. Power consumption increases with increasing frictional pressure drop. Thus, the assigned pressure drop should not be any larger than necessary for adequate control. Example 8.1 illustrates the procedure for valve sizing.

Example 8.1 Valve Sizing

What size valve will be required to control the flow rate of 50 gal/min (0.169 m^3/min) of brine ($\eta = 1.2$), if the frictional pressure drop in the system, excluding the valve, is 15 psi (1.03 bar)? Assume a linear valve.

Q (design) = 1.3 (50) = 65 gal/min (0.246 m^3/min).

For adequate process control, the pressure drop across the valve for a linear valve is

$$\frac{(\Delta p)_V}{\Delta H' + (\Delta p)_V} = 0.33$$

$(\Delta p)_V = (0.33 / 0.67) (15) = 7.388$ psi (0.510 bar)

Substituting into Equation 8.7, the valve size is

$$C_V = 65 \left(\frac{1.2}{7.388} \right)^{1/2} = 26.20$$

or, after rounding, $C_V = 26$. Now, a valve can be selected from a manufacturer's catalog.

PIPE SIZING

Pipe sizing consists of determining the diameter, length, and wall thickness.

Pipe Length

Determining pipe length for a flow system is a simple problem. After locating all equipment, the length of pipe is automatically determined. Piping is nearly always connected from one process unit to another by making ninety-degree turns. Occasionally, a forty-five degree turn is needed.

Pipe Diameter

Threaded piping is available in 12 in (30.5 cm) or smaller, but is usually used in sizes 2 in (5.08 cm) and smaller because fabrication costs increase rapidly above 2 in (5.08 cm) [1]. Threaded piping is used mostly for utilities and welded piping for process piping [31].The inside diameter of a pipe could be calculated by optimizing pumping and piping costs. As the inside diameter of the pipe increases, the liquid velocity decreases, and the cost of pumping decreases. This occurs because the frictional pressure loss decreases with a decrease in liquid velocity. On the other hand, as the pipe diameter increases, its weight increases, and the installed cost of the piping increases. As illustrated in Figure 8.13, the pipe diameter selected is at the total minimum cost. For most purposes, such as rough or preliminary designs, and for small installations, this calculation is not necessary. Rules-of-thumb are sufficient. Ludwig [13] lists velocities for several liquids and pipe materials. From this list, a velocity of 6 ft/s (1.83 m/s) for the discharge side of the pump seems to be a reasonable average value. After specifying the volumetric flow rate of the liquid and selecting a liquid velocity, calculate the inside diameter of the pipe. Piping is only available in standard diameters, which does not exactly correspond to either the inside or outside diameter of a pipe, as shown in Table 8.2A, where the pipe dimensions are in inches. In Table 8.2B the dimensions are in millimeters. Other technical factors may change the suggested velocity of 6 ft/s (1.83 m/s). For example, if a liquid contains suspended particles, the liquid velocity must be reduced to prevent erosion. For clear fluids, expect erosion above 10 ft/s (3.05 m/s) [31]. On the other hand, to prevent suspended particles from settling or deposits from forming on the pipe wall requires increasing the liquid velocity. A liquid velocity is selected to balance these opposing factors as determined by experience.

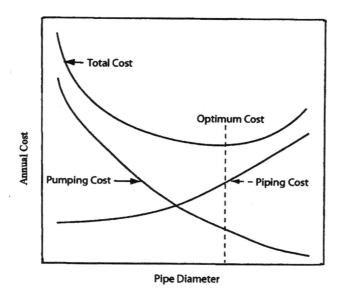

Figure 8.13 Optimum pipe diameter.

Table 8.2A Dimensions of American Standard Pipe in Inches – Schedule 40 Pipe

Nominal Pipe Size	Inside Diameter	Outside Diameter	Threads[a] per in	Bottom Diameter of Thread[c]	Threaded Length	Thread Engaged[b]	Pipe-Tap Drill Size
1/8	0.269	0.405	27	0.334	0.412	5/16	21/64
1/4	0.364	0.540	18	0.433	0.625	7/16	27/64
3/8	0.493	0.675	18	0.568	0.630	7/16	9/16
1/2	0.622	0.840	14	0.701	0.819	9/16	11/16
3/4	0.824	1.050	14	0.911	0.913	9/16	29/32
1	1.049	1.315	11-1/2	1.144	1.03	11/16	1-1/8
1-1/4	1.380	1.660	11-1/2	1.488	1.06	11/16	1-15/16
1-1/2	1.610	1.900	11-1/2	1.727	1.07	11/16	1-23/32
2	2.067	2.375	11-1/2	2.199	1.10	3/4	2-3/16
2 ½	2.469	2.875	8	2.619	1.64	1-1/16	2-9/16
3	3.068	3.500	8	3.241	1.70	1-1/8	3-3/16
3 ½	3.548	4.000	8	3.738	1.75	1-3/16	3-11/16
4	4.026	4.500	8	4.234	1.80	1-3/16	4-3/16
5	5.047	5.563	8	5.290	1.91	1-5/16	5-1/4
6	6.065	6.625	8	6.346	2.01	1-3/8	6-5/16

a) Threads are tapered.
b) Length of threaded section screwed into fitting, which is determined by experience.
c) At the end of the pipe.

Source: Adapted from Ref. 14.

Table 8.2B Dimensions of American Standard Pipe in Millimeters – Schedule 40 Pipe

Nominal Pipe Size in	Inside Diameter mm	Outside Diameter mm	Threads[a] per mm	Bottom Diameter of Thread[c]	Threaded Length mm	Thread Engaged[b] mm	Pipe-Tap Drill Size in
1/8	6.83	10.3	1.06	8.48	10.5	7.94	21/64
1/4	9.25	13.7	0.709	11.0	15.9	11.11	27/64
3/8	12.5	17.1	0.709	14.4	16.0	11.11	9/16
1/2	15.8	21.3	0.551	17.8	20.8	14.3	11/16
3/4	20.9	26.7	0.551	23.1	23.2	14.3	29/32
1	26.6	33.4	0.453	29.0	26.2	17.5	1-1/8
1 ¼	35.1	42.2	0.453	37.8	26.9	17.5	1-15/16
1 ½	40.9	48.3	0.453	43.9	27.2	17.5	1-23/32
2	52.5	60.3	0.453	55.9	27.9	19.1	2-3/16
2 ½	62.7	73.0	0.315	66.5	41.7	27.0	2-9/16
3	77.9	88.9	0.315	82.3	43.2	28.6	3-3/16
3 ½	90.1	102	0.315	94.9	44.5	30.2	3-11/16
4	102	114	0.315	108	45.7	30.2	4-3/16
5	127	141	0.315	134	48.5	33.3	5-1/4
6	154	168	0.315	161	51.1	34.9	6-5/16

a) Threads are tapered.
b) Length of threaded section screwed into a fitting, which is determined by experience.
c) At the end of the pipe. mm

Pipe Wall Thickness

The pipe wall thickness and hence its strength is determined by the schedule number. The schedule number is defined by

Schedule Number = 1000 (p/S)   (8.8)

where p is the internal pressure in lb_F/in^2 gage and S is the allowable stress in lb_F/in^2 for the pipe material. Tables 8.2A and 8.2.B lists the pipe wall thickness for Schedule 40 pipe up to six-inch pipe sizes. The Chemical Engineering Handbook [1] contains dimensions for other pipe sizes and schedule numbers. Frank [31] recommends using Schedule 40 for carbon steel pipe and Schedule 10 for carbon steel alloys at moderate pressures. The allowable pressure should be checked using the ASME (American Institute of Mechanical Engineers) code.

FLOW METERING

Figure 8.14 shows some commonly used flow meters. Dolenc [23] reviews these flow-meter types in addition to other types. The meters in Figure 8.14 are divided into two classes: the variable-head meters, which are the orifice, venturi,

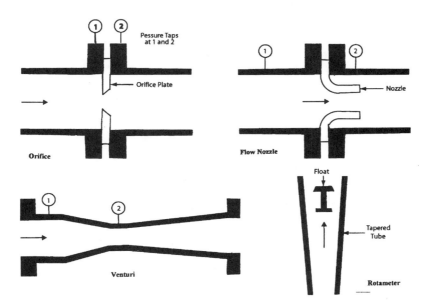

Figure 8.14 Examples of flow meters.

flow nozzle, and the variable-area meter, which is the rotameter. Head is equivalent to pressure. It is the height that the flowing liquid must be elevated to give the required pressure. For the variable-head meter, the flow rate is obtained by measuring the pressure drop across the meter, which varies with the flow rate. For the rotameter, the position of the float determines the flow rate.

Variable-Head Meters

To size a variable-head meter, we must calculate the orifice, venturi throat or nozzle diameter. Using Bernoulli's equation we can derive a relationship between the flow rate, the pressure drop across the meter, and the orifice diameter.

Because the change in elevation and the work done is zero, Equation 8.2 becomes

$$\frac{v_2^2}{2\,\alpha_2\,g_C} - \frac{v_1^2}{2\,\alpha_1\,g_C} + \frac{p_2 - p_1}{\rho} + E = 0 \tag{8..9}$$

The friction loss term, E, can be related to the downstream velocity, v_2, by

$$E = K\,\frac{v_2^2}{2\,g_C} \tag{8.10}$$

where K, the friction loss factor, is experimentally determined.

From the conservation of mass for an incompressible fluid flowing through the orifice we find that

$$v_1 = v_2 = v \tag{8.11}$$

Substituting Equations 8.10 and 8.11 into Equation 8.9 and solving for the fluid velocity in the pipe, we find that

$$v = \left(\frac{2\,g_C\,(p_1 - p_2)\,/\,\rho}{1\,/\,\alpha_2 - 1\,/\,\alpha_1 + K}\right)^{1/2} \tag{8.12}$$

Bird et al. [6] showed that $\alpha_1 \approx 1$ and $1/\alpha_2 \approx (A_0/A)^2$ for an orifice meter. Substitute these values into Equation 8.12. Then, multiply each side of Equation 8.12 by the cross-sectional area of the pipe to obtain the volumetric flow rate. Also, for frictionless flow, K = 0. Thus, Equation 8.12 becomes

$$Q = A_O \left(\frac{2 \, g_C \, (p_1 - p_2) / \rho}{[1 - (A_O / A)^2]} \right)^{1/2} \qquad (8.13)$$

This formula is the same for frictionless flow through the venturi and nozzle meters.

To account for friction and the approximate values of α used, multiplied Equation 8.13 by a discharge coefficient, C_D.

$$Q = C_D A_O \left(\frac{2 \, g_C \, (p_1 - p_2) / \rho}{[1 - (A_O / A)^2]} \right)^{1/2} \qquad (8.14)$$

The discharge coefficient is a function of the meter type and Reynolds number.

Using the orifice meter as an example, Example 8.2 illustrates the sizing procedure. Calculating the orifice diameter requires assigning the pressure drop across the orifice.

Example 8.2 Orifice-Meter Sizing

Size an orifice meter to meter 70 gal/min (0.265 m³/min) of acetone at 15 °C. The pipe size is a two-inch Schedule 40 pipe. The viscosity of acetone is 0.337 cp $(3.37 \times 10^{-4}$ Pa-s), and its specific gravity is 0.792.

To size an orifice meter requires calculating the orifice diameter from Equation 8.14. After dividing and multiplying Equation 8.14 by A, substituting A= π $D^2/4$, and letting $\beta = D_O / D$, where D_O is the orifice diameter and D the inside pipe diameter, we obtain

$$Q = C_D \frac{\pi D^2}{4} \beta^2 \left[\frac{2 \, g_C \, (p_1 - p_2) / \rho}{(1 - \beta^4)} \right]^{1/2}$$

Because $C_D = f$ (Re), first calculate the Reynolds number in the pipe. From Table 8.2A, the inside diameter of a Schedule 40, two-inch pipe is 2.067 in (5.25 cm).

$$v = \frac{4 \, Q}{\pi \, D^2} = \frac{4}{\pi} \frac{70.0 \text{ gal/min}}{7.48 \text{ gal/ft}^3} \frac{1}{60 \text{ s/min}} \frac{1}{(2.067/12)^2 \text{ ft}^2} = 6.692 \text{ ft/s (2.04 m/s)}$$

$\mu = 0.337$ cp $(6.72 \times 10^{-4} \text{ lb}_M/\text{ft-s-cp}) = 2.265 \times 10^{-4} \text{ lb}_M/\text{ft-s} (3.37 \times 10^{-4} \text{ Pa-s})$

$\rho = 0.792 \, (62.4) = 49.42 \text{ lb}_M/\text{ft}^3$

$$Re = \frac{\rho\, D\, v}{\mu} = \frac{49.42}{2.265 \times 10^{-4}} \frac{lb_M/ft^3}{lb_M/ft\text{-}s} \frac{2.067 \text{ in}}{12 \text{ in/ft}} \frac{6.692 \text{ ft/s}}{1} = 2.52 \times 10^5$$

$$Q = \frac{70 \text{ gal/min}}{60 \text{ s/min}} \frac{1}{7.48 \text{ gal/ft}^3} = 0.1560 \text{ ft}^3/\text{s} \ (4.42 \text{ m}^3/\text{s})$$

Select 50 in (127 cm) of water as the pressure drop across the orifice. The pressure drop in force per unit area is related to the pressure drop in terms of a liquid height by

$$p_1 - p_2 = \frac{g}{g_C} \rho_w\, \Delta z = \frac{32.17}{32.17} \frac{ft/s^2}{lb_M\, ft/s^2\text{-}lb_F}\ 62.4 \frac{lb_M}{ft^3} \frac{50 \text{ in}}{12 \text{ in/ft}}$$

$$p_1 - p_2 = 260.0 \text{ lb}_F/ft^2 \ (12.45 \text{ kPa})$$

After substituting the values of Q, D, $p_1 - p_2$, ρ, and g_C into the first equation above, we obtain

$$0.156 = C_D\, (\pi/4)\, (2.0667/12)^2\, \beta^2 \left(\frac{2\,(32.17)\,(260.0)}{0.792\,(62.4)\,(1 - \beta^4)} \right)^{1/2}$$

Considine [16] gives equations for the orifice coefficient, C_D, for several ways of measuring pressure drop across the orifice. For corner pressure taps, shown in Figure 8.2.1, the equation is

$$C_D = 0.5959 + 0.0312\, \beta^{2.1} - 0.184\, \beta^{8.0} + 91.71 \beta^{2.5}/Re^{0.75}$$

Solving these two equations simultaneously for C_D and β using Polymath [27], the orifice coefficient, $C_D = 0.6035$ and β = 0.6690. Thus, the orifice diameter,

$$D_O = 0.6690\,(2.067) = 1.383 \text{ in} \ (3.51 \text{ cm}).$$

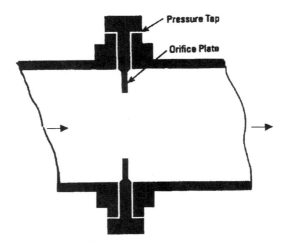

Figure 8.2.1 An orifice meter with corner pressure taps. Adapted from Ref. 26 with permission.

Variable-Area Meters

To size a rotameter requires calculating the volumetric flow rate of a standard fluid at standard conditions. Most manufacturers calibrate rotameters using a stainless-steel float and water at a standard temperature for liquids and air at a standard temperature and pressure for gases. For other fluids, float materials, and operating conditions, the flow rate must be converted to an equivalent flow rate of water or air. To derive a formula for making this conversion, Bernoulli's equation is applied across the float shown in Figure 8.15 to give Equation 8.9.

Because a rotameter tube is tapered, the annular flow area varies with position of the float, as shown in Figure 8.15. The conservation of mass for an incompressible fluid becomes

$$A_1 v_1 = A_2 v_2 = A_0 v_0 \tag{8.15}$$

where the subscript 1 refers to the entrance of the rotameter, 2 the exit of the meter, and o to the annular area between the float and tube.

The friction loss term,

$$E = \frac{K v_0^2}{2 g_C} \tag{8.16}$$

After substituting v_1, v_2, and E from Equations 8.15 and 8.16 into Equation 8.9 and solving for the fluid velocity in the annular area surrounding the float, we find that

$$v_O = \left[\frac{2 g_C (p_1 - p_2) / \rho}{[(A_O/ A_2)^2 / \alpha_2] - [(A_O / A_1)^2 / \alpha_1]} \right]^{1/2} \qquad (8.17)$$

By multiplying Equation 8.17 by the annular area, A_O, the volumetric flow rate,

$$Q = A_O \left[\frac{2 g_C (p_1 - p_2) / \rho}{[(A_O/ A_2)^2 / \alpha_2] - [(A_O / A_1)^2 / \alpha_1]} \right]^{1/2} \qquad (8.18)$$

For any flow rate, the float is kept at a stationary position in the fluid by the drag and buoyant forces acting upwards and the gravitational force acting downward. The force balance is

$$F_G = F_D - F_B \qquad (8.19)$$

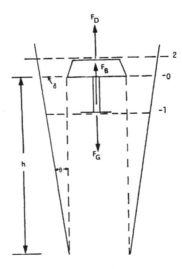

Figure 8.15 Geometry of a rotameter tube and float.

The drag force across the float is defined as equal to the product of the drag coefficient, C_D, the pressure drop across the float, $p_1 - p_2$, and a characteristic area for the float, A_F. After substituting this definition and expressions for the gravitational and buoyant forces into Equation 8.19, the force balance becomes

$$C_D (p_1 - p_2) A_F = (V_F \rho_F - V_F \rho) (g/g_C) \tag{8.20}$$

where the subscript F refers to the float.

Next, substitute $(p_1 - p_2)$ from Equation 8.20 into Equation 8.18. Then, the volumetric flow rate,

$$Q = A_O C_R \left(\frac{V_F}{A_F} \frac{\rho_F - \rho}{\rho} \right)^{1/2} \tag{8.21}$$

where the rotameter coefficient, C_R, is defined by

$$C_R = \left[\frac{2 g/C_D}{[(A_O/A_2)^2/\alpha_2] - [(A_O/A_1)^2/\alpha_1] + K} \right]^{1/2} \tag{8.22}$$

As Figure 8.15 shows, the annular flow area between the tube and float is

$$A_O = \frac{\pi (D_F + 2\delta)^2}{4} - \frac{\pi D_F^2}{4} \tag{8.23}$$

Expanding Equation 8.23 and dropping the term that contains δ^2, which is small, then $A_O = \pi \delta D_F$. From the geometry of the meter, $\delta = h \tan \theta$, as Figure 8.15 shows. Substituting these relations into Equation 8.21, the volumetric flow rate becomes

$$Q = \pi C_R D_F h (\tan \theta) \left(\frac{V_F}{A_F} \frac{\rho_F - \rho}{\rho} \right)^{1/2} \tag{8.24}$$

If C_R does not vary with float position, which is usually the case for float diameters of one-half inch or greater, then the volumetric flow rate is directly proportional to h.

To size a rotameter, we must convert the flow rate to an equivalent flow rate of water or air. The flow rate of the metered fluid is given by Equation 8.24. For the same meter at the same float position, the flow rate of the standard fluid is given by

$$Q_S = \pi C_{RS} D_F h (\tan \theta) \left(\frac{V_{FS}}{A_F} \frac{\rho_F - \rho_S}{\rho_S} \right)^{1/2} \tag{8.25}$$

where the subscript, s, refers to the standard fluid, water or air.

By dividing Equation 8.25 by Equation 8.24, the flow rate of the standard fluid, at standard conditions, in terms of the flow rate of the actual fluid, at actual conditions, is given by

$$Q_S = Q \left(\frac{\rho_{FS} - \rho_S}{\rho_F - \rho} \frac{\rho}{\rho_S} \right)^{1/2} \tag{8.26}$$

provided the rotameter coefficient is independent of the fluid being metered, i.e., $C_R \approx C_{RS}$. Equation 8.26 may be used for either liquid or gases, but for gases it may be simplified because $\rho_S \ll \rho_F$ and $\rho \ll \rho_F$. Thus, Equation 8.26 reduces to

$$Q_S = Q \frac{\rho_{FS}}{\rho_F} \frac{\rho}{\rho_S} \tag{8.27}$$

If the ideal gas law is obeyed, then $\rho = M P / R T$. Substituting this equation into Equation 8.27, the flow rate at standard conditions,

$$Q_S = Q \frac{\rho_{FS}}{\rho_F} \left(\frac{P}{P_S} \frac{M}{M_S} \frac{T_S}{T} \right)^{1/2} \tag{8.28}$$

where M is the molecular weight of the gas.

Thus, sizing rotameters requires using either Equation 8.26 or Equation 8.28 to calculate the flow rate of the standard fluid. Then use Table 8.3, supplied by a manufacturer, to select a rotameter. The procedure for sizing a rotameter is illustrated in Example 8.3.

Example 8.3 Rotameter Sizing

Find the rotameter size required to meter 1.5 gal/min (5.68×10^{-3} m^3/min) of carbon tetrachloride at 20 °C (68 °F).

To select a rotameter from Table 8.3 first calculate an equivalent flow rate of water from Equation 8.26. In Table 8.3 stainless steel floats are used. The density of stainless steel is 8.02 g/cc (501 lb/ft^3) and the density of carbon tetrachloride is 1.60 g/cc (99.9 lb/ft^3). After substituting numerical values into Equation 8.26, the volumetric flow rate of water,

$$Q_S = 1.50 \left[\frac{(8.02 - 1.00)}{(8.02 - 1.60)} \frac{1.60}{1.00} \right]^{1/2} = 1.984 \text{ gal/min } (7.51 \times 10^{-3} \text{ m}^3/\text{min})$$

Therefore, from Table 8.3 select a ½ inch rotameter having a maximum flow rate of 2.44 gal/min (9.24 l/min). This rotameter size is somewhat larger than needed, allowing for a safety factor.

Table 8.3 Commercial Rotameter Sizes

Tube Size	Maximum Flow gpm H₂O Equiv.	scfm air Equiv.	Tube Number	Float Number (316 sst)	Total Δ P (See Note 1)	V.I.C. (See Note 2)	psia Critical (See Note 3)
1/2"	0.267	1.10	FP-1/2-17-G-10	1/2-GUSVT-40A	1.2	2.9	5.5
	0.328	1.35	FP-1/2-21-G-10	1/2-GUSVT-40A	1.4	2.9	3.5
	0.442	1.82	FP-1/2-27-G-10	1/2-GUSVT-40A	2.0	2.9	2.7
	0.480	1.92	FP-1/2-17-G-10	1/2-GSVT-45A	3.5	5.1	17.9
	0.600	2.47	FP-1/2-21-G-10	1/2-GSVT-45A	4.6	5.1	11.5
	0.619	2.55	FP-1/2-35-G-10	1/2-GUSVT-40A	3.1	2.9	2.0
	0.670	2.76	FP-1/2-17-G-10	1/2-GSVT-44A	6.4	7.1	33.4
	0.690	2.85	FP-1/2-17-G-10	1/2-GSVT-48A	7.3	7.6	39.0
	0.810	3.35	FP-1/2-27-G-10	1/2-GSVT-45A	6.8	5.1	8.4
	0.830	3.42	FP-1/2-21-G-10	1/2-GSVT-44A	7.7	7.1	33.8
	0.880	3.62	FP-1/2-21-G-10	1/2-GSVT-48A	8.0	7.6	24.6
	0.885	3.65	FP-1/2-17-G-10	1/2-GNSVT-48A	8.2	1.1	19.8
	1.10	4.52	FP-1/2-21-G-10	1/2-GNSVT-48A	9.9	1.1	20.0
	1.12	4.60	FP-1/2-27-G-10	1/2-GSVT-44A	12.3	7.1	16.2
	1.15	4.74	FP-1/2-35-G-10	1/2-GSVT-45A	8.2	5.1	8.5
	1.19	4.90	FP-1/2-27-G-10	1/2-GSVT-48A	13.7	7.6	18.6
	1.44	5.93	FP-1/2-27-G-10	1/2-GNSVT-48A	15.8	1.1	16.5
	1.56	6.43	FP-1/2-35-G-10	1/2-GSVT-44A	14.8	7.1	16.5
	1.66	6.85	FP-1/2-35-G-10	1/2-GSVT-48A	17.2	7.6	18.8
	2.00⁴	8.24⁴	FP-1/2-50-G-9	1/2-GSVT-45A	12.0	5.1	4.0
	2.76⁴	11.4⁴	FP-1/2-50-G-9	1/2-GSVT-44A	31.0	7.1	7.7
	2.90⁴	12.0⁴	FP-1/2-50-G-9	1/2-GSVT-48A	35.2	7.6	8.9
	3.52⁴	14.5⁴	FP-1/2-50-G-9	1/2-GNSVT-48A	52.0	1.1	8.8
3/4"	1.96	8.08	FP-3/4-21-G-10	3/4-GSVGT-54A	5.3	10.4	13.9
	2.49	10.2	FP-3/4-21-G-10	3/4-GNSVGT-54A	6.8	1.6	13.9
	2.66	11.0	FP-3/4-27-G-10	3/4-GSVGT-59A	7.0	14.1	28.7
	2.70	11.1	FP-3/4-27-G-10	3/4-GSVGT-54A	7.7	10.4	9.6
	3.37	13.9	FP-3/4-21-G-10	3/4-GNSVGT-59A	11.5	2.1	26.3
	3.55	14.6	FP-3/4-27-G-10	3/4-GNSVGT-54A	11.5	1.6	9.6
	3.67	15.1	FP-3/4-27-G-10	3/4-GSVGT-59A	13.7	14.1	19.8
	4.80	19.8	FP-3/4-27-G-10	3/4-GNSVGT-59A	20.5	2.1	19.8
1"	4.25	17.5	FP-1-27-G-10	1-GSVGT-64A	12.9	14.8	11.5
	4.82	19.9	FP-1-27-G-10	1-GSVGT-68A	18.7	16.9	15.6
	5.63	23.2	FP-1-27-G-10	1-GNSVGT-64A	20.7	2.2	11.3
	6.00	24.7	FP-1-35-G-10	1-GSVGT-64A	24.6	14.8	6.8
	6.46	26.6	FP-1-27-G-10	1-GNSVGT-68A	32.5	2.5	15.6
	6.80	28.0	FP-1-35-G-10	1-GSVGT-68A	37.0	16.9	8.9
	7.62	31.4	FP-1-27-G-10	1-GNSVGT-69A	75.0	1.5	22.2
	7.84	32.4	FP-1-35-G-10	1-GNSVGT-64A	37.7	2.2	6.8
	9.00	37.0	FP-1-35-G-10	1-GNSVGT-68A	62.8	2.5	8.9
	9.50	39.2	FP-1-35-G-10	1-GSVGT-69A	65.3	8.5	13.4
	11.0	45.3	FP-1-35-G-10	1-GNSVGT-69A	112	1.5	13.4
1½"	13.2	54.4	FP-1½-27-G-10	1½-GSVGT-87A	9.5	27.6	15.4
	14.6	60.0	FP-1½-27-G-10	1½-GSVGT-86A	13.5	31.0	22.0
	17.6	72.0	FP-1½-27-G-10	1½-GNSVGT-87A	12.8	4.20	15.4
	18.6	76.5	FP-1½-27-G-10	1½-GNSVGT-86A	15.2	4.80	22.0
2"	24.0	99.0	FP-2-27-G-10	2-GSVGT-97A	24.0	26.5	16.4
	30.6	126.0	FP-2-27-G-10	2-GNSVGT-97A	32.0	3.0	16.4
	31.6	130.0	FP-2-27-G-10	2-GSVGT-98A	34.0	18.5	21.2
	36.1	149.0	FP-2-27-G-10	2-GNSVGT-98A	45.0	3.30	21.2

NOTES:
1. Pressure drop is total pressure loss across the meter at 100% flow rate in inches of water column.
2. Meter is unaffected by viscosity when the value of cps/$\sqrt{\rho}$ (using operating density in g/cc and viscosity in centipoises) is less than V.I.C. (viscosity immunity ceiling). V.I.C. is applicable to liquids only; all gas flows fall below Viscosity Immunity Ceiling.
3. Meters not recommended for gas service where pressure is below minimum shown. For such applications use low pressure drop capacity table. A flow throttling valve close coupled to meter outlet is recommended for all gas applications.
4. Not available with metal scale. Specify percent scale or direct reading scale on tube.

Source:Ref.. 17.

PUMP SIZING AND SELECTION

Chapter 5 considered pump types and their evaluation and selection. After selecting a pump type, the next step is to size the pump. This requires calculating the
flow rate and the pressure rise across the pump or the pump head. The net positive
suction head (NPSH), is also important, particularly for centrifugal pumps. NPSH
is the difference between the total pressure and the vapor pressure of the fluid at
the pump inlet. NPSH will be discussed later.

Pump Head

Apply Bernoulli's equation over the whole flow system to develop an expression
for the pump head. After rearranging Equation 8.2, to obtain the suction and discharge heads we find that

$$-\frac{g_C}{g}W = \frac{v_2^2}{2g} + z_2 + \frac{g_C\,p_2}{g\,\rho} + \frac{g_C}{g}E_D - \left(\frac{v_1^2}{2g} + z_1 + \frac{g_C\,p_1}{g\,\rho} - \frac{g_C}{g}E_S\right) + \frac{g_C}{g}E_P$$

(8.29)

where the subscript 1 refers to the suction end of the flow system, 2 to the discharge end of the flow system, S the suction line, and D the discharge line. The
units of each term in Equation 8.29 are in feet of liquid, called head. Engineers call
the first term to the right of the equal sign in Equation 8.29, velocity head; the
second term, elevation head; the third term, pressure head; and the fourth term,
friction head. The frictional loses consists of three terms. These are: the friction
losses in the discharge piping, $H_{FD,}$ the friction losses in the suction piping, H_{FS},
and the friction losses in the pump, E_P. The friction head in the pump is accounted
for in the pump efficiency. Therefore, from Equation 8.29,

$$-\frac{g_C}{g}(W + E_P) = (H_D - H_S)$$

(8.30)

where H_D and H_S is the sum of the velocity, elevation, pressure, and friction heads.
The difference between the discharge and suction heads is sometimes called the
total dynamic head.

Because the work done on the system is negative, $W = -W_P$, where W_P is the
pump work. Substituting W_P into Equation 8.30, we find that

$$\frac{g_C}{g}(W_P - E_P) = (H_D - H_S)$$

(8.31)

The pump efficiency, η_P, is defined by

$$\eta_P = \frac{W_P - E_P}{W_P} \tag{8.32}$$

Substitute Equation 8.32 into Equation 8.31, to obtain

$$\frac{g_C}{g} \eta_P W_P = (H_D - H_S) \tag{8.33}$$

Next, divide the suction and discharge friction heads into two parts. One part consists of the piping losses, and the other part consists of fittings losses. The friction-head loses,

$$H_F = \sum_i \left(4f \frac{L}{R_H} \frac{v^2}{2g} \right)_i + \sum_j \left(K \frac{v^2}{2g} \right)_j \tag{8.34}$$

where the subscript i refers to various pipe diameters and j to various fittings.

The hydraulic radius, R_H, defined as the cross-sectional area for flow divided by the wetted perimeter, is discussed by Bird et al. [6]. For flow in circular conduits, the hydraulic radius equals D/4.

The friction factor, f, depends on the Reynolds number and the relative roughness, ε/D. Table 8.4 contains roughness factors, ε, for several pipe materials. Surface roughness is very irregular and non-uniform. Thus, ε for any pipe material is an average value. Figure 8.16 is a plot of the friction factor as a function of Reynolds number with the relative roughness as a parameter.

For pipe fittings and other resistances, we can calculate the frictional losses using the friction loss factor, K, in Equation 8.34. Figures 8.17 to 8.20 and Table 8.5 contains factors for several fittings, flow meters, and valves. Sometimes, friction losses of fittings are accounted for by using equivalent lengths of straight piping. The equivalent length is that length of straight piping that will give the same frictional pressure loss as the fitting [18]. In this case, the equivalent lengths of piping are added to the straight lengths of piping and are substituted into the first term of Equation 8.34. Frictional losses for new piping can be predicted to roughly ± 25% for fittings and ± 10% for piping [19].

After calculating the head, then calculate the power supplied to pump by the shaft of the pump driver, i.e., the brake horsepower, which is given by

$$P_p = \frac{m W_P}{550} = \frac{m g (H_D - H_S)}{550 g_C \eta_P} \tag{8.35}$$

Cavitation

If the pressure in a flowing liquid falls below its vapor pressure, the liquid will vaporize. If vapor bubbles form on the suction side of the pump, the bubbles will move with the stream and will subsequently collapse in a region of high pressure. This phenomenon is called cavitation. Dissolved gases in the fluid, such as air in water, could also form bubbles. The collapsing vapor or gas bubbles subject the pump surfaces to tremendous shock. The energy involved in the shock is explosive enough to flake off small bits of metal and in time the pump will become pitted. Cavitation also results in a loss of energy. Immediate clues of cavitation are reduced flow rate, loss of head, pumping in spurts, and excessive noise and vibration.

Table 8.4 Pipe- Roughness Factors

Pipe Material	Roughness Factor, ε 1×10^{-6} ft
Riveted Steel[a]	3,000 - 30,000
Concrete[a]	1,000 - 10,000
Wood Stave[a]	600 - 30,000
Cast Iron[a]	850
Galvanized Iron[a]	500
Asphalted Cast Iron[a]	400
Steel or Wrought Iron[a]	150
Tubing[a]	5
Hard Plastic[b]	0.17 - 0.83
Glass[b]	0.17 - 0.83
Electropolished Stainless[b]	0.17 - 0.83
Mechanically-polished Stainless[b]	0.33 - 1.3
New-unpolished Stainless[b]	1.3 - 8.3
New Copper or Brass[b]	1.3 - 8.3
Rubber[b]	2.7 - 10
Seamless Carbon Steel[b]	10 - 42
Corrugated Steel[b]	>170
Tuberculated Iron Pipe[b]	42 - 170

a) Source: Reference 1
b) Source: Reference 24

Cavitation will not occur as long as the pressure at the suction side of the pump is sufficiently high. The suction pressure required to avoid cavitation depends on the pump design and is specified by the pump manufacturer. The term manufacturers use to describe the pressure required is NPSH (net-positive-suction-head). NPSH is defined as the difference between the pressure head and the head corresponding to the liquid vapor pressure at the pump inlet, i.e.,

$$NPSH = H_{Pi} - H_{VPi} \qquad (8.36)$$

The required NPSH, $(NPSH)_R$, is specified by the pump manufacturer, and the available NPSH, $(NPSH)_A$, is determined by the design of the pump suction piping. To prevent cavitation the available NPSH must be equal to or greater than the required NPSH.

$$(NPSH)_A \geq (NPSH)_R \qquad (8.37)$$

The vapor-pressure head of the liquid in Equation 8.36 is calculated by converting the vapor pressure into head in feet. The pressure head at the pump inlet is calculated by applying Bernoulli's Equation between the surface of the liquid at

Table 8.5 Friction-Loss Factors for Flow Meters

Meter Type	Friction-Loss Factor[a], K
Disk	7.0
Piston	15.0
Rotary (star-shaped Disk)	10.0
Turbine	6.0
	Pressure Drop[b]
Orifice	50 in H_2O
Flow Totalizer	4.0 psi
Rotameter	3.0 psi
Flow Tube	psi

a) Source: Reference 1
b) Source: Reference 25

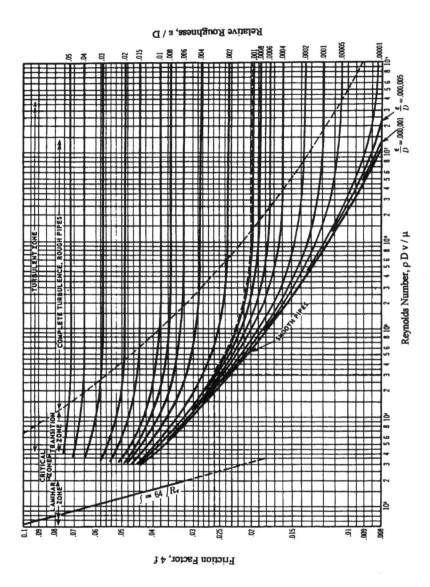

Figure 8.16 Friction-factor chart. From Ref. 18 with permission.

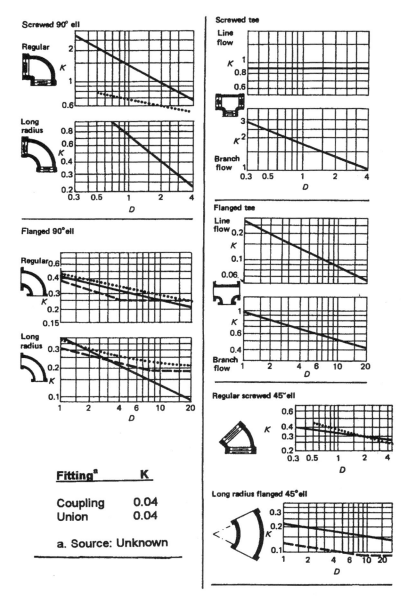

Figure 8.17 Friction-loss factors for pipe fittings. From Ref. 19 with permission.

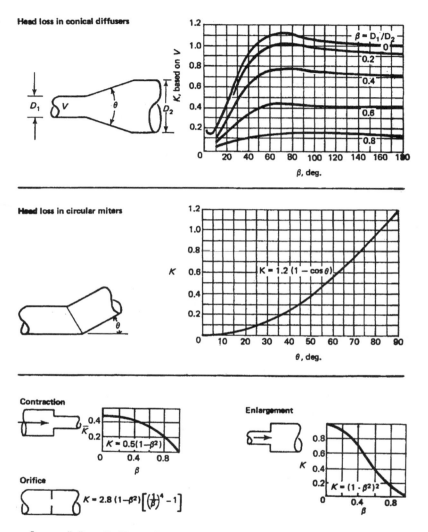

Figure 8.18 Friction-loss factors for pipe transitions. From Ref. 19 with permission.

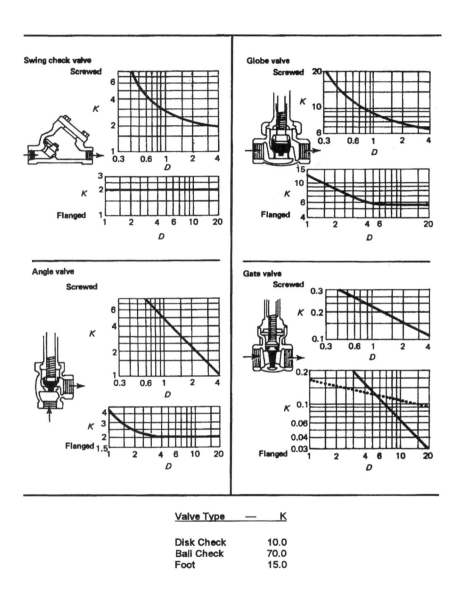

Valve Type	—	K
Disk Check		10.0
Ball Check		70.0
Foot		15.0

Figure 8.19 Friction-loss factors for valves. From Ref. 19 with permission.

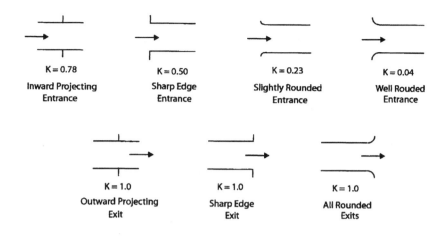

Figure 8.20 Friction-loss factors for tank entrances and exits. From Ref. 18.

the inlet to the flow system, point 1, and the pump inlet, point i. Solving for pressure head at the pump inlet,

$$H_{Pi} = H_{P1} - (H_{Vi} - H_{V1}) - (H_{Zi} - H_{Z1}) - H_{FS} \tag{8.38}$$

First, drop the velocity head term in Equation 8.38 because it is small. Then, substitute the pressure head, H_{Pi}, from Equation 8.38 into Equation 8.36 to obtain an equation for calculating $(NPSH)_A$.

$$(NPSH)_A = H_{P1} + H_{Z1} - H_{Zi} - H_{FS} - H_{VPi} \tag{8.39}$$

The risk of cavitation is great when the $(NPSH)_A$ is small. To keep $(NPSH)_A$ large, as Equation 8.39 shows, the inlet pressure to the system, H_{P1}, should be as large as possible. Also, keep the liquid level above the centerline of the pump, $H_{Z1} - H_{Zi}$, the friction losses low, H_{FS}, and the vapor-pressure head, H_{VPi}, low, by keeping the inlet temperature low.

Centrifugal Pump Selection

To select a centrifugal pump size we must examine pump characteristic curves, which are plots of head versus flow rate. Reference 8.21 discusses the factors influencing the selection of a centrifugal pump. The characteristic curves in Figure 8.21 are given for impeller sizes ranging from 7 to 9 ½ in (17.8 to 24.1 cm). The curves intersect the ordinate and gradually curve downward as the flow rate increases. Also, the characteristic curves intersect the efficiency curves at several flow rates. The intersection of a characteristic curve with the horsepower curves (dashed lines) gives the brake horsepower at several flow rates. Finally, the lower curve is the required NPSH for the pumps. The best operating point is the point where the efficiency is a maximum. Thus, for the 9 ½ in (24.1 cm) impeller in Figure 8.21, the maximum efficiency is 84 % at a head of 72 ft (21.9 m).

When selecting a pump, get as close to the maximum-efficiency point as possible. For example, if the flow system requires a pump to deliver 1250 gpm (4.73 m³/min) at 52 ft (15.8 m) of head, the pump with a 9 ½ in (24.1 cm) impeller – shown in Figure 8.21 – will deliver 58 ft (17.7 m) of head at 1250 gpm (4.73 m³/min) with an efficiency of slightly less than 80%. You will, however, be operating too far to the right of the maximum-efficiency point, near the end of the characteristic curve, where the pump efficiency is low and the required NPSH high. Also, the pump will be noisy, and there is little flexibility if you need to increase the flow rate. On the other hand, by selecting an operating point too far to the left of the maximum efficiency point, the load on the bearings and seals will be large, reducing their life. Operating slightly to the left of the maximum efficiency point, where the efficiency is still high, is recommended [21]. Thus, no pump in Figure 8.21 is suitable. If the flow rate, however, is 900 gpm (3.41 m³/min) and the head 55 ft (15.2 m), the point will be located between the 8 and 8 ½ in (20.3 and 21.6 cm) diameter impellers in Figure 8.21. Then, select the pump with the 8 ½ in (21.6 cm) impeller diameter.

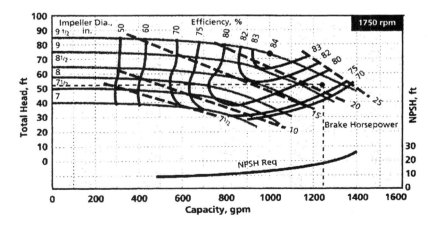

Figure 8.21 Characteristic curves for centrifugal pumps. From Ref. 28.

Example 8.4 Centrifugal-Pump Sizing

In the flow system in Figure 8.4.1, the centrifugal pump is delivering water at 100 gal/min (0.379 m³/min and 70 °F (21.1 °C) into a boiler operating at a pressure of 35 psig (2.41 barg). The water levels in the feed tank and boiler are constant. The approximate liquid velocity is 3 ft/s (0.914 m/s) in the suction side of the pump and 6 ft/s (1.83 m/s) in the discharge side of the pump. Because corrosion is expected to be negligible, carbon steel pipe will be used. Assume a frictional pressure drop of 5 psi across the heat exchanger. Find the suction and discharge pipe sizes, the head the pump must deliver, the brake horsepower, the electric-motor horsepower, and the control-valve size, i.e., the valve coefficient. Assume a linear valve. Also, select an impeller size from the characteristic curves shown in Figure 8.4.2, and determine if cavitation will occur in the pump.

Data
At 70 °F (21.1 °C):

Density of water	62.3 lb$_M$/ft³ (998 kg/m³)
Vapor pressure	0.3631 psia (0.0250 bara)
Viscosity of water	0.982 cp (9.82x10⁻⁴ Pa-s)

First, define the flow system, i.e., show the entrance and exit points. These are points one and two in Figure 8.2.1. These points are selected because the pressures

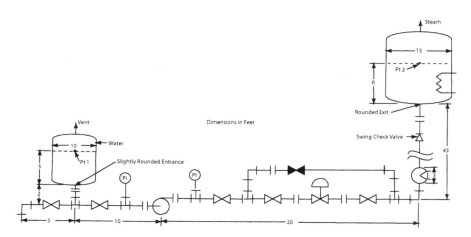

Figure 8.4.1 Boiler flow system.

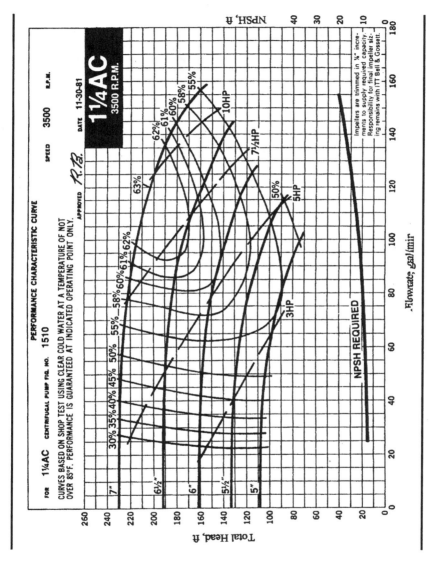

Figure 8.4.2 Characteristic curves for centrifugal pumps. (Source Ref. 28 with permission).

are known. The pump will supply the work to transfer the water from the liquid surface in the feed tank to the liquid surface in the boiler. From Equation 8.33, calculate the pump work as the difference of the pump discharge and suction heads. Both heads are the sum of the velocity, elevation, pressure, and friction heads. Divide the calculation into two parts. First, calculate the suction head, and then calculate the discharge head.

The pump head,

$$H_D - H_S = \frac{v_2^2}{2\,\alpha_2\,g} + z_2 + \frac{g_C\,p_2}{g\,\rho} + \frac{g_C}{g}\,E_D - \left(\frac{v_1^2}{2\,\alpha_2\,g} + z_1 + \frac{g_C\,p_1}{g\,\rho} - \frac{g_C}{g}\,E_S \right)$$

The friction pressure losses are calculated by summing up the pipe and fittings losses.

$$H_F = \frac{g_C}{g}\,E = \sum_i \left(4\,f\,\frac{L}{D}\,\frac{v^2}{2\,g} \right)_i + \sum_j \left(K\,\frac{v^2}{2\,g} \right)_j$$

SUCTION SIDE OF THE PUMP

Pipe Sizing

First, calculate a preliminary cross-sectional area for the inside of the pipe, which is

$$A = \frac{Q}{v} = 100\,\frac{gal}{min}\,\frac{1}{7.481}\,\frac{ft^3}{gal}\,\frac{1\,min}{60\,s}\,\frac{1\,s}{3\,ft} = 0.07426\,ft^2$$

The calculated pipe diameter (3.69) does not correspond to any standard pipe size, as shown in Table 8.2A. We could select either a 3 ½ or 4 in nominal pipe size. To keep the (NPSH)$_A$ as high as possible, select a 4 in pipe. If we select 3 ½ in pipe, (NPSH)$_A$ would be reduced because of an increase in the frictional pressure drop. Because the pipe size is greater than two inches, select welded piping. We will then use flanges to connect piping to valves and other equipment.

For the 4 inch pipe in Table 8.2A,

D = 4.026/12 = 0.3355 ft (0.102 m), and

A= (3.142/4) (0.3355)^2 = 0.08842 ft^2 (8.21x10^{-3} m^2)

Thus, the actual liquid velocity,

$$v = 100 \; \frac{gal}{min} \; \frac{1 \; ft^3}{7.481 \; gal} \; \frac{1 \; min}{60 \;\; s} \; \frac{1}{0.08842 \; ft^2} = 2.520 \; ft/s \; (0.768 \; m/s)$$

Velocity Head

The velocity head is usually negligible. This is shown in the following calculation.

$$A = \pi \, D^2/4 = (3.142 / 4) \, (10.0)^2 = 78.55 \; ft^2$$

The velocity in the feed tank is given by

$$v = 100 \; \frac{gal}{min} \; \frac{1 \; ft^3}{7.481 \; gal} \; \frac{1 \; min}{60 \;\; s} \; \frac{1}{78.55 \; ft^2} = 2.836 \times 10^{-3} \; ft/s \; (8.64 \times 10^{-4} \; m/s)$$

If the flow in the feed tank is laminar, then $\alpha = \frac{1}{2}$, but if the flow is turbulent, then $\alpha \approx 1$. Next, calculate the Reynolds number.

$$Re = \frac{\rho \, v \, D}{\mu}$$

$$\mu = 0.982 \; cp \; (6.72 \times 10^{-4}) \; \frac{lb_M}{ft\text{-}s\text{-}cp} = 6.60 \times 10^{-4} \; \frac{lb_M}{ft\text{-}s} \; (9.82 \times 10^{-4} \; Pa\text{-}s)$$

$$Re = \frac{\dfrac{62.3 \; lb_M}{1 \; ft^3} \; \dfrac{2.836 \times 10^{-3} \; ft}{1 \;\;\; s} \; \dfrac{10.0 \; ft}{1}}{6.60 \times 10^{-4} \; lb_M/ft\text{-}s} = 267.7$$

Because the Reynolds number is less than 2100, the flow in the feed tank is laminar, and $\alpha \approx \frac{1}{2}$.

The velocity head in the feed tank is

$$H_V = \frac{\alpha \, v^2}{2 \, g} = \frac{(2.836 \times 10^{-3})^2 \; ft^2/s^2}{2 \, (2) \, (32.2) \;\;\; ft/s^2} = 6.244 \times 10^{-8} \; ft \; (1.90 \times 10^{-8} \; m)$$

which is negligible. The velocity head can usually be neglected at the outset in both the suction and discharge sides of the pump.

Pressure Head

The pressure head is given by

$$H_P = \frac{g_C \, p}{g \, \rho} = \frac{\dfrac{32.2 \; lb_M\text{-ft}}{1 \; lb_F\text{-}s^2} \; \dfrac{14.7 \; lb_F}{1 \; in^2} \; \dfrac{144 \; in^2}{1 \; ft^2}}{\dfrac{32.2 \; ft}{1 \; s^2} \; \dfrac{62.3 \; lb_M}{1 \; ft^3}} = 34.0 \; ft \; (10.3 \; m)$$

This is a useful number to remember. It means that atmospheric pressure can support a column of water 34 ft (10.3 m) high.

Friction Head

Obtain the friction factor from Figure 8.16 after calculating the Reynolds number and the relative roughness, ε/D, for the pipe. From Table 8.4, the roughness factor for steel pipe, $\varepsilon = 1.5 \times 10^{-4}$ ft $(4.57 \times 10^{-5}$ m$)$.

$$\frac{\varepsilon}{D} = \frac{1.5 \times 10^{-4} \; ft}{0.3357 \; ft} = 0.4468 \times 10^{-3}$$

$$Re = \frac{\rho \, v \, D}{\mu} = \frac{62.3 \, (2.52) \, (0.3355)}{6.60 \times 10^{-4}} = 7.981 \times 10^4$$

From Figure 8.16, $4f = 0.021$.

The computations for the friction head, H_{FS}, in the suction side of the pump are completed in the Table 8.4.1.

Table 8.4.1 Summary of Flow-System Design Computations

	Suction	Discharge
Flow Rate (gal/min)[a]	100	100
Density (lb/ft^3)[b]	62.3	62.3
Pipe ID (ft)[c]	0.336	0.256
Pipe Length (ft)[c]	12.0	60.0
Velocity (ft/s)[d]	2.84	4.34
Viscosity (lb$_M$/ft-s)[d]	6.60×10^{-4}	6.60×10^{-4}
Reynolds Number	8.0×10^4	1.05×10^5
Relative Roughness	5.89×10^{-4}	5.89×10^{-4}
Friction Factor, 4f	0.0205	0.021

	Suction (4 in pipe)			Discharge (3 in pipe)		
Fitting	No.	K	Total K	No.	K	Total K
Gate Valves	1	0.16	0.16	3	0.20	0.60
Check Valves				1	2.00	2.00
Tees, Branch	1	0.67	0.67			
Tees, Line	1	0.13	0.13	3	0.16	0.48
90° Elbows				1	0.33	0.33
Flanges[e]	2	0.04	0.08	4	0.04	0.16
Entrance	1	0.23	0.23			
Exit				1	1.00	1.0
Total (ΣK)			1.27			4.57

	Suction	Discharge
$\Sigma(4\,f\,L/D)$	0.732	4.92
$1.1[\Sigma(4\,f\,L/D)]$	8.05	5.41
$1.25\,\Sigma K$	1.59	5.71
$v^2/2g$ (ft)[d]	0.125	0.292
H_F' (ft)[d]	0.299	3.25[f]
H_Z (ft)[d]	7.0	49.0
H_P (ft)[d]	34.0	115.0

a) To convert to m^3/min multiply by 3.785×10^{-3}.
b) To convert to kg/m^3 multiply by 16.019.
c) To convert to m multiply by 0.3048.
d) To convert to m/s multiply by 0.3048.
e) No data are available. Use a union to approximate a flanged connection.
f) Frictional losses do not include the control valve and heat exchanger.

Net Positive Suction Head (NPSH)

To prevent cavitation in the pump $(NPSH)_A$ must be greater than $(NPSH)_R$. The design of the suction piping and not the discharge piping determines $(NPSH)_A$. The pump manufacturer specifies the $(NPSH)_R$.

$$(NPSH)_A = H_{Zi} + H_{Pi} - H_{FS} - H_{VPi}$$

where the subscript i refers to the inlet of the pump.

Table 8.6 contains values for the first three terms. Next, calculate the vapor-pressure head

$$H_{VPi} = \frac{g_C \, p_V}{g \, \rho} = \frac{32.2 \, (0.363)}{32.2 \, (62.3)} \quad 144 = 0.8393 \text{ ft} (0.256 \text{ m})$$

$$(NPSH)_A = 7.00 + 34.0 - 0.290 - 0.8393 = 39.9 \text{ ft} (12.2 \text{ m})$$

From Figure 8.4.2, $(NPSH)_R = 22$ ft (6.71 m) at 100 gal/min (0.379 m³/min). Because $(NPSH)_A > (NPSH)_R$ the pump will not cavitate.

DISCHARGE SIDE OF THE PUMP

Pipe Sizing

Calculate a preliminary area.

$$A = \frac{Q}{v} = 100 \; \frac{\text{gal}}{\text{min}} \; \frac{1}{7.481} \; \frac{\text{ft}^3}{\text{gal}} \; \frac{1}{60} \; \frac{\text{min}}{\text{s}} \; \frac{1 \text{ s}}{6 \text{ ft}} = 0.03713 \text{ ft}^2 (34.5 \text{ cm}^2)$$

$$D = [\,(4 / 3.142)\,(0.03713)\,]^{1/2} = 0.2174 \text{ ft} (2.609 \text{ in}, 6.63 \text{ cm})$$

From Table 8.2A select a 3 in pipe. Therefore, D = 3.068 in (0.2557 ft, 7.79 cm) and A = 0.05136 ft² (47.7 cm²). Because the pipe size is greater than 2 in (5.08 cm), the discharge piping is also welded, requiring flanged connections to equipment.

Now, the actual water velocity,

$$v = \frac{100}{7.481} \; \frac{1}{60.0} \; \frac{1}{0.05136} = 4.338 \text{ ft/sec} (1.32 \text{ m/s})$$

Velocity Head

We showed that the velocity head in the feed tank is negligible. The velocity head in the boiler will also be negligible. We will not repeat the calculation for the boiler.

Pressure Head

$$H_{P2} = \frac{g_C\, P_2}{g\, \rho} = \frac{32.2}{32.2} \frac{49.7}{62.3} \; 144 = 114.9 \text{ ft } (35.0 \text{ m})$$

Friction Head

$$\frac{\varepsilon}{D} = \frac{1.5 \times 10^{-4}}{0.2557} = 5.886 \times 10^{-4}$$

$$Re = \frac{62.3\,(4.338)\,(0.2557)}{6.60 \times 10^{-4}} = 1.047 \times 10^{5}$$

From Figure 8.16, $4f = 0.205$. The computations for friction head on the discharge side of the pump, H_{FD}, are completed in Table 8.4.1.

Valve Pressure Drop

So far, the frictional head does not include the frictional pressure drop across the control valve. To insure good process control, the designer specifies the pressure drop across the valve. The pressure drop should be about 33% of the frictional pressure drop for a linear valve.

$$\frac{(\Delta H)_V}{(\Delta H)_F' + (\Delta H)_V} = 0.33$$

where $(\Delta H)_F'$ is the frictional pressure drop in the flow system, excluding the frictional pressure drop across the control valve. Now, convert the 5 psi drop across the heat exchanger to head.

$$(\Delta H)_E = \frac{g_C \, \Delta p}{g \, \rho} = \frac{32.2 \, (144) \, (5)}{32.2 \, (62.3)} = 11.56 \text{ ft } (\, 3.52 \text{ m})$$

$$(\Delta H)'_F = 3.25 + 11.56 + 0.299 = 15.11 \text{ ft } (\, 4.61 \text{ m})$$

$$(\Delta H)_V = (0.33 \, / \, 0.67) \, (15.11) = 7.442 \text{ ft } (\, 2.26 \text{ m})$$

Pump Size

Next, calculate the total head.

$$\Delta H = 49 + 115.0 + 3.25 + 11.56 + 7.442 - (7 + 34.0 - 0.29) = 145.5 \text{ ft } (44.3 \text{ m})$$

The point at 100 gal/min (0.379 m^3/min) and 145.5 ft (44.3 m) in Figure 8.4.2 is slightly above the characteristic curve for the 6 in (1.52 cm) impeller diameter. Select the pump with a 6 ½ in (16.5 cm) impeller diameter. Pump casings can be fitted with several impeller diameters. Thus, in the future the pump size can be expanded from the 6 ½ in (16.5 cm) to the 7 in (17.8 cm) impeller diameter.

Valve Size

Now, calculate the actual pressure drop across the control valve. From Figure 8.4.2 and at 100 gal/min for the 6 ½ in impeller, the total available head is 173 ft (52.7 m). Therefore,

$$\Delta H = 49 + 115.0 + 3.25 + 11.56 + (\Delta H)_V - (7 + 34.0 - 0.29) = 173.0 \text{ (52.7 m)}$$

$$(\Delta H)_V = 34.90 \text{ ft } (10.6 \text{ m})$$

$$(\Delta p)_V = \frac{g \, \rho \, (\Delta H)_V}{g_C} = \frac{32.2 \, (62.3) \, (34.90)}{32.2 \, (144)} = 15.10 \text{ lb}_F/\text{in}^2 \text{ (1.04 bar)}$$

Size the valve for a flow rate greater than the anticipated operating flow rate to allow for some flexibility.

Q (design) = 1.3 (100) = 130 gal/min (0.492 m^3/min)

For incompressible flow, the valve size (C_V) is given by

$$C_V = Q \, [\, \eta \, / \, (\Delta p)_V \,]^{1/2} = 130 \, (1.0 \, / \, 15.10)^{1/2} = 33.45 \text{ gal/min-(psi)}^{1/2}$$

Round off C_V to 33.

Table 8.4.2 Summary of Flow-System Design

Item	Specification	Specification
Pipe Size		
Discharge	3 in	7.62 cm
Suction	4 in	10.2 cm
Pipe Connections	Flanged	
Pipe Material	Carbon Steel	
Pipe Construction	Welded	
Schedule No.	40	
Flow Rate	100 gpm	0.379 m³/min
Total Head	173 ft	52.7 m
Valve Size (C_V)	33	
Motor Power	10 hp	7.46 kW
(NPSH)$_R$	22 ft	6.71 m
(NPSH)$_A$	39.9 ft	12.2 m
Impeller Diameter	6 ½ in	15.2 cm

Pump Power

The mass flow rate, m, in Equation 8.35 equals ρ Q.

$$m = \frac{62.3\ \text{lb}_M}{1\ \text{ft}^3}\ \frac{100\ \text{gal}}{1\ \text{min}}\ \frac{1\ \text{ft}^3}{7.48\ \text{gal}} = 832.8\ \text{lb}_M/\text{min} \ (378\ \text{kg/min})$$

Use Equation 8.35 to calculate the shaft or brake horsepower. The pump efficiency, taken from Figure 8.4, is 62%.

$$P_P = \frac{\dfrac{832.8\ \text{lb}_M}{1}\ \dfrac{1\ \text{min}}{\text{min}\ 60\ \text{s}}\ \dfrac{32.2\ \text{ft}}{1\ \text{s}^2}\ \dfrac{173\ \text{ft}}{1}}{\dfrac{550\ \text{ft-lb}_F}{1\ \text{s-hp}}\ \dfrac{32.2\ \text{ft-lb}_M}{1\ \text{s}^2\text{-lb}_F}\ \dfrac{0.62}{1}} = 7.042\ \text{hp}\ (5.25\ \text{kW})$$

which is the pump or brake horsepower. The pump horsepower is also plotted in Figure 8.4.2.

Now, calculate the electric-motor horsepower. For an electric motor, the efficiency is about 88%. Therefore, the motor horsepower is 8.002 hp (5.97 kW). The next standard-size electric motor is 10 hp (7.46 kW), which results in a safety factor of 25.0% .

NOMENCLATURE

English

A	area
C_D	discharge coefficient or drag coefficent
C_R	rotameter coefficient
C_V	valve coefficient
D	diameter
E	friction loss
F	friction factor
F	force
G	acceleration of gravity
g_C	conversion factor
h	height
H	head
K	friction loss factor
L	length
$(NPSH)_A$	available net-positive-suction-head
$(NPSH)_R$	required net-positive-suction-head
m	mass flow rate
M	molecular weight
p	pressure
P	power
Q	volumetric flow rate

Re	Reynolds number
R_H	hydraulic radius
S	allowable stress
T	absolute temperature
v	average velocity
V	volume
W	work
z	elevation

Greek

α	kinetic energy correction factor
β	ratio of the orifice to pipe diameter
δ	annular thickness
ε	roughness
η	specific gravity or efficiency
θ	rotameter-tube taper
μ	viscosity
ρ	density

Subscripts

B	brake or buoyant
D	discharge side of the pump or drag
F	float or friction
G	gravity
h	height
i	pump inlet
O	orifice
p	pressure

P	Pump
R	rotameter
Re	Reynolds group
S	suction side of the pump or standard
v	vapor or velocity
vp	vapor pressure
w	water
z	elevation

REFERENCES

1. Perry, R.H., and Green, D.V., eds., Chemical Engineers Handbook, 7th ed., McGraw-Hill Book Co., New York, NY, 1973.
2. Ciancia, J., Valves in the Chemical Process Industries, Chem. Eng., $\underline{72}$, 18, 97, 1965.
3. Brochure, Pressure Regulators, Reducing – Pilot Operated, 11 Series, Masoneilan, Houston, TX, 1994.
4. Fontana, M.G., Greene, N.D., Corrosion Engineering, 3rd ed., McGraw-Hill Book Co., New York, NY, 1986
5. Craig, B.D., Anderson, D.S., eds., Handbook of Corrosion Data,2nd ed., ASM International, Metals Park, OH, 1989.
6. Bird, R.B., Stewart, W.E. Lightfoot, E.N., Transport Pheonmena, John Wiley & Sons, New York, NY, 1960.
7. Booth, R.G., Epstein, N., Kinetic Energy and Momentum Factors for Rough Pipes, Can. J. of Chem. Eng., $\underline{47}$, 10, 515, 1969.
8. Driskell, L.R., Practical Guide to Control Valve Sizing, Instr.Techn., $\underline{14}$, p.47, 1967.
9. Chalfin, S., Specifying Control Valves, Chem. Eng., $\underline{81,}$ 21, 105, 1974.
10. Boger, H.W., Flow Characteristics for Control Valve Installations, ISA Jour., $\underline{13}$, 11, 50, 1966.
11. Moore, R.L., Flow Characteristics of Valves, I.S.A. Handbook of Control Valves, J.W. Hutchison, Ed., Instrument Society of America, Pittsburgh, PA, 1971.
12. Forman, E.R., Fundamentals of Process Control, Part 2, Chem. Eng., $\underline{72}$, 13, 127, 1965.
13. Ludwig, E.E., Fluid Flow Fundamentals, Chem. Eng., $\underline{67}$, 12, 122, 1960.
14. Masek, J.A., Metallic Piping, Chem. Eng., $\underline{67}$, 13, 215, 1968.
15. Maintenance Instructions, Control Valve Series 20/25 000, Kammer Valves Inc., Pittsburgh, PA, 2002.

16. Considine, D.M., ed., Process Instruments and Control Handbook, 3rd ed., McGraw-Hill, New York, NY, 1985.
17. Specification Sheet (l0A3500), Extruded Body Indicator Flowrator Meters, Fischer & Porter Inc., Warminster, PA, 1982.
18. Anonymous, Flow of Fluids Through Valves, Fittings and Pipe, Technical Paper No. 410, Crane Co., New York, NY, 1982.
19. Simpson, L.L., Sizing Piping for Process Plants, Chem. Eng., 75, 13, 192, 1968.
20. Anonymous, Flow Equations for Sizing Control Valves, ISA-S75.01, Instrument Society of America, Research Triangle Park, NC, 1995.
21. Anonymous, How to Select and Size the Right Centrifugal Pumps, Tech Talk, 12, 2, 1, ITT Fluid Handling, Morton Grove, IL, 1997.
22. Anonymous, Steam Trapping Guide, Technical Bulletin No. T507, Sarco Company, Allentown, PA, 1967.
23. Dolenc, J. W., Choose the Right Flow Meter, Chem. Eng. Progr., 92, 1, 22, 1996.
24. Tverberg, J.C., Effect of Surface Roughness on Fluid Friction, Flow Control, 1, 8, 11, 1995.
25. Grossel, S., Personal Communication, Hofmann-LaRoche, Clifton, NJ, Oct. 19, 1983.
26. Khandelwal, P.K., Gupta, V., Make the Most of Orifice Meters, Chem. Eng. Prog., 89, 5, 32, 1993.
27. Shacham, M., Cutlip, M. B., Polymath, Version 4.0, CACHE Corp., Austin, TX, 1996.
28. Base Mounted Centrifugal Pump Performance Curves, Curve Booklet B-260E, Bell & Gossett ITT, Fluid Handling Division, Morton Grove, IL, 1987.
29. Sandler, H.J., Lukiewicz, E.T., Practical Process Engineer, McGraw-Hill, New York, NY, 1987.
30. Kern, R., Useful Properties of Fluids, Chem. Eng., 81, 27, 1974.
31. Frank, O., Personal Communication, Consulting Engineer, Convent Station, NJ, Jan. 2002.
32. Merrick, R.C., A Guide to Selecting Manual Valves, Chem. Eng., 93, 17, 52, 1986.

Appendix: SI Units and Conversion Factors

Reproduced with permission of the American Institute of Chemical Engineers. Copyright 1977 AIChE. All rights reserved.

AIChE Goes Metric

Beginning in 1979, the International System of Units (SI) will be used in all Institute publications, meeting papers, and course texts.

J. Y. Oldshue, Mixing Equipment Co., Inc., Rochester, N.Y.

Schedules for AIChE entering into metric conversion using SI were determined by the AIChE Council at their March, 1977, meeting in Houston, Tex., based on recommendations from the Metrication Committee. The key point is that every paper submitted for presentation in an AIChE meeting, or submitted for publication in an AIChE journal, or any new course text submitted for presentation at an AIChE-sponsored course after January 1, 1979, must use SI units. Other units, such as Centimeter-Gram-Second (CGS) Metric, or English, may be used in addition, although this practice is discouraged.

On the accompanying pages is a guide to SI, including tables of conversion, which will be made available in quantity to all AIChE committees and divisions that need it.

SI is somewhat different than the CGS system, in use for many years, which has often been called the Metric System. SI is a system adopted internationally by the General Conference of Weights and Measures. Among some of the principles are the use of the kilogram for mass only, and the use of newton for force or weight.

Pressure is expressed in terms of newtons per square meter, and is given the name, pascal. The pascal is a very small unit, and the kilopascal is suggested as the most common unit for pressure.

The main feature of SI is in the fact that it is coherent, which means that no conversion factors are needed when using basic or derived SI units. Any exception to the SI units destroys the coherency of the system, and is not really a step forward in usefulness.

The third column of Table 1 shows the metric units that may be used for an indefinite period of time with SI. These include the minute, hour, year, and liter. The fourth column contains units that are accepted for a limited period of time, probably on the order of five to 10 years, although this duration has not been established by the Institute. And finally, the fifth column lists those units that are definitely outside SI, and which will not be allowed in AIChE publications.

In the opinion of the Metrication Committee, there is no longer any question about eventual conversion to the metric system, and to SI in particular. The only question really is, when and how? AIChE is following the practice being instituted by many technical societies; we are not either leading or trailing significantly at present.

On the lighter side, the magnitude of the newton is about the weight of an apple. If we were to grind up that apple and spread it out over one square meter, we would have a pressure of one pascal, which may give a better feeling for the small size of that particular unit. Your Chairman of the Metrication Committee is approximately 2 meters tall, which was not a requirement, but can serve as a benchmark.

The Metrication Committee plans to submit a series of articles to *CEP* at two or three month intervals that will deal with various aspects of metric conversion. These are planned to include a typical process flow diagram in SI, a consideration of hard vs. soft conversion, consideration of conversion of various physical properties into SI, case histories of conversion in various industries and companies, and a description of the working of the International Standards Organizations.

Every AIChE committee and division has a member on the committee who acts as its liaison. Please feel free to call upon us for any assistance or information on conversion.

The Council resolution adopted National Bureau of Standards special publication 330, 1974 edition, entitled, "International System of Units (SI)." This is a translation of the proceedings of the last General Conference of Weights and Measures, which set up the present rules of SI. In the last several months, there have been several American National Standards Institute publications on metric practices. The AIChE Committee is looking into adopting some of these or other publications, or preparing a separate, more detailed guide, if needed, on metric practice. In particular, the Institute of Electrical & Electronics Engineers' document, ANSI-210.1-19xx is accepted.

Table 1. Acceptable and unacceptable metric units.

| Quantity | SI Unit | AIChE Recommendations | | |
		Accepted Alternate*	Temporary Alternate**	To Be Avoided†
Time	second	year		
		day		
		hour		
Pressure	pascal		bar, atmosphere	kg force / m²
Energy	joule	—	—	calorie, kilowatt-hr.
Force	newton	—	—	dyne, kilogram
Mass	kilogram	ton	—	
Volume	m³	liter		
Viscosity	pascal-second		—	poise¶

These units are to be incorporated into a one page document similar to that published in *CEP*, May, 1971. Units will be added where appropriate and modifications made in accordance with this table.
*Table VIII NBS 330   **Table X NBS 330   †Table XII NBS 330
¶To be avoided because they were formerly used with the CGS system and are not part of SI.

In addition, the American Metric Council has published an editorial guide that contains much information for authors, editors, secretaries, and other people involved in publication. This is available through the American National Metric Council, 1625 Massachusetts Ave. N.W., Washington, D. C. 20036.

Note: Reprints of this article and guide will be made available to AIChE groups at no charge for Institute business purposes. Individuals interested in copies for their personal use may obtain the reprints for $1.50 prepaid. In either case write: Publications Dept., AIChE, 345 E. 47 St., New York, N.Y., 10017.

A Word About the Guide

This guide for the use of SI units, originally published in the May, 1971, issue of CEP, has been updated and expanded slightly since then to conform to present practices. This material was prepared by Evan Buck, staff engineer, Union Carbide Corp., South Charleston, W. Va., a member of the Metrication Committee.

Oldshue

Buck

Abbreviated Guide for Use of the SI

These tables summarize the SI unit system adopted by the AIChE Council on March 19, 1977, for use within the AIChE after January 1, 1979. This unit system is based on that documented in the National Bureau of Standards (NBS) Special Publication 330, 1974 edition, titled "The International System of Units (SI)," with the following modifications:

1. The "year" as a time unit has been added.
2. The symbol "L" rather than "l" is to be used as the abbreviation for liter, which avoids possible confusion with the numeral "1."
3. The prefixes "peta" (10^{15}) and "exa" (10^{18}) have been added.

Items 2 and 3 have been adopted by the NBS subsequent to the appearance of Publication 330.

SI Base Units

Quantity	Name	Symbol
length	meter	m
mass	kilogram	kg
time	second	s
electric current	ampere	A
thermodynamic temperature	kelvin	K
amount of substance	mole	mol
luminous intensity	candela	cd

SI Supplementary Units

Quantity	SI unit Name	Symbol
plane angle	radian	rad
solid angle	steradian	sr

Examples of SI Derived Units Expressed in Terms of Base Units

Quantity	SI Unit Name	Symb
area	square meter	m^2
volume	cubic meter	m^3
speed, velocity	meter per second	m/s
acceleration	meter per second squared	m/s^2
kinematic viscosity	square meter per second	m^2/s
wave number	1 per meter	m^{-1}
density, mass density	kilogram per cubic meter	kg/m^3
current density	ampere per square meter	A/m^2
magnetic field strength	ampere per meter	A/m
concentration (of amount of substance)	mole per cubic meter	mol/m
activity (radioactive)	1 per second	s^{-1}
specific volume	cubic meter per kilogram	m^3/kg
luminance	candela per square meter	cd/m^2
angular velocity	radian per second	rad/s
angular acceleration	radian per second squared	rad/s^2

SI Derived Units With Special Names

Quantity	SI Unit Name	Symbol	Expression in terms of other units
frequency	hertz	Hz	s^{-1}
force	newton	N	$kg \cdot m/s2$
pressure, stress	pascal	Pa	N/m^2
energy, work, quantity of heat	joule	J	$N \cdot m$
power, radiant flux	watt	W	J/s
quantity of electricity, electric charge	coulomb	C	$A \cdot s$
electric potential, voltage, potential difference, electromotive force	volt	V	W/A
capacitance	farad	F	C/V
electric resistance	ohm	Ω	V/A
conductance	siemens	S	A/V
magnetic flux	weber	Wb	$V \cdot s$
magnetic flux density	tesla	T	Wb/m^2
inductance	henry	H	Wb/A
luminous flux	lumen	lm	$cd \cdot sr$
illuminance	lux	lx	$cd \cdot sr/m^2$

Examples of SI Derived Units Expressed by Means of Special Names

Quantity	SI Unit Name	SI Unit Symbol
dynamic viscosity	pascal-second	Pa · s
moment of force	meter-newton	N · m
surface tension	newton per meter	N/m
heat flux density, irradiance	watt per square meter	W/m^2
heat capacity, entropy	joule per kelvin	J/K
specific heat capacity, specific entropy	joule per kilogram-kelvin	J/(kg · K)
specific energy	joule per kilogram	J/kg
thermal conductivity	watt per meter-kelvin	W/(m · K)
energy density	joule per cubic meter	J/m^3
electric field strength	volt per meter	V/m
electric charge density	coulomb per cubic meter	C/m^3
electric flux density	coulomb per square meter	C/m^2
permittivity	farad per meter	F/m
permeability	henry per meter	H/m
molar energy	joule per mole	J/mol
molar entropy, molar heat capacity	joule per mole-kelvin	J/(mol · K)
radiant intensity	watt per steradian	W/sr
radiance	watt per square meter-steradian	$W \cdot m^{-2} \cdot sr^{-1}$

Units in Use With the International System

Name	Symbol	Value in SI Units
minute	min	1 min = 60 s
hour	h	1 h = 60 min = 3600 s
day	d	1 d = 24 h = 86400 s
year	yr	1 yr ≈ 365 d
degree	°	1° = (π/180) rad
minute	′	1′ = (1/60)° = (π/10800) rad
second	″	1″ = (1/60)′ = (π/648000) rad
liter	L	1 L = 1 dm^3 = 10^{-3} m^3
ton	t	1 t = 10^3 kg
nautical mile		1 nautical mile = 1852 m
knot		1 nautical mile per hour = (1852/3600) m/s
ångström	Å	1 Å = 0.1 nm = 10^{-10} m
are	a	1 a = 1 dam^2 = 10^2 m^2
hectare	ha	1 ha = 1 hm^2 = 10^4 m^2
barn	b	1 b = 100 fm^2 = 10^{-28} m^2
bar	bar	1 bar = 0.1 MPa = 10^5 Pa
standard atmosphere	atm	1 atm = 101325 Pa
gal	Gal	1 Gal = 1 cm/s^2 = 10^{-2} m/s^2
curie	Ci	1 Ci = 3.7 × 10^{10} s^{-1}
röntgen	R	1 R = 2.58 × 10^{-4} C/kg
rad	rad	1 rad = 10^{-2} J/kg

Note: *In addition to the thermodynamic temperature (symbol T), expressed in kelvins, use is also made of Celsius temperature (symbol t) defined by the equation*

$$t = T - T_0$$

where T_0 = 273.15 K by definition. The Celsius temperature is expressed in degrees Celsius (symbol °C). The unit "degree Celsius" is thus equal to the unit "kelvin," and an interval or a difference of Celsius temperature may also be expressed in degrees Celsius.

SI Prefixes

Factor	Prefix	Symbol	Factor	Prefix	Symbol
10^{18}	exa	E	10^{-1}	deci	d
10^{15}	peta	P	10^{-2}	centi	c
10^{12}	tera	T	10^{-3}	milli	m
10^9	giga	G	10^{-6}	micro	μ
10^6	mega	M	10^{-9}	nano	n
10^3	kilo	k	10^{-12}	pico	p
10^2	hecto	h	10^{-15}	femto	f
10^1	deka	da	10^{-18}	atto	a

Directions for Use

SI symbols are not capitalized unless the unit is derived from a proper name; e.g., Hz for H. R. Hertz. Unabbreviated units are not capitalized; e.g., hertz, newton, kelvin. Only E, P, T, G, *and* M *prefixes are capitalized.*

Except at the end of a sentence, SI units are not to be followed by periods.

With derived unit abbreviations, use center dot to denote multiplication and a slash for division; e.g., newton-second/meter^2 = N·s/m^2.

Conversion Factors to SI for Selected Quantities

*An asterisk after the seventh decimal place indicates the conversion factor is exact and all subsequent digits are zero.

To convert from	To	Multiply by
barrel (for petroleum, 42 gal)	meter^3 (m^3)	1.5898729 E − 01
British thermal unit (Btu, International Table)	joule (J)	1.0550559 E + 03

Continued

Continued from page 137

To convert from	To	Multiply by	
Btu/lbm-deg F (heat capacity)	joule/kilogram-kelvin (J/kg·K)	4.1868000*	E + 03
Btu/hour	watt (W)	2.9307107	E − 01
Btu/second	watt (W)	1.0550559	E + 03
Btu/ft²-hr-deg F (heat transfer coefficient)	joule/meter²-second-kelvin (J/m²·s·K)	5.6782633	E + 00
Btu/ft²-hour (heat flux)	joule/meter²-second (J/m²·s)	3.1545907	E + 00
Btu/ft-hr-deg F (thermal conductivity)	joule/meter-second-kelvin (J/m·s·K)	1.7307347	E + 00
calorie (International Table)	joule (J)	4.1868000*	E + 00
cal/g·deg C	joule/kilogram-kelvin (J/kg·K)	4.1868000*	E + 03
centimeter	meter (m)	1.0000000*	E − 02
centimeter of mercury (0°C)	pascal (Pa)	1.3332237	E + 03
centimeter of water (4°C)	pascal (Pa)	9.80638	E + 01
centipoise	pascal-second (Pa·s)	1.0000000*	E − 03
centistoke	meter²/second (m²/s)	1.0000000*	E − 06
degree Fahrenheit (°F)	kelvin (K)	$t_K = (t_F + 459.67)/1$	
degree Rankine (°R)	kelvin (K)	$t_K = t_R/1.8$	
dyne	newton (N)	1.0000000*	E − 05
erg	joule (J)	1.0000000*	E − 07
farad (International of 1948)	farad (F)	9.99505	E − 01
fluid ounce (U.S.)	meter³ (m³)	2.9573530	E − 05
foot	meter (m)	3.0480000*	E − 01
foot (U.S. Survey)	meter (m)	3.0480061	E − 01
foot of water (39.2°F)	pascal (Pa)	2.98898	E + 03
foot²	meter² (m²)	9.2903040*	E − 02
foot/second²	meter/second² (m/s²)	3.0480000*	E − 01
foot²/hour	meter²/second (m²/s)	2.5806400*	E − 05
foot-pound-force	joule (J)	1.3558179	E + 00
foot²/second	meter²/second (m²/s)	9.2903040*	E − 02
foot³	meter³ (m³)	2.8316847	E − 02
gallon (U.S. liquid)	meter³ (m³)	3.7854118	E − 03
gram	kilogram (kg)	1.0000000*	E − 03
horsepower (550 ft·lbf/s)	watt (W)	7.4569987	E + 02
inch	meter (m)	2.5400000*	E − 02
inch of mercury (60°F)	pascal (Pa)	3.37685	E + 03
inch of water (60°F)	pascal (Pa)	2.48843	E + 02
inch²	meter² (m²)	6.4516000*	E − 04
inch³	meter³ (m³)	1.6387064*	E − 05
kilocalorie	joule (J)	4.1868000*	E + 03
kilogram-force (kgf)	newton (N)	9.8066500*	E + 00
micron	meter (m)	1.0000000*	E − 06
mil	meter (m)	2.5400000*	E − 05
mile (U.S. Statute)	meter (m)	1.6093440*	E + 03
mile/hour	meter/second (m/s)	4.4704000*	E − 01
millimeter of mercury (0°C)	pascal (Pa)	1.3332237	E + 02
ohm (International of 1948)	ohm (Ω)	1.000495	E + 00
ounce-mass (avoirdupois)	kilogram (kg)	2.8349523	E − 02
ounce (U.S. fluid)	meter³ (m³)	2.9573530	E − 05
pint (U.S. liquid)	meter³ (m³)	4.7317647	E − 04
poise (absolute viscosity)	pascal-second (Pa·s)	1.0000000*	E − 01
poundal	newton (N)	1.3825495	E − 01
pound-force (lbf avoirdupois)	newton (N)	4.4482216	E + 00
pound-force-second/ft²	pascal-second (Pa·s)	4.7880258	E + 01
pound-mass (lbm avoirdupois)	kilogram (kg)	4.5359237*	E − 01
pound-mass/foot³	kilogram/meter³ (kg/m³)	1.6018463	E + 01
pound-mass/foot-second	pascal-second (Pa·s)	1.4881639	E + 00
psi	pascal (Pa)	6.8947573	E + 03
quart (U.S. liquid)	meter³ (m³)	9.4635295	E − 04
slug	kilogram (kg)	1.4593903	E + 01
stoke (kinematic viscosity)	meter²/second (m²/s)	1.0000000*	E − 04
ton (long, 2240 lbm)	kilogram (kg)	1.0160469	E + 03
ton (short, 2000 lbm)	kilogram (kg)	9.0718474*	E + 02
torr (mm Hg, 0°C)	pascal (Pa)	1.3332237	E + 02
volt (International of 1948)	volt (absolute) (V)	1.000330	E + 00
watt (International of 1948)	watt (W)	1.000165	E + 00
watt-hour	joule (J)	3.6000000*	E + 03
yard	meter (m)	9.1440000*	E − 01

Index